Design of Foundations for Offshore Wind Turbines

Design of Foundations for Offshore Wind Turbines

Subhamoy Bhattacharya
University of Surrey
UK

This edition first published 2019

Registered Offices
John Wiley & Sons, Inc., 111 River Street, Hoboken, NJ 07030, USA
John Wiley & Sons Ltd, The Atrium, Southern Gate, Chichester, West Sussex, PO19 8SQ, UK

Editorial Office
The Atrium, Southern Gate, Chichester, West Sussex, PO19 8SQ, UK

For details of our global editorial offices, customer services, and more information about Wiley products visit us at www.wiley.com.

Wiley also publishes its books in a variety of electronic formats and by print-on-demand. Some content that appears in standard print versions of this book may not be available in other formats.

Library of Congress Cataloging-in-Publication Data

Names: Bhattacharya, Subhamoy, author.
Title: Design of foundations for offshore wind turbines / Subhamoy Bhattacharya, University of Surrey.
Description: Hoboken, NJ, USA : Wiley, 2019. | Includes bibliographical references and index. |
Identifiers: LCCN 2018046931 (print) | LCCN 2018047955 (ebook) | ISBN 9781119128151 (Adobe PDF) | ISBN 9781119128144 (ePub) | ISBN 9781119128120 (hardcover)
Subjects: LCSH: Offshore wind power plants–Design and construction. | Offshore structures–Foundations.
Classification: LCC TK1541 (ebook) | LCC TK1541 .B44 2019 (print) | DDC 621.31/2136–dc23
LC record available at https://lccn.loc.gov/2018046931

Cover Design: Wiley
Cover Image: © silkwayrain/Getty Images

Set in 10/12pt WarnockPro by SPi Global, Chennai, India

Contents

Preface

The offshore wind power industry is expanding at a rapid pace in Europe and Asia and has the potential to solve many issues: clean air, clean energy, and energy security for fossil-fuel-starved countries (e.g. Japan, India). Furthermore, the cost of offshore wind has reduced drastically over the last five years, and it is speculated that this will become the cheapest form of energy in the industrialised world. Foundation selection for these offshore structures plays an important role in the overall concept design for offshore wind farms, as there are large financial implications attached to the choices made. Typically, foundations cost 16–34% of the overall costs, depending on the location and size of the wind farm. This book provides an overview of the civil engineering aspects of these significant infrastructure projects and then focuses on the foundation design.

The industry started by following the design of offshore oil and gas (O&G) structures. This book shows the differences in the design of foundations of two types of offshore structures: O&G and offshore wind turbines (OWTs). It is now widely acknowledged that OWT structures are unique in their features. The most important difference with respect to O&G installation structures is dynamic sensitivity – i.e. natural frequencies of these structures are very close to the forcing frequencies from wave, rotor frequency (1P), and blade frequency (2P/3P). This book aims to distil the knowledge gained through research and describe the different calculations methods for the benefit of practicing engineers and engineering students carrying out projects in this area.

The book is aimed at an international readership of practising engineers within the renewable energy industry and offshore O&G engineers who are considering entering the industry. This would also be helpful for engineering students (undergraduate research and design project students), PhD and EngD researchers, and project managers. The book is intended to provide a deeper understanding of the foundation issues for these structures, and references are provided for further reading. The book is not aimed for repeating materials readily available in other textbooks; its goal is to provide the incremental knowledge in this area required for carrying out the design. There are excellent textbooks in the area of offshore geotechnics, and those teaching materials are not repeated here.

This book is shaped and developed based on the author's teaching and research over the past decade. Some or all of this textbook material has been delivered to a wide range of engineering students and professional engineers: fourth-year MEng course on 'Offshore Foundations to Engineering Science Students' at University of Oxford (UK); masters-level course titled 'Engineering for Offshore Wind and Marine Power' delivered to civil, mechanical, and aerospace engineering students of University of Bristol (UK);

MSc courses titled 'Energy Geotechnics' and 'Renewable Energy Systems Engineering' at University of Surrey (UK); specialised lecture series titled 'Foundation Design for Offshore Wind Turbines' at Zhejiang University (China) and Qingdao University of Technology (China); Global Initiative for Academic Network (GIAN) course in Indian Institute of Technology (IIT, Bhubaneswar); and ISWT (International Summer and Winter Term) course in Indian Institute of Technology (Kharagpur).

To obtain feedback on the material from the industry, a two-day CPD (continuing professional development) course titled 'Design of Foundations for Offshore Wind Turbines' was developed, where one chapter constitutes one lecture and was delivered to a wide range of participants from Europe and Asia in different European and Asian cities. The author is thankful to Professor Purnendu Das (ASRANET: **A**dvanced **S**tructural **R**eliability **A**nalysis **NET**work) for making many of the arrangements for the CPD course. Some of the CPD courses was delivered in conjunction with CENER (Spain) and GH-GL (Garrad Hassan). Many of the discussions presented in the book are based on interactions with the participants of the courses. The author acknowledges these organisations: Atkins, DnV, Motts MacDonald, Eon, GeoSea, London Offshore Consultants, Scottish Power, Parkwind, Technalia, CENER, China Energy, Beaurea Veritas, RES offshore, EDF, Iberdrola, RWE Innogy and Ramboll.

The author would also like to thank the European Commission for the appointment to position of technical expert for reviewing FP7: INNWIND project, which is at the forefront of technological development.

Many of the materials presented in this book is the research work of my past and current students, which includes undergraduate, masters, and PhD students. Special mention goes to my PhD students: Georgios Nikitas, Saleh Jalbi, Piyush Mohanty, Pradeep Dammalla, Aleem Mohammad, Dr. James Cox, Dr. Laszlo Arany, Dr. Yuliqing Yu, Dr. Domenico Lombardi, Dr. Suresh Dash, Dr. Mehdi Rouholamin, and Dr. Masoud Shadlou. I also acknowledge the review of chapters by my former MSc student Carlos Molina Messa (ex-GEO, Copenhagen and now Ramboll).

I would like to acknowledge my collaborators and colleagues (current and former) for various discussions: Mr. Nathan Vimalan, Mr. Julian Garnsey, Mr. Chris Thomas, Mr. Daniel Birtminn, Dr. Liang Cui, Dr. Ying Wang, Dr. S. Szyniszewski, Dr. Barnali Ghosh, Dr. Nick Nikitas, Dr. Nick Alexander, Dr. John Macdonald, Dr. Erdin Ibrahim, Prof. Marios Chryssanthapoulos, Prof. Chris Martin, Prof. Byron Byrne, Prof. John Hogan, Prof. David Muir Wood, Prof. George Mylonakis, Prof. Colin Taylor, Prof. Bouzid Djillali, Prof. Lizhong Wang.

The book would not be possible without discussions with my professors, mentors, and well-wishers: Professor Malcolm Bolton and Professor Gopal Madabhushi (University of Cambridge); Professor Guy Houlsby and Prof Harvey Burd (University of Oxford); Professor Mark Randolph (University of Western Australia); and Mr. Tom Aldridge, Mr. Pat Power, and Mr. Tim Carrington (Fugro Geoconsulting).

I wish to record my appreciation and experience gained from my first academic job at the University of Oxford as a departmental lecturer (started in 2005) and junior research fellow at Somerville College (Oxford University), where I developed an interest in this subject (under the leadership of Prof Houlsby), sitting around Wallis table in the Jenkins building at tea time. These experiences provided the initial understanding of the subject. The discussions with Christian Leblanc (DONG and now WoodThilsed) while he was doing his PhD experiments at Oxford provided me with an understanding of design challenges and issues. The professional experience gained while working at Fugro Geoconsulting on some of the exciting offshore projects of ACG (Azeri Chirag Gunashli), anchor piles for Alvheim FPSO, and foundation-related issues with the Judy and Munro platform were very helpful in developing the materials for the book.

The author would like to thank his family members for supporting all his ambitions. Last, but not least, the author would like to thank the copy-editing team and Wiley team, especially Cheryl Ferguson, Kingsly Jemima, Sathishwaran Pathbanabhan, Anne Hunt Special thanks to Gustava Sanchez (Scottish Power), Dr Laszlo Arany (Atkins) Dr Matthijs Soede (DG, RTD, European Commission), George Nikitas (Univ of Surrey) and Dr Paul Harper (Univ of Bristol) for various illustration, used in this book.

This is a new area, and the technology is developing very fast. Offshore turbines are being sited not only in deeper waters and further offshore but also in seismic and typhoon zones. Much of the information and methodology presented is expected to be outdated in the next few years and the book will need a new edition. There could also be errors and omissions in the book, and I would like to know about them. Please email me at Subhamoy.Bhattacharya@gmail.com. The comments will be duly acknowledged in the next edition.

London, 10 September 2018

About the Companion Website

This book is accompanied by a companion website:

www.wiley.com/go/bhattacharya/offshorewindturbines

The website includes:

(1) Teaching materials:
  (a) Powerpoint slides of each chapter showing the summary and key learning points
  (b) Tutorials questions from each chapter to reinforcing the understanding
  (c) Solved example problems in conjunction with Chapter 6
  (d) Videos of some concepts
(2) Expanded biography of the author
(3) Reviews of the book by some experts

Scan this QR code to visit the companion website

1

Overview of a Wind Farm and Wind Turbine Structure

Learning Objectives

The aim of this chapter is to provide an overview of the power generation from wind and features of a wind turbine structure. The overall layout of a wind farm is also discussed to appreciate the multidisciplinary nature of the subject. The fundamental concepts and understanding of other disciplines and fields not directly related to foundations but are necessary to carry out the foundation design are also described with references for further study. The chapter also provides description of different types of foundations that are being used and planned to be used.

After you read this chapter, you will be able to: (i) appreciate the complexity and multidisciplinary nature of the design; (ii) get an overview of the subject; (iii) differentiate between oil and gas (O&G) structure and offshore wind turbine structure.

The chapters of the book are arranged in the following way: It starts with a system-level understanding (overall wind farm – Chapter 1) and then to component level (foundations design – Chapters 2 and 3) and finally to the element level (soil behaviour, provided in Chapter 4). Chapter 5 discusses the different methods of analyses and Chapter 6 provides some example applications.

1.1 Harvesting Wind Energy

Offshore wind power generation has established itself as a source of reliable energy rather than a symbolism of sustainability. It has been reported by National Grid of the United Kingdom (UK) that on 19 October 2014, 24% of the electricity supply in the United Kingdom was provided by offshore wind farms due to an unexpected fire in Didcot power station and when few of the nuclear power stations were offline due to maintenance and technical issues. Furthermore, National Grid also reported that on 21 October 2014, UK wind farms generated 14.2% of the electricity, which is more than the electricity generated by its nuclear power station (13.2%) for a 24-hour period.

Before the details of engineering of these systems are discussed, it is considered useful to discuss the sustainability of wind resources as it is often noted that wind doesn't blow all the time. Wind, essentially atmospheric air in motion, is a secondary source of energy and is dependent on the sun. The electromagnetic radiation of the Sun

Design of Foundations for Offshore Wind Turbines, First Edition. Subhamoy Bhattacharya.

Companion website: www.wiley.com/go/bhattacharya/offshorewindturbines

unevenly heats the Earth's surface and creates a temperature gradient in the air, thereby also developing a density and pressure difference. The disparity in differential heating of the surface of the Earth is also a result of specific heat and absorption capacity of sand, clay, intermediate and mixed soils, rocks, water, and other materials. This also results in differential heating of air in different regions and at different rates. The physical process or mechanism that governs the air flow is convection. Common examples are land and sea breezes in coastal regions. The direction and velocity of wind are partly influenced by the rotation of the Earth and topography of the Earth's surface, and thus coastal areas are attractive locations for harvesting wind power. This above discussion shows the sustainability of the wind resource as it is related to the Sun and Earth's motion.

In 2017, Europe was the global leader for offshore wind energy, with the United Kingdom leading the field. This is partially due to the aspirations and policies of the European Union to reduce its greenhouse emissions from the 1990 levels by 20% by the year 2020 and then a further reduction of 80–95% by 2050. There is also an initiative in Europe to make its energy system clean, secure, and efficient.

Offshore wind farming is considered to be one of the most reliable ways to produce clean green energy for five reasons:

1. The average wind speed over sea is generally higher and more consistent than onshore, making the offshore wind farming more efficient.
2. The noise and vibrations from the wind turbines will have minimum impact on human beings due to their distance from land.
3. Large capacity can be installed offshore in comparison to an equivalent onshore wind farm. The reasons are that heavier wind turbine generators (WTGs) or towers can be easily transported and installed using sea routes. In contrast, transporting these large and heavy structures/components during construction will substantially disrupt the daily life for people who live in the vicinity of the wind farm due to blockage of roads.
4. Wave and current loading can be harvested alongside wind through the use of hybrid systems.
5. Wind turbine technology is relatively more mature than other forms of renewables.

1.2 Current Scenario

Currently, the United Kingdom is leading in offshore wind harvesting (currently generating around 3.6 GW). However, Denmark was the first country to build an offshore wind farm 2.5 km off the Danish coast at Vindeby. Figure 1.1a shows the cumulative offshore wind power capacity by country in 2013 and Figure 1.1b displays the evolution of global offshore wind power capacity from 1993 to 2013. Construction of large-scale offshore wind farms are on the rise – due to initiatives in many countries such as Germany, Spain, Portugal, South Korea, China, and Japan. The growth is further enhanced possibly due to diminishing public confidence following the 2011 Fukushima Dai-ichi nuclear power plant (NPP) incident. Figure 1.2a shows the planned offshore wind farm development in the UK waters and Figure 1.2b shows some of the wind farms in Europe. Asian countries such as China, Taiwan, Japan, and South Korea are also fast progressing; see Figure 1.2c.

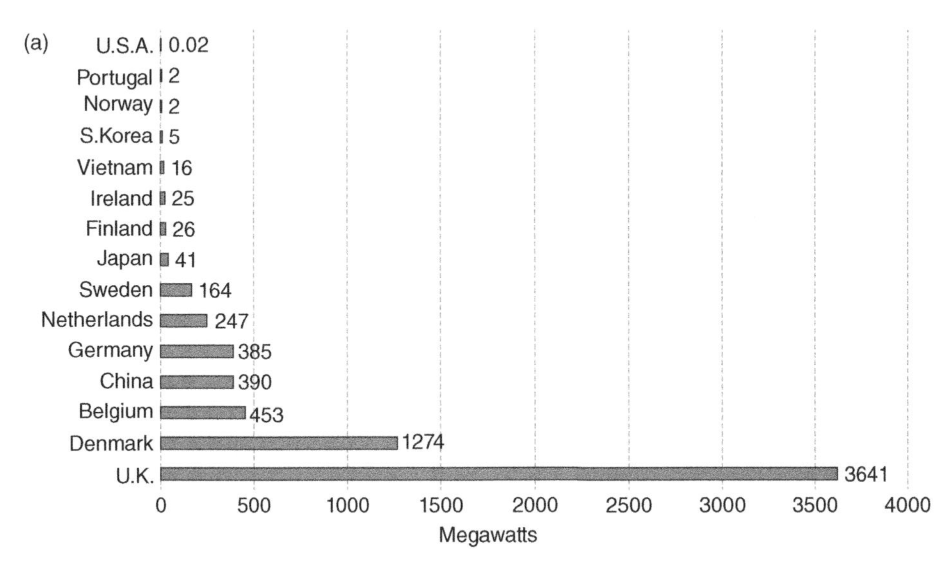

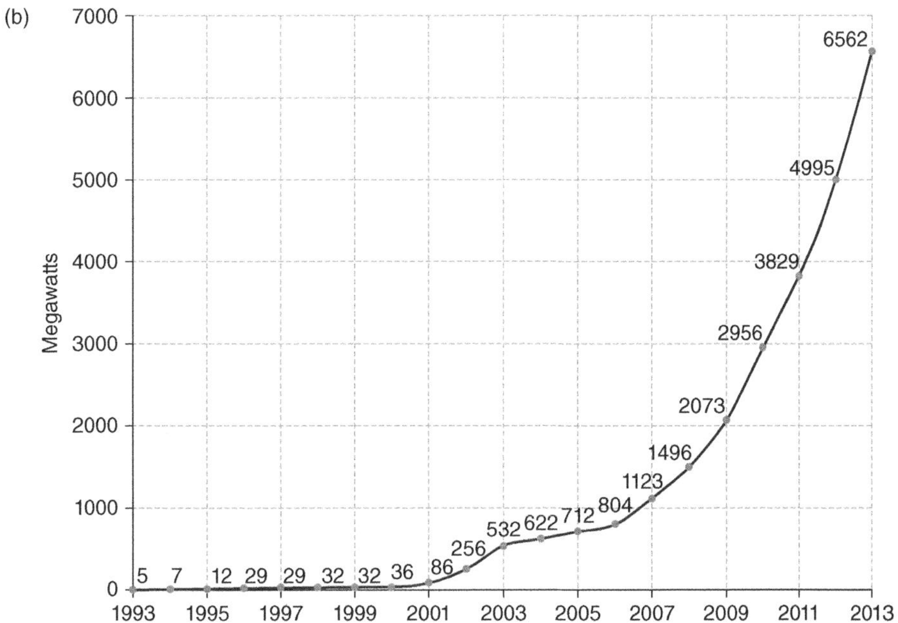

Figure 1.1 (a) Offshore wind power capacity (cumulative) by country in 2013 (Source: E.W.E.A.) and (b) evolution of cumulative global offshore wind power capacity for 1993–2013 (Source: E.W.E.A.).

ASIDE

Energy challenge: With the discovery of shale natural gas (fracking) and lower oil prices, it is predicted that reliance of oil (often termed as Oil Age) may be ending. With the increasing use of electric cars and wind turbines, it may be argued that this move toward low-carbon energy is irreversible and quite similar to the transition from the *Stone Age* to the *Bronze Age.*

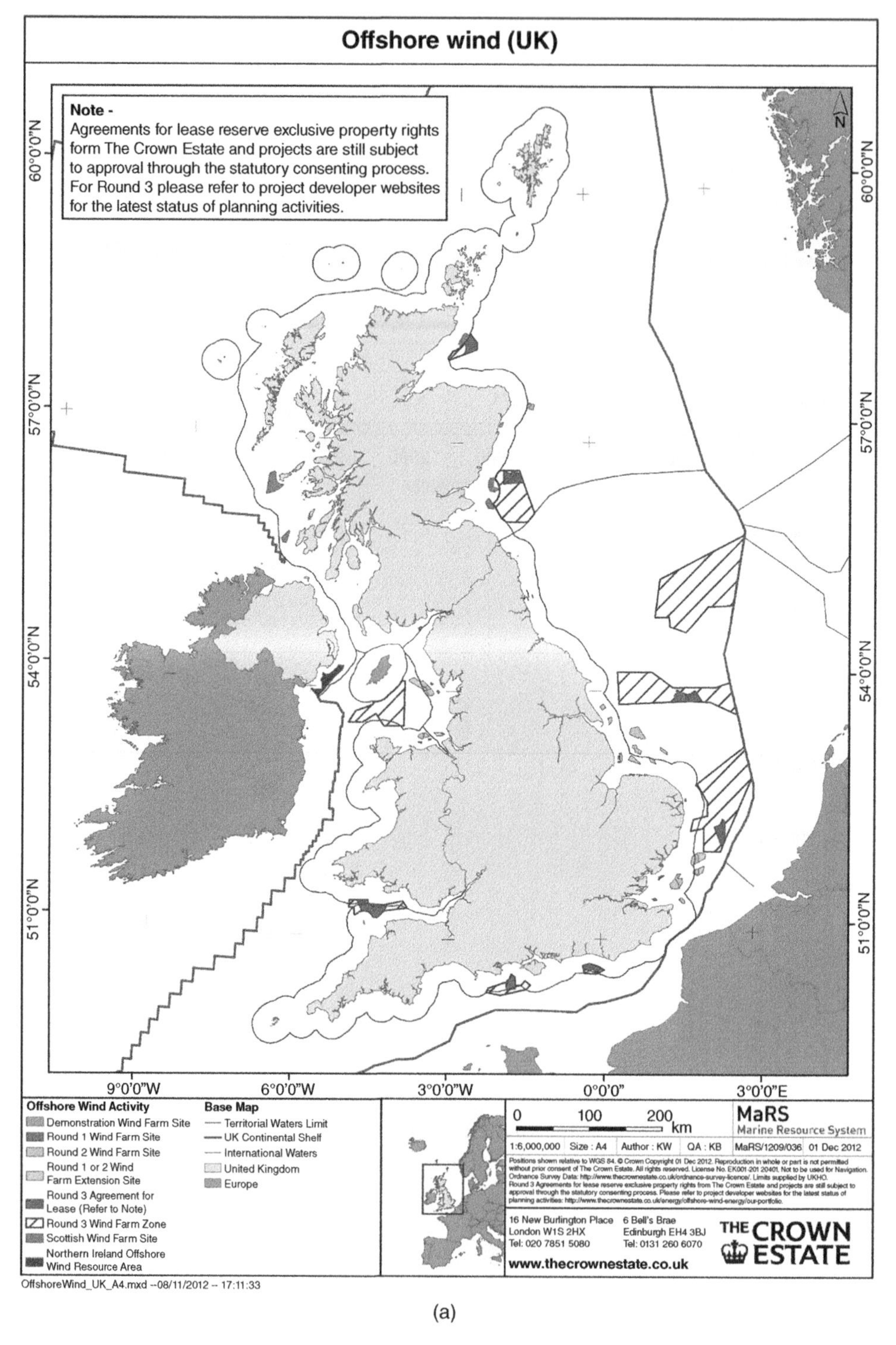

(a)

Figure 1.2 (a) Offshore wind farms around the United Kingdom; (b) wind farms in Europe; and (c) developments in China, Korea, Japan, Taiwan.

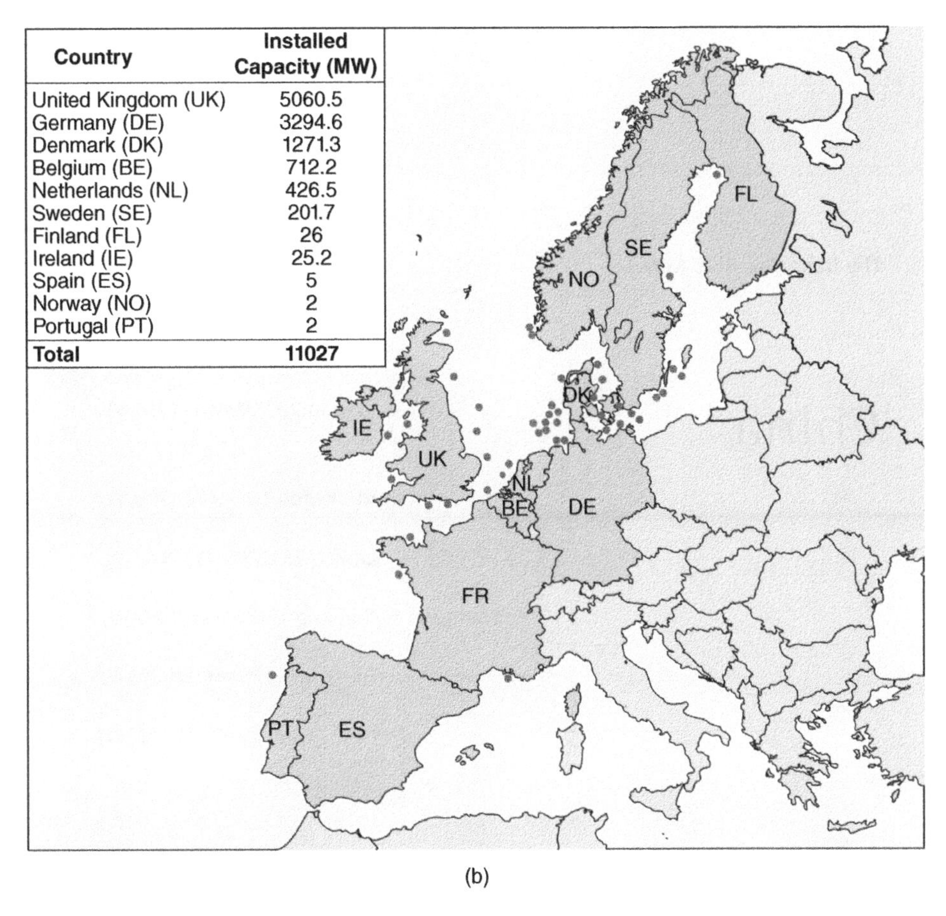

Country	Installed Capacity (MW)
United Kingdom (UK)	5060.5
Germany (DE)	3294.6
Denmark (DK)	1271.3
Belgium (BE)	712.2
Netherlands (NL)	426.5
Sweden (SE)	201.7
Finland (FL)	26
Ireland (IE)	25.2
Spain (ES)	5
Norway (NO)	2
Portugal (PT)	2
Total	**11027**

(b)

Figure 1.2 (*Continued*)

1.2.1 Case Study: Fukushima Nuclear Plant and Near-Shore Wind Farms during the 2011 Tohoku Earthquake

A devastating earthquake of moment magnitude $M_w 9.0$ struck the Tohoku and Kanto regions of Japan on 11 March at 2 :46 p.m., which also triggered a tsunami (see Figure 1.3 for the location of the earthquake and the operating wind farms). The earthquake and the associated effects such as liquefaction and tsunami caused great economic loss, loss of life, and tremendous damage to structures and national infrastructures but very little damage to the wind farms. Extensive damage was also caused by the massive tsunami in many cities and towns along the coast. Figure 1.4a shows photographs of a wind farm at Kamisu (Hasaki) after the earthquake and Figure 1.4b shows the collapse of pile-supported building at Onagawa. At many locations (e.g. Natori, Oofunato, and Onagawa), tsunami heights exceeded 10 m, and sea walls and other coastal defence systems failed to prevent the disaster.

The earthquake and its associated effects (i.e. tsunami) also initiated the crisis of the Fukushima Dai-ichi nuclear power plant. The tsunami, which arrived around 50 minutes following the initial earthquake, was 14 m high, which overwhelmed the 10 m high

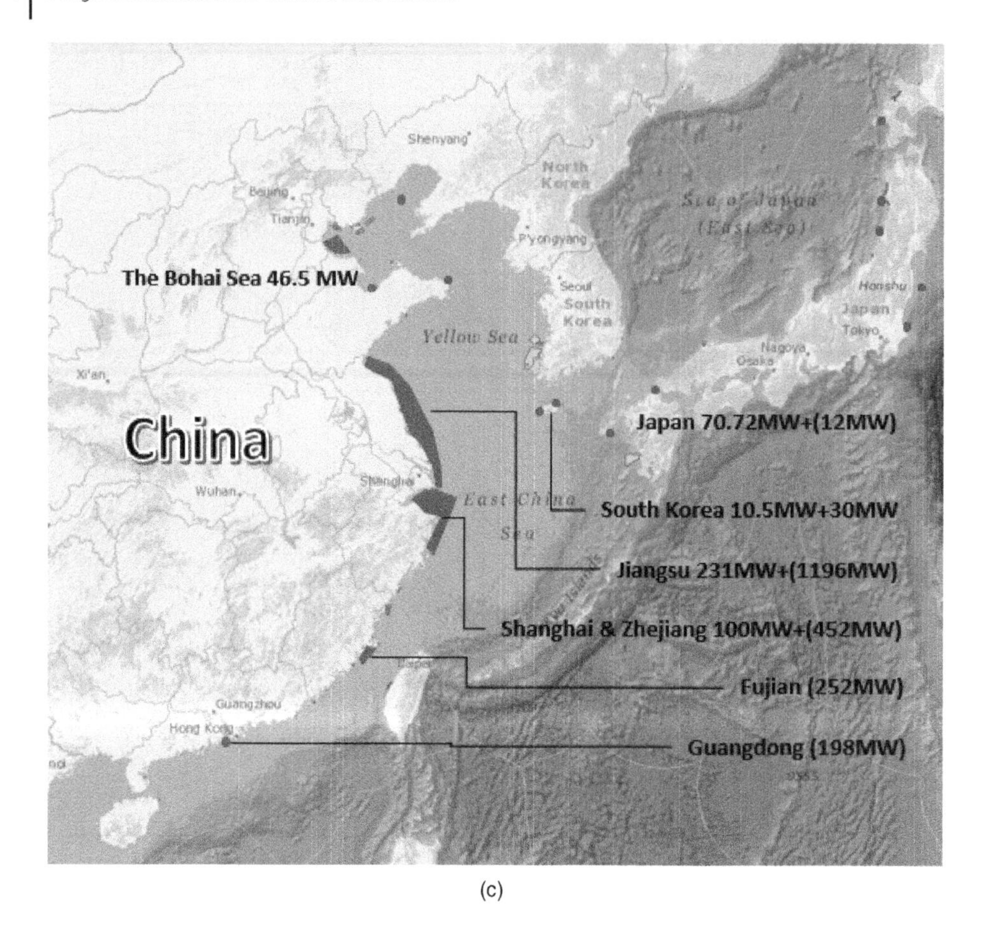

(c)

Figure 1.2 *(Continued)*

plant sea walls, flooding the emergency generator rooms and causing power failure to the active cooling system. Limited emergency battery power ran out on 12 March and subsequently led to the reactor heating up and melting down, which released harmful radioactive materials into the atmosphere. Power failure also meant that many of the safety control systems were not operational. The release of radioactive materials caused a large-scale evacuation of over 300 000 people, and the clean-up costs are expected to be in the tens of billions of dollars. On the other hand, following/during the earthquake, the wind turbines were automatically shut down (like all escalators or lifts), and following an inspection they were restarted.

1.2.2 Why Did the Wind Farms Survive?

Recorded ground acceleration time-series data in two directions (north–south [NS] and east–west [EW]) at the Kamisu and Hiyama wind farms (FKSH 19 and IBRH20) are presented in Figure 1.5 in frequency domain. The dominant period ranges of the recorded ground motions at the wind farm sites were around 0.07–1.0 seconds and the period

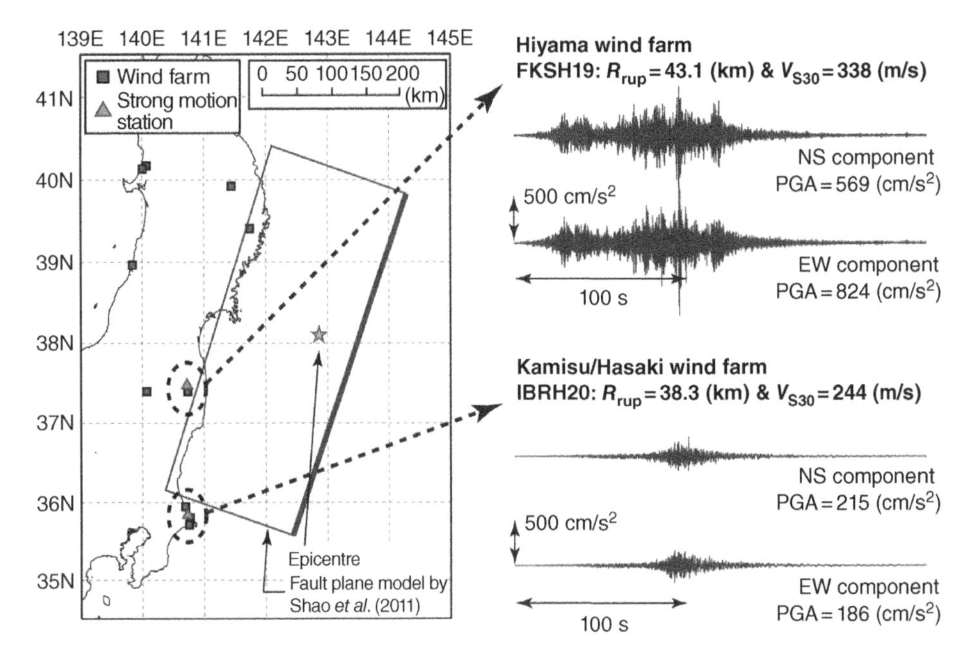

Figure 1.3 Details of the 2011 Tohoku earthquake and locations of the wind farms.

Figure 1.4 (a) Photograph of the Kamisu (Hasaki) wind farm following the 2011 Tohoku earthquake; and (b) collapse of the pile-supported building following the same earthquake.

of offshore wind turbine systems are in the range of 3.0 seconds. Due to nonoverlapping, these structures will not get tuned and as a result, they are relatively insensitive to earthquake shaking. However, earthquake-induced effects such as liquefaction may cause some damages. Further details can be found in Bhattacharya and Goda (2016)

ASIDE

One may argue that had there been a few offshore wind turbines operating, the disaster might have been averted or the scale of damages could have certainly been reduced. The wind turbines could have run the emergency cooling system and prevented the reactor meltdown. In this context, it is interesting to note that there are plans to replace

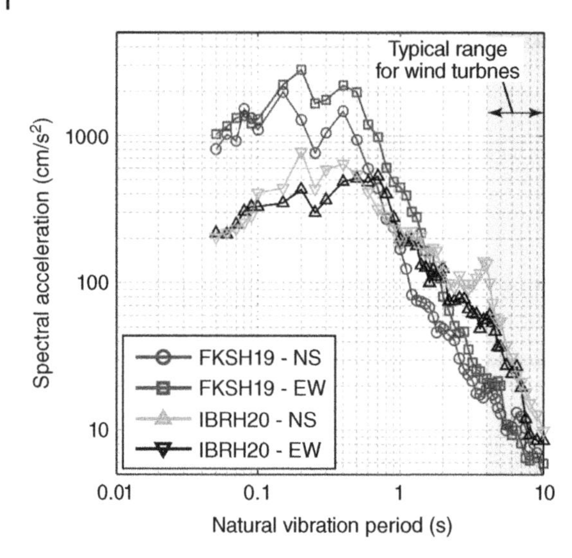

Figure 1.5 Power spectra of the earthquake and natural frequency of wind turbines.

the Fukushima NPP by a floating wind farm. The project is in advanced stages whereby 2 MW semi-sub-floating turbine is under operation for few years. An innovative 7 MW oil-pressure-drive type wind turbine on a three-column semi-sub floater has recently been tested.

1.3 Components of Wind Turbine Installation

The majority of wind turbines or wind energy converters conform to a generic arrangement, typically characterised by a three-bladed turbine driving a horizontally mounted generator. To ease the understanding, the general terminology and components of a wind turbine are shown in Figures 1.6 and 1.7. Typically, a turbine manufacturer supplies

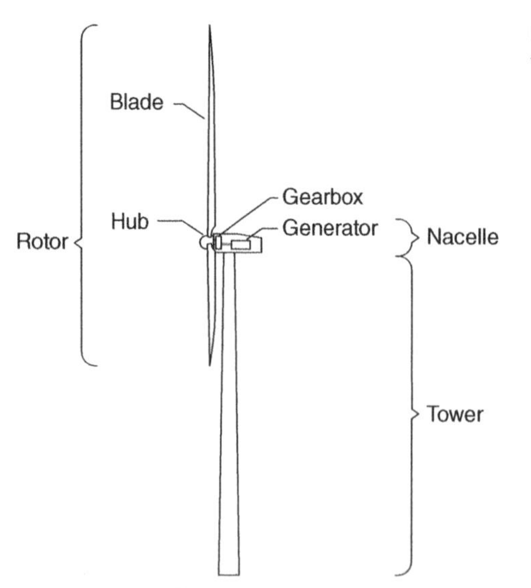

Figure 1.6 RNA (rotor-nacelle assembly) and the tower.

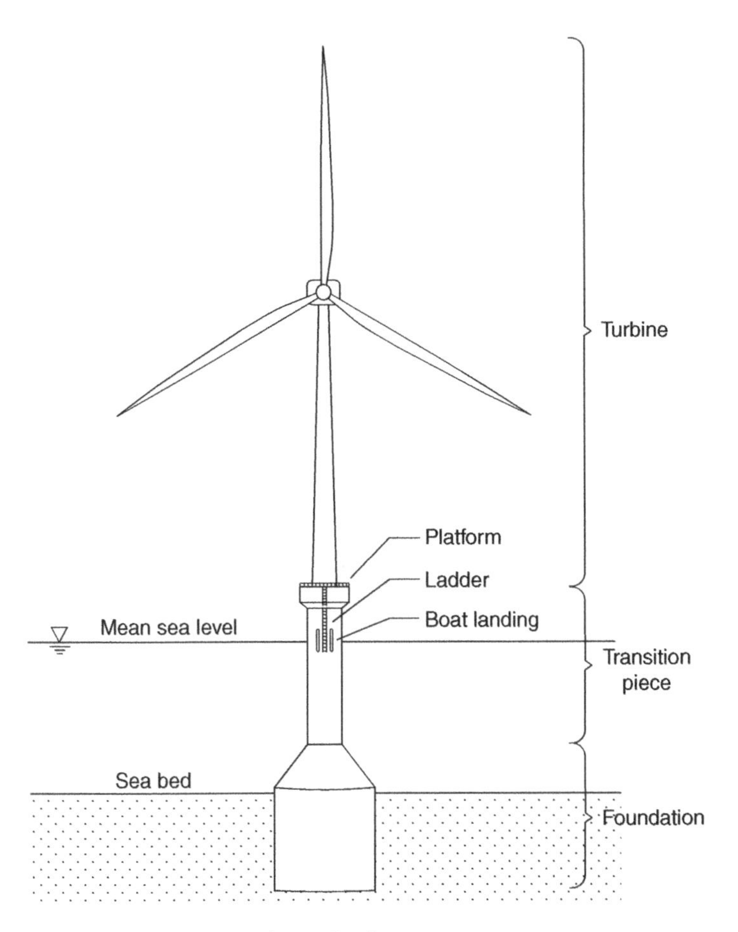

Figure 1.7 Components of a wind turbine structure.

rotor-nacelle assembly (RNA) assembly and the tower, i.e. the components shown in Figure 1.6. The working principle is very simple: essentially, the kinetic energy of the flowing wind is converted into rotational kinetic energy in the turbine and then to electrical energy through a generator. Figure 1.8 displays the components inside the nacelle for a typical turbine.

Readers are referred to the specialised book for details of turbines, such as Burton et al. (2011) and Jameison (2018).

The nacelle, as shown in Figure 1.6, is mounted on the top of the tower and can have different shapes and sizes depending on the turbines. The nacelle contains the generator, which is driven by the high-speed shaft. The high-speed shaft is usually connected to the low-speed shaft by a gearbox. The low-speed shaft goes out of the nacelle and the rotor hub is placed on it. The blades are connected to the rotor hub. The low-speed shaft rotates with the turbine blades and the typical speed is about 20 revolutions per minute (20 RPM). A typical gearbox has a speed ratio of about 1 : 100, and the high-speed shaft drives the generator.

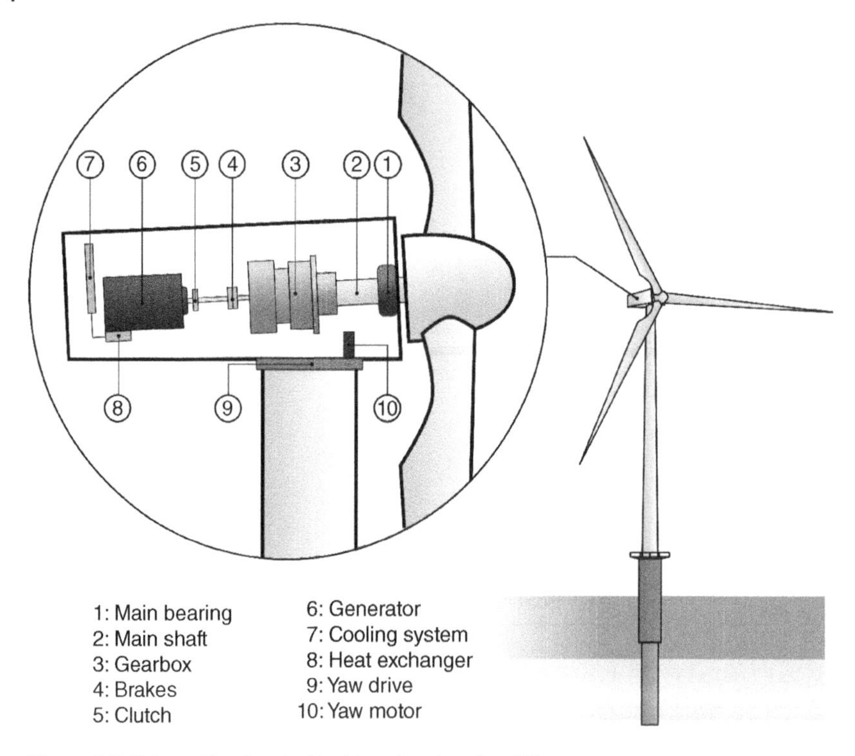

Figure 1.8 Schematic of a wind turbine showing the different components.

For economic viability of a site, it is necessary to estimate the expected power and energy output of each turbine. The wind power capture can be estimated using Eq. (1.1):

$$P = \frac{1}{2} C_p \rho A U^3 \tag{1.1}$$

where

C_p the power coefficient,

ρ the density of air,

A the area of the rotor swept area $A = \frac{\pi}{4} D^2$ (with D being the rotor diameter),

U the wind speed

Based on the relationship, it is clear that for a given swept area and for a particular wind speed and air density, there are two possible ways of increasing the output power:

1. increasing the power coefficient C_p
2. extending the rotor swept area (designing wind turbines with large rotor diameter) thereby increasing A.

ASIDE

It is becoming obvious that it is more convenient to increase the rotor diameter than to invest in a more efficient blade design. The swept area is proportional to the second power of the rotor diameter. This increase in rotor diameter, however, requires taller towers and larger nacelle and components, which pose many challenges in the design.

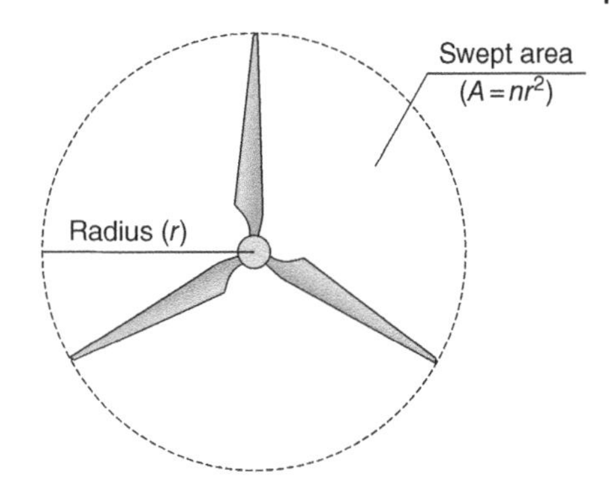

Figure 1.9 Power output for the example.

It may be noted that the 8 MW wind turbine has a rotor of 164 m – i.e. approximately 82 m long blades.

1.3.1 Betz Law: A Note on C_p

Betz Limit or Betz law, based on Betz (1919), states that no wind turbine can convert more than 16/27 (59.3%) of the kinetic energy of the wind into mechanical energy through turning a rotor, and therefore, the theoretical maximum power coefficient ($C_{p,\max}$) is 0.59. However, wind turbines cannot operate at this maximum limit, and this value depends on turbine type, number of blades, and the speed of the rotor. The value of C_p for the best designed wind turbines is in the range 0.35–0.45.

Example 1.1 ***Power Output Problem***
For a wind turbine site, the average wind speed at the hub height is 10 m s^{-1}. If the length of the blade is 60 m, and the density of air is 1.223 kg m^{-3}, find the power output. Assume the power coefficient is 0.4.

The swept area is circle of 120 m diameter; see Figure 1.9. Following Eq. (1.1), the power is given by:

$$P = \frac{1}{2}C_p\rho AU^3 = \frac{1}{2} \times 0.4 \times 1.223\frac{kg}{m^3} \times \frac{\pi}{4} \times 120^2 m^2 \times 10^3 \frac{m^3}{s^3} = 2.76MW$$

1.4 Control Actions of Wind Turbine and Other Details

Wind speeds vary with time, and following Eq. (1.1), it is clear that this will cause a fluctuation in the power generation. This may pose a particular challenge to the power supplied to the electricity grid, and this is known as PTO (power take-off) issues. Control system are in place in the RNA, and the main purpose is to have steady power by ensuring that the rotor turns at a constant rate. There are also issues such as changing the direction of wind, variation of loads in the hub due to unsteady blade aerodynamics, blade flapping, etc., which must be controlled to reduce the fatigue stresses.

The nacelle contains an anemometer to measure wind speeds. At the cut-in wind speed (typically 4 m s^{-1}), the wind turbine starts producing power. At a certain wind speed, the rated power and rotational speed are reached. Typically, wind turbines reach the highest efficiency at the designed wind speed between 12 and 16 m s^{-1}. At this wind speed, the power output reaches its rated capacity. Above this wind speed, the power output of the rotor is limited to the rated capacity and is carried out by various means: stall regulation (constant rotational speed i.e. RPM) or pitch regulation.

Figure 1.10 shows schematically a wind turbine model, along with the definition of the tilt, yaw, and pitch. If the horizontal wind speed is perpendicular to the rotor plane, the wind turbine is in optimal position. The angle of the wind speed with the plane of the axis of the tower and the low speed shaft is called the yaw angle (θ_{yaw}). The turbine has a mechanism that tries to rotate itself in the optimal direction, so that the rotor is perpendicular to the wind. This control action is known as yawing.

Some wind turbines are capable of tilting motion, which means that the angle of the low-speed shaft with the horizontal direction changes. The angle between the horizontal direction and the low-speed shaft is called the tilt angle (θ_{tilt}). This tilt angle can be a deflection due to wind loading or the result of a control action called tilting.

When a wind turbine is producing power, usually a constant rotational speed is required. This can be done in multiple ways; the most common are yaw and pitch control. In pitch-controlled wind turbines, the pitch angle (which is the angle of the blades around the axis that runs from the blade root to the blade tip) can be changed. With the change in the pitch, the angle of attack on the aerofoil profiles of the blades changes, causing a change in the lift force and therefore the rotational speed and power output. This control action is called pitching. The actual pitch angle is denoted by θ_{pitch} or θ_p.

The turbine blades are not perpendicular to the low-speed shaft; usually, there is a small cone angle (θ_{cone}). This coning of the blades provides more stability against the wind, and also has an effect called centrifugal relief. This means that the blades are bent downwind by the wind loading on the blades, but the rotation of the turbine causes a centrifugal load, which is opposite to the load by the wind.

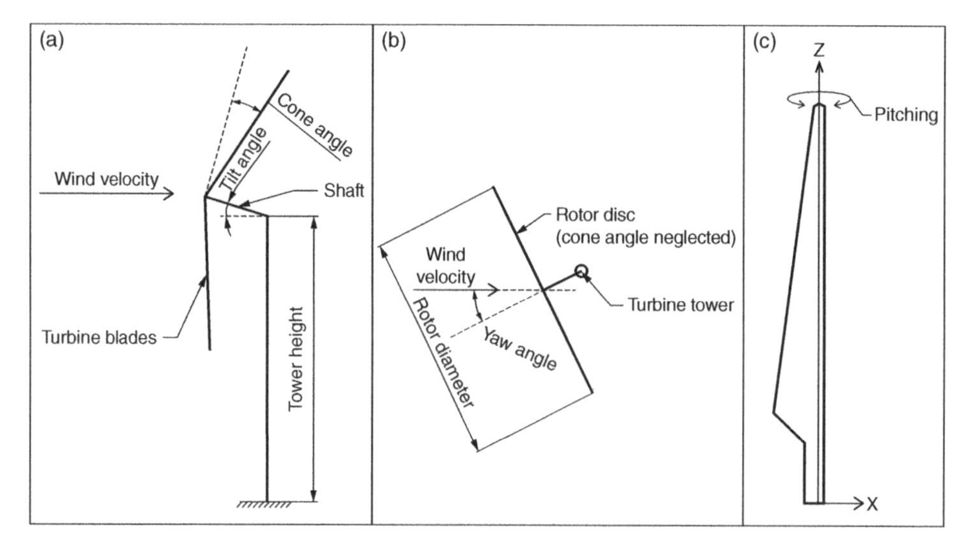

Figure 1.10 Simple wind turbine model for aerodynamic calculations; (a) tilt; (b) yaw; and (c) pitch.

When considering blade loads, a thrust force and a tangential force are defined on the blades. The thrust force is the force acting in the direction normal to the rotor and the tangential force is the one acting in the rotor plane. The flapwise direction is the direction perpendicular to the chord of the aerofoil, and edgewise direction is parallel to the chord of the aerofoil cross section.

It must be mentioned that the tilt angle is fixed for a particular wind turbine installation. For example, the tilt angle for proposed 10 MW SeaTitan is 5°. Blades are very flexible (can be idealised as a cantilever) and will vibrate in flapwise and edgewise direction. The tilt is allowed so as to avoid the blades hitting the tower.

Considering energy extraction from wind, there is a wind speed below which wind turbines cannot be in operation, and there is a wind speed above which the wind turbine must be shut down to avoid serious damage to the blades and machinery. Between these levels lies the operational range of the wind turbine known as 1P range. It is to be noted that this range strongly depends on the size of the turbine, the method of regulation (yaw or pitch regulated), and the parameters of the wind turbine blades.

The definitions of a few terms often required in the design stage follow:

Start-up speed: The wind speed at which the rotor and blade assembly begins to rotate.

Cut-in speed: The wind speed at which the wind turbine starts generating usable power. Typically, 10–15 km h^{-1} (2.8–4.1 m s^{-1}).

Rated speed: The minimum wind speed at which the wind turbine generates its rated power.

Cut-out speed: The wind speed at which the wind turbine ceases power generation and shuts down because of safety reasons.

1P range: This is the rotor frequency range of the turbine. For example, following Table 1.1, Vestas V164-8.0 MW turbine has an operating range of 4.8–12.1 RPM. Essentially, it means the turbine will operate within this range in its entire life cycle. At low wind speed duration, it is expected to operate at 4.8 RPM, and when the wind speed is high, it will operate at rated RPM – i.e. the maximum RPM of 12.1.

Structural design speed: A wind turbine is designed to survive very high wind speeds with 50 years mean recurrence interval without any damage.

There are four ways of shutting down the wind turbine:

1. Use of automatic break when the wind speed sensor measures the cut-out wind speed
2. Pitching the blades to spill the wind

Table 1.1 Details of the various turbines showing the cut-in and rated frequencies.

Turbine make and details	Rating (MW)	Cut in (rpm)	Rated (rpm)
Vestas V 164-8.0 MW	8	4.8	12.1
Siemens SWT-6.0-154	6	5	11
RE power 6 M	6.15	7.7	12.1
RE power 5 M	5.075	6.9	12.1
Vestas V120	4.5	9.9	14.9
Vestas V90	3	8.6	18.4
Sinovel SL3000/90	3	7.5	17.6

3. Use of spoilers mounted on the blade to increase drag and reduce the speed
4. Yawed out of wind (turning the blades sideways to the wind)

1.4.1 Power Curves for a Turbine

Based on the typical wind speeds, a wind turbine manufacturer can provide the so-called power curves for their turbines. Power curve is essentially the power production of the turbines plotted with respect to wind speed. Power curves and rotational speed curves are available for a wind turbine and provided by the manufacturer. Typical examples are given in Figures 1.11–1.12 for a 2.5 and 6.2 MW turbine, respectively.

Figure 1.13 shows a photograph of overhang of a turbine.

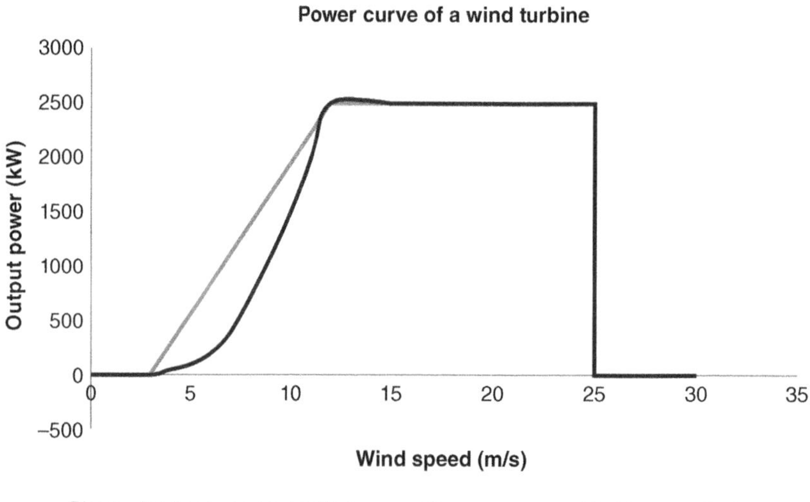

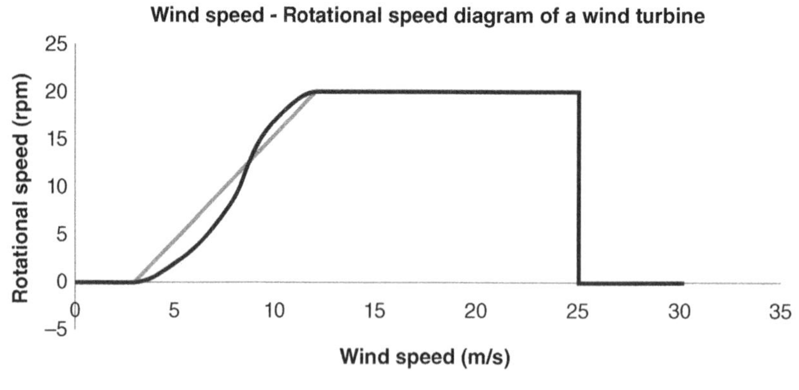

Figure 1.11 Power curve and rotational speed of a wind turbine.

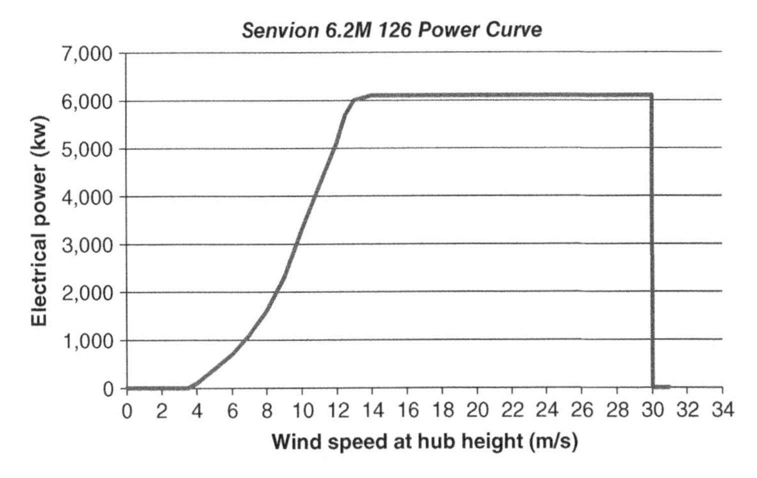

Figure 1.12 Power curve for a 6.2 MW turbine.

Figure 1.13 Showing the dimensions and the rotor overhang. [Photo Courtesy: VESTAS].

1.4.2 What Are the Requirements of a Foundation Engineer from the Turbine Specification?

The foundation designer needs cut-in and cut-out RPM in order to decide the target natural frequency of the whole system. For example, if we consider 8 MW turbine, the 1P range is 4.8–12.1 RPM. If this is converted to Hz, the frequency range is (4.8/60) to (12.1/60), which is 0.08–0.201 Hz. It is therefore advisable to avoid the global natural frequency of the whole system in this range – otherwise resonance-related effects will reduce the service life.

1.4.3 Classification of Turbines

For the purpose relevant to civil engineering design, wind turbines can be classified (simplistically) into three types:

Table 1.2 Typical weight of rotor and nacelle of different WTG.[a)]

Turbine	Sinovel 3.0 MW (SL3000/113)	3.6 MW Siemens	4 MW Siemens	6 MW Siemens	8 MW Vestas
Rotor diameter	113 m	120 m	130 m	154 m	164 m
Nacelle weight	120 t	140 t	140 t	360 t	390 t
Rotor weight	41.5 t	100 t	100 t	81 t	105 t

a) The values presented may change with improved design. Recently, the 8 MW with similar blade length is rated upwards to 9.5 MW.

Table 1.3 Wind farms where gravity-based foundation has been used.

Year of commission	Project	Type of foundation	Depth of water in metres (m)	Distance from shore (km)
2001	Middelgrunden – 40 MW project	Gravity	3–5	2
2003	Nysted 1 (Rodsand I)	Gravity	6–10	10.8
2009	Thornton Bank phase 1	Gravity	12–27	26

1. *Having a gearbox.* The main purpose of a gearbox is to amplify the slow-moving blades. For example, Vestas V164-8.0 MW has an operational range of 4.8–12.1 RPM. Assuming a speed ratio of 1 : 100, the blade rotation of 12 RPM can therefore be amplified 100 times to 1200 RPM, which would necessitate a small-size generator.
2. *Direct drive with no gear box.* This option is attractive as this eliminates the failure of gearbox, which reduces the operational cost (OPEX) cost. For example, in 10 MW Direct Drive SeaTitan Wind Turbines, the cut-in wind speed is 4 m s^{-1} and the cut-out is 30 m s^{-1}. The rated power is generated at wind speed of 11.5 m s^{-1} at 10 RPM.
3. Hybrid, which is a mix of two systems having limited step gear box.

The 1P frequency range is important, as it imparts vibration to the system. For a gearbox-wind turbines, one has to consider the vibration for the whole operational range (4.8–12.1RPM for V164-8.0 MW turbines). On the other hand, for direct drive – only the particular RPM has to be considered.

Size of a turbine: With rated power, the size of a turbine also increases. For example, the size of Sinovel SL3000/113 size is 12.5 long × 5.0 m wide × 6.6 m height. On the other hand, the size of the 8 MW turbine is 24 m long × 12 m wide × 7.5 m high. Table 1.2 provides rotor diameter and masses of several turbines.

Figure 1.14 shows relevant details from the manufacturer catalogue that are required for design.

1.5 Foundation Types

Foundations constitute the most important design consideration and often determines the financial viability of a project. Typically, foundations cost 25–34% of the whole

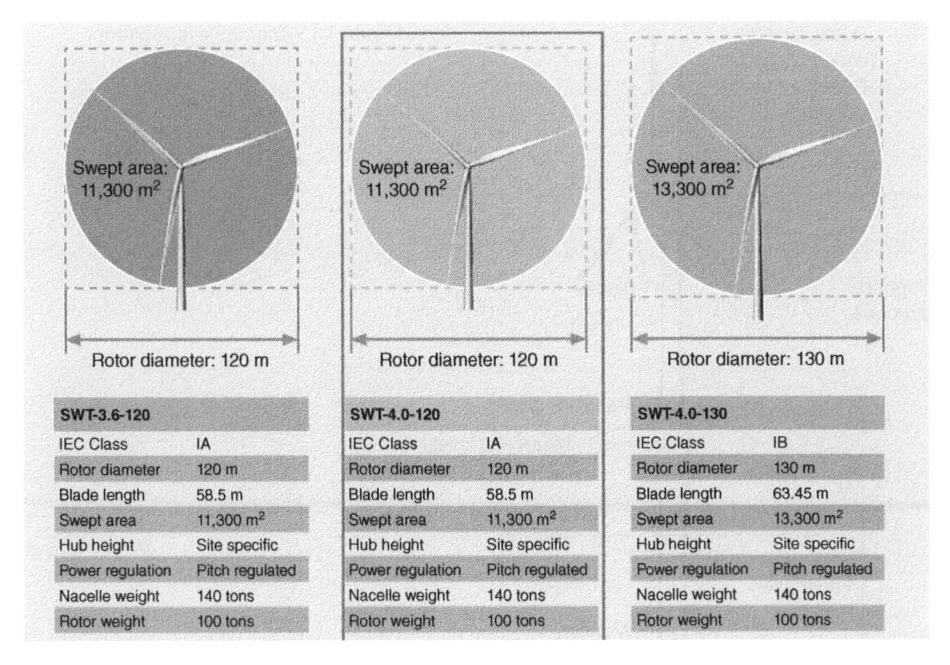

SWT-3.6-120	
IEC Class	IA
Rotor diameter	120 m
Blade length	58.5 m
Swept area	11,300 m^2
Hub height	Site specific
Power regulation	Pitch regulated
Nacelle weight	140 tons
Rotor weight	100 tons

SWT-4.0-120	
IEC Class	IA
Rotor diameter	120 m
Blade length	58.5 m
Swept area	11,300 m^2
Hub height	Site specific
Power regulation	Pitch regulated
Nacelle weight	140 tons
Rotor weight	100 tons

SWT-4.0-130	
IEC Class	IB
Rotor diameter	130 m
Blade length	63.45 m
Swept area	13,300 m^2
Hub height	Site specific
Power regulation	Pitch regulated
Nacelle weight	140 tons
Rotor weight	100 tons

Figure 1.14 Data for turbines required for design purposes (Source: Reproduced with permission from Siemens Brochure).

project and there are attempts to get the costs down. Many aspects must be considered while choosing and designing the foundation for a particular site. They include: ease to install under most weather conditions, varying seabed conditions, aspects of installation including vessels and equipment required, and local regulations concerning the environment (noise). Figure 1.15 shows a schematic diagram of a wind turbines supported on a large-diameter column inserted deep into the ground (known as monopile). This is the most used foundation so far in the offshore wind industry due to its simplicity.

Figure 1.16 shows the various types of foundations commonly used today for different depths of water. Monopiles (Figure 1.16c), gravity-based foundations (Figure 1.16b), and suction caissons (Figure 1.16a) are currently being used or considered for water depths of about 30 m. For water depth between 30 and 60 m, jackets or seabed frame structures supported on piles or caissons are either used or planned. A floating system is being considered for deeper waters, typically more than 60 m. However, selection of foundations depends on seabed, site conditions, turbine and loading characteristics, and economics – not always on the water depth.

The substructure can be classified into two types, as described in the flowchart (Figure 1.17):

1. Grounded system or fixed structure where the structure is anchored to the seabed. Grounded system can be further subdivided in two types in the terminology of conventional foundation/geotechnical engineering: shallow foundation (gravity-based solutions and suction caisson) and deep foundation.
2. Floating system where the system is allowed to float and is anchored to the seabed by a mooring system. Floating systems have certain ecological advantages in the sense

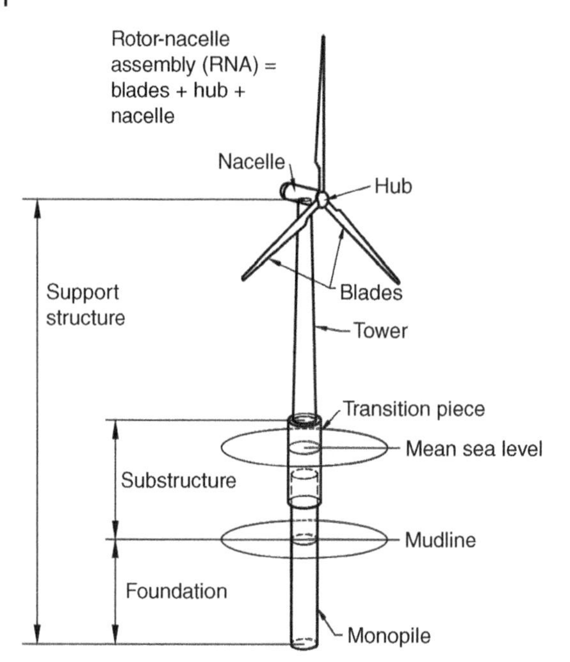

Figure 1.15 Monopile foundation.

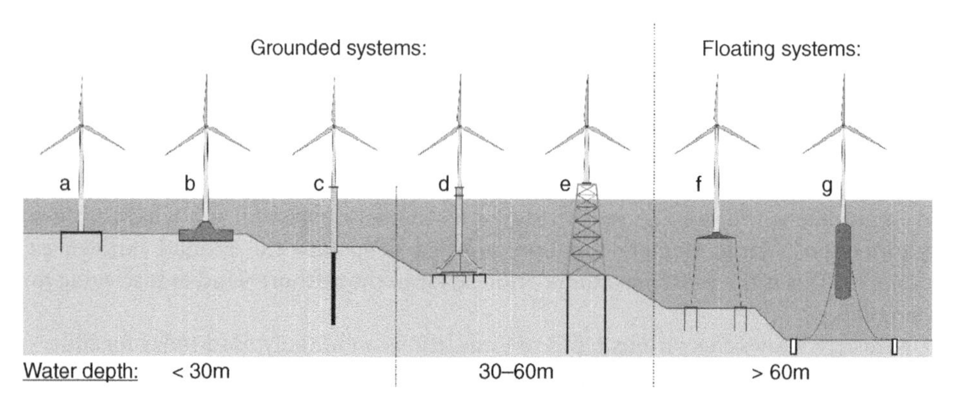

Figure 1.16 (a) Bucket/suction caisson: (b) gravity-based; (c) monopile; (d) tripod on bucket/suction caisson; (e) jacket/lattice structure; (f) tension leg platform; and (g) spar buoy floating concept.

that the foundations leave a very low seabed footprint, they are easy to decommission, and maintenance is easier as the system can be de-anchored and floated out to the harbour.

1.5.1 Gravity-Based Foundation System

The gravity-based foundation is designed to avoid uplift or overturning, i.e. no tensile load between the support structure and the seabed. This is achieved by providing adequate dead load to stabilise the structure under the action of overturning moments. If the dead loads from the support structure and the superstructure (tower + RNA) is not sufficient, additional ballast will be necessary. The ballast consists of rock, iron ore, or

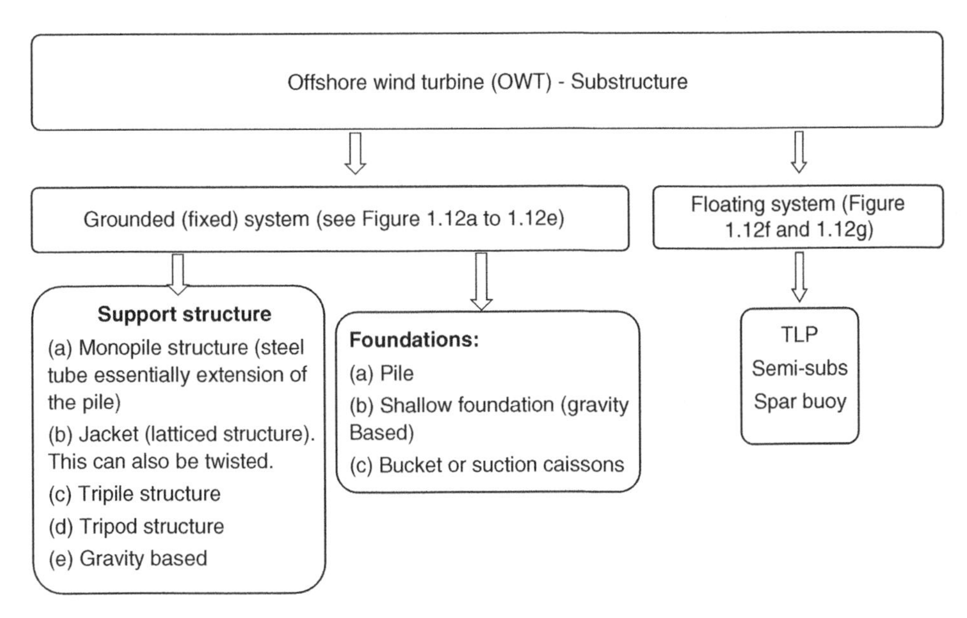

Figure 1.17 Flowchart classifying the substructure.

concrete. Installation of these foundations often requires seabed preparation to avoid inclination. The gravity-based structures in most cases are constructed *in-situ* concrete or with precast concrete units. The gravity-based concept can be classified into the two types, depending on the method of transportation and installation:

1. *Crane-free solution, also known as 'float-out and sink' solution.* These types of foundations will be floated (either self-buoyant or with some mechanism) and towed to the offshore site. At the site, the foundation will be filled with ballast, causing it to sink to the seabed. This can be an attractive solution for sites having very hard or rocky soil conditions. This operation does not require a crane and thus is known as a crane-free solution.
2. *Crane-assisted solution.* In these types, the foundation does have the capacity to float and is therefore towed to the site on-board a vessel. The foundation is then lowered to the seabed using cranes. An example is the Thornton Bank, shown in Figure 1.18b, where the shape of the gravity-based substructure is compared to a champagne bottle. Table 1.3 provides some examples of gravity-based foundations.

1.5.1.1 Suction Caissons or Suction Buckets

Suction buckets (sometimes referred to as suction caissons) are similar in appearance to a gravity-based foundation but with long skirts around the perimeter. Essentially, they are hybrid foundations taking design aspects from both shallow and pile foundation arrangements. A caisson consists of a ridged circular lid with a thin tubular skirt of finite length extending below, giving it the appearance of a bucket. Typically, such foundations will have a diameter-to-length ratio (D/Z) of around 1, making them significantly shorter than a pile but deeper than a shallow foundation. A sketch of a suction caisson with terminology can be seen in Figure 1.19a. Suction caissons themselves

Figure 1.18 (a) Example of GBS (transportation for Karehamn wind farm – Sweden); Courtesy: Jan DE Nul Group. (b) GBS from Thornton Bank project; (c) GBS – Strabag concept; and (d) foundation for Middelgrunden wind farm (Denmark) – shallow gravity-based foundation.

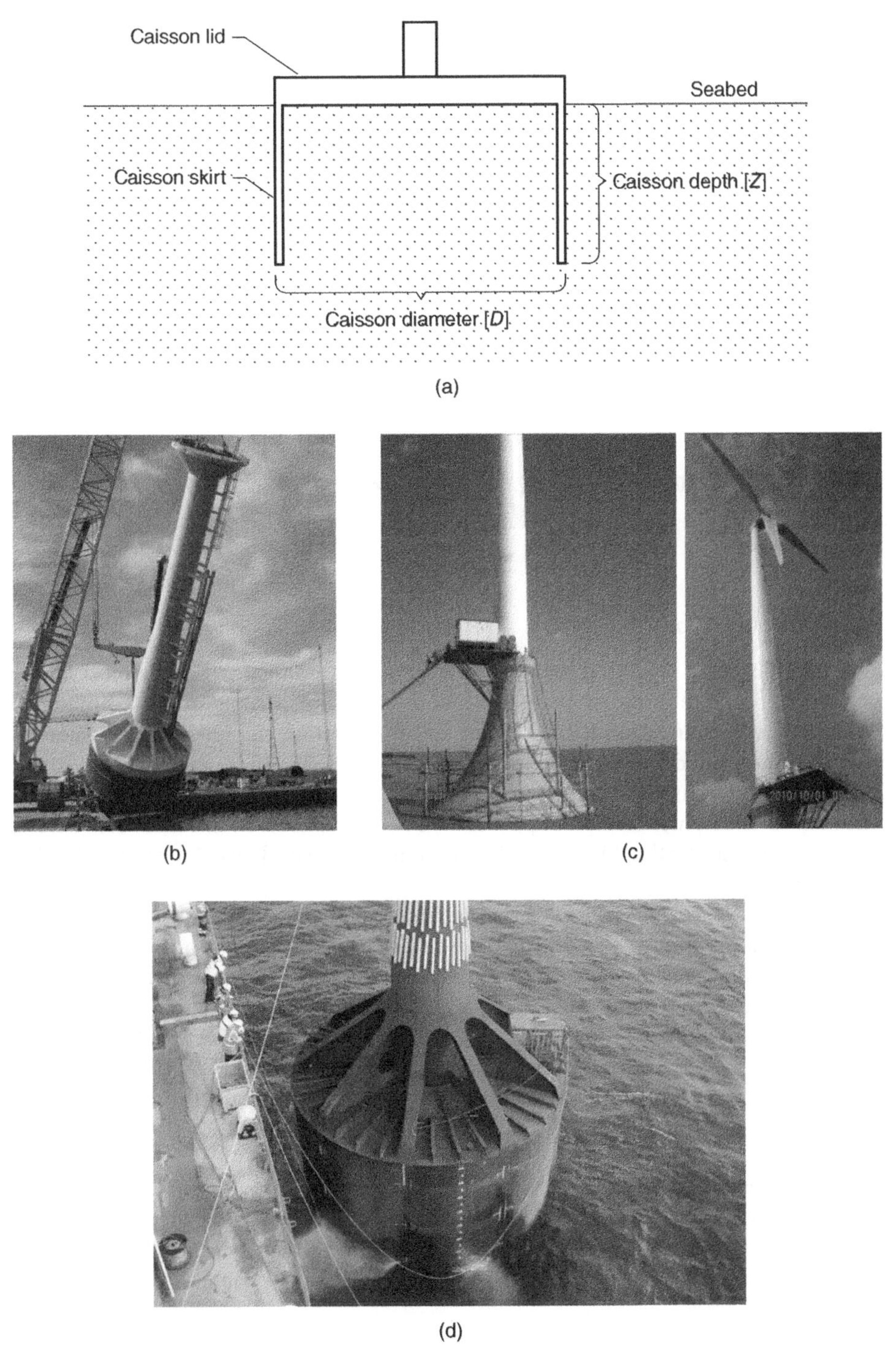

Figure 1.19 (a) Typical layout of a suction caisson foundation; (b) Horns Rev 2 being manoeuvred from the fabrication site; (c) Qidong Sea offshore wind turbine; and (d) installation of the Dogger Bank met mast caisson, OffshoreWIND.biz, (2013). Photo Courtesy: Prof Lizhong Wang (Zhejiang University).

are a fairly recent development in the offshore industry. Caissons first came into use around 30 years ago as a foundation structure for offshore oil and gas production platforms.

1.5.1.2 Case Study: Use of Bucket Foundation in the Qidong Sea (Jiangsu Province, China)

The first suction bucket supporting offshore wind turbine was installed in the Qidong Sea area, Jiangsu province (China), in October 2010, see Figure 1.19c. The whole structure was designed to support a 2.5 MW wind turbine in approximately 6 m of water. The bucket is 30 m in diameter and 7 m deep. Fabrication of the caisson took around two months and the entire foundation structure was towed to site and installed with the aid of gas jetting around the caisson skirt. By constructing the caisson out of concrete, the fabrication cost of the structure was reduced in comparison to that of a similar steel design.

1.5.1.3 Dogger Bank Met Mast Supported on Suction Caisson

Suction caissons were used to support met mast at the Dogger Bank offshore wind farm; see Figure 1.19d.

1.5.2 Pile Foundations

Single large-diameter steel tubular piles, also known as monopiles, are the most common form of foundation for supporting offshore wind turbines. Figure 1.20 shows the monopile type of foundation that is essentially a large steel pile (3–7 m in diameter) driven into the seabed with typical penetration depth of 25–40 m. A steel tube, commonly called the transition piece (TP), is connected to the steel pile and the tower is attached to it. The transition piece supports the boat landings and ladders used for entering the turbine. Currently, this type of foundation is extensively used for water depths up to 25–30 m.

These foundations can be reliability driven into the seabed using a steam or hydraulically driven hammer, and the practice is standardised due to offshore oil and gas industry. The handling and driving of these foundations require the use of either floating vessels or jack-up, which must be equipped with large cranes, suite of hammers, and drilling equipment. If the ground profile at the site contains stiff clay or rock, drive-drill-drive procedures may need to be adopted. Pile-driving results in noise and vibrations. Therefore, the turbine (nacelle and rotor) is always installed after the piling is carried
out.

Often a group of small-diameter piles can be used to support a wind turbine. Small-diameter piles are also used to support a jacket, which, in turn, supports a tower and the WTG; see Figure 1.20b. Figure 1.20c shows the Shanghai Donghai Bay project where a group of piles is used – known as HRPC (high-rise pile cap).

ASIDE

Noise emission of conventional pile driving may often exceed acceptable levels in certain locations. Vibro driving method is suitable in such cases.

Figure 1.20 (a) Large diameter monopile; (b) WTG structure supported on a jacket. Jackup vessel is used for installation; (c) a group of piles to support a wind turbine; and (d) photo of 1.20c and (e) installation of the turbine (Chinese case study). Photo Courtesy Figure 1.20a, b): Dr Matthijs Soede [European Commission].

1.5.3 Seabed Frame or Jacket Supported on Pile or Caissons

Often, a seabed frame or a jacket supported on piles or caissons can act as a support structure and can be classified as multipods. Multipods have more than one point of

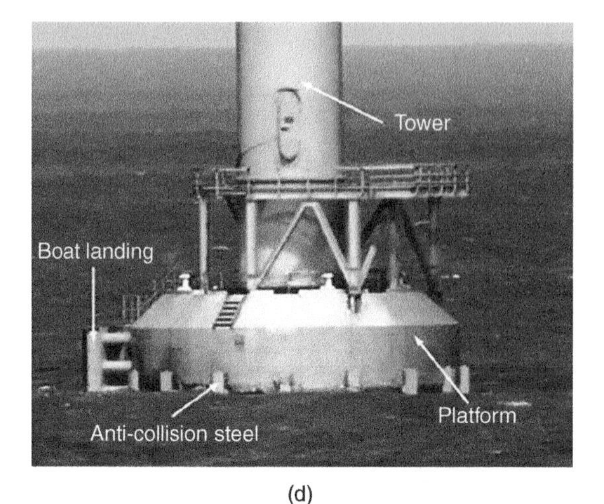

(d)

(e)

Figure 1.20 *(Continued)*

contact between the foundation and the soil. Soil-embedded elements may include flexible piles, gravity bases, and suction caissons.

Example 1.2 ***Jacket Supported on Multiple Suction Caissons*** Jackets or seabed frames supported on multiple shallow foundations are currently being installed to support offshore wind turbines in deep waters ranging between 23 and 60 m; see, for example, Borkum Riffgrund 1 (Germany, water depth 23–29 m), Alpha Ventus Offshore (Germany, water depth 28–30 m), and Aberdeen Offshore wind farm (Scotland, water depth 20–30 m). The jackets are typically designed as three- or four-legged and are supported on shallow foundations (suction caissons or slender piles). The height of the jacket currently in use is between 30 and 35 m, where it is governed by water depth and wave height. However, it is expected that future offshore developments will see jacket heights up to 65 m to support larger turbines (12–20 MW) in deeper waters.

Figures 1.21–1.25 show a schematic of a three-legged jacket inspired by some recent offshore developments.

1.5.4 Floating Turbine System

The floating system can be classified into three main types (see Figure 1.26):

1. *Mooring stabilised TLP (tension leg platform) concept*. This type of system is stabilised with tensioned mooring and anchored to the seabed for buoyancy and stability; see Figure 1.26a.
2. *Ballast stabilised Spar buoy concept with or without motion control stabiliser*. This type of system will have a relatively deep cylindrical base providing the ballast, whereby the lower part of the structure is much heavier that the upper part. This

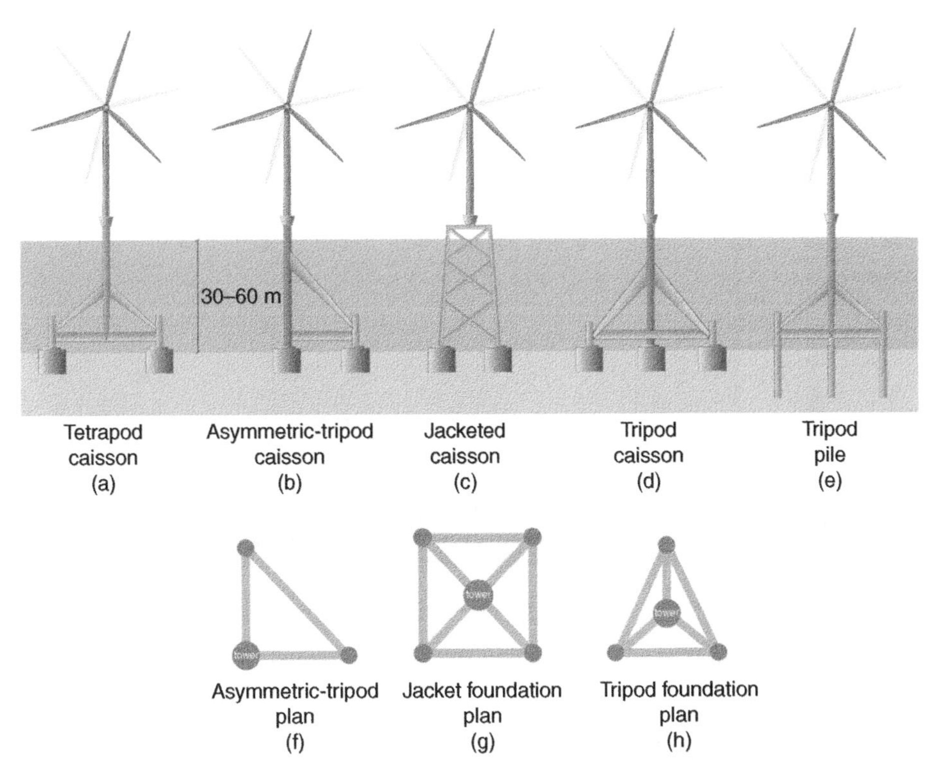

Figure 1.21 Multipod foundations and various proposed and existing multi-foundation arrangements to support WTG; (a) tetrapod substructure supported by four suction caisson foundations; (b) asymmetric tripod substructure supported by three suction caissons foundations; (c) jacket substructure supported by four suction caisson foundations; (d) symmetric tripod substructure supported by three suction caisson foundations; (e) tri-pile substructure and foundation; (f) plan view of an asymmetric tripod substructure; (g) plan view of a jacket substructure; and (h) plan view of a symmetric tripod substructure.

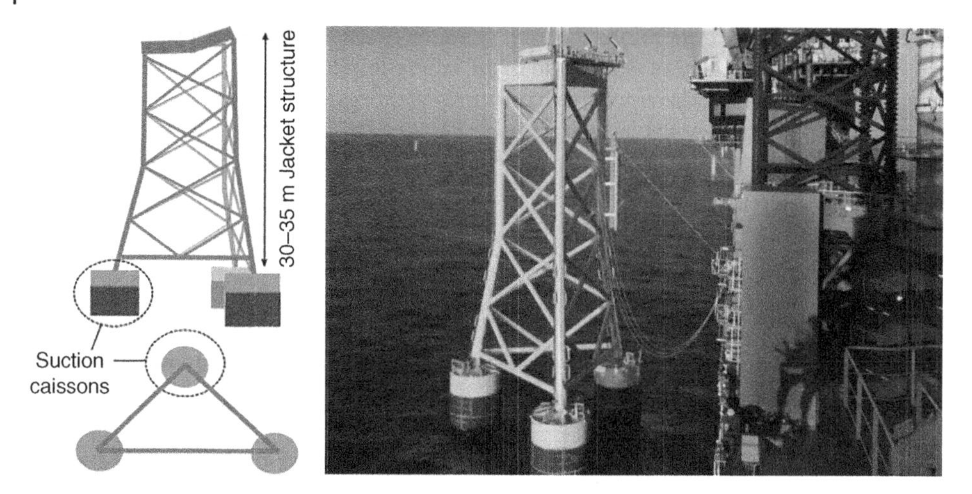

Figure 1.22 Schematic of a three-legged jacket supported on a suction caisson.

Figure 1.23 Different type of support structure for a wind farm in China.

would raise the centre of buoyancy about the center of gravity of the system. While these are simple structures having a low capex cost, they need a deeper draft (i.e. deeper water) and are not feasible in shallow water. Motion stabilisers can be used to reduce the overall tilt of the system, see Figure 1.26b.

3. *Buoyancy stabilised semi-submersible.* This concept is a combination of ballasting and tensioning principle and consumes lot of steel; see Figure 1.26c.

There are varieties of anchors that can be used to moor the floating system, and they can be classified into surface anchors and embedded anchors. An example of surface anchors is a large, heavy box containing rocks or iron ore, and the holding capacity depends on the weight of the anchor itself and the friction between the base of the

Figure 1.24 Support structure.

anchor and the seabed. On the other hand, examples of embedded anchors are anchor piles, such as shown in Figure 1.26c. Figure 1.26b,c are floating wind turbine concepts suitable for deeper waters with Figures 1.26a and 1.27a showing the floating concept (semi-sub) implemented in a wind farm in Japan (offshore Fukushima). By contrast, Figure 1.27b shows the Hywind concept (spar concept). Figure 1.27c is a TLP concept developed by GICON.

1.6 Foundations in the Future

Foundations typically cost 15–35% of an overall offshore wind farm project, and in order to reduce the LCOE, new innovative foundations are constantly being researched. However, before any new type of foundation can actually be used in a project, a thorough technology review is often carried out to derisk it.

The European Commission defines this thorough technology readiness level (TRL) numbering, starting from 1 to 9; see Table 1.4 for different stages of the process. One of the early works that needs to be carried out is technology validation in the laboratory environment (TRL 4). In this context of foundations, it would mean carrying out tests to verify various performance criteria, including the long-term performance. It must be realised that it is very expensive and operationally challenging to validate in a relevant environment, i.e. in an offshore environment, and therefore, laboratory-based evaluation has to be robust so as to justify the next stages of investment.

Figure 1.25 Jacket in Belgium water. [Photo Courtesy: Dr Matthias.]

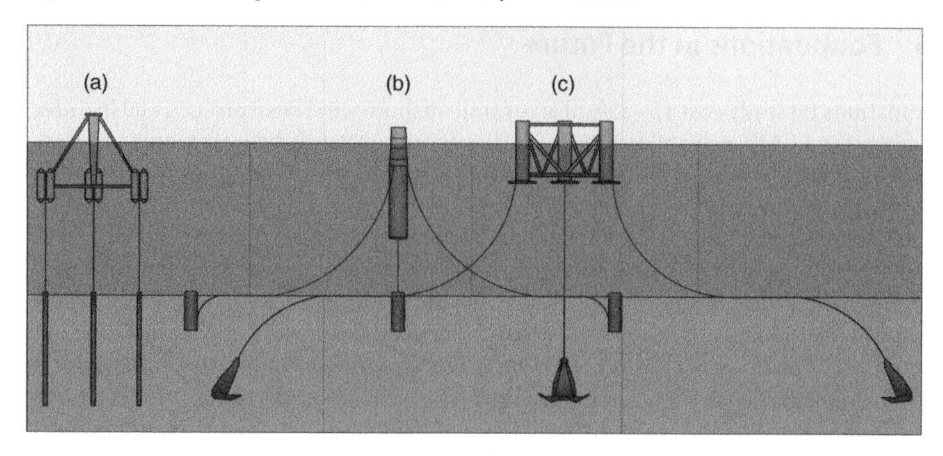

Figure 1.26 Three main types of floating system to support WTG (wind turbine generator). [Photo Courtesy: Dr Lazlo Arany]

Figure 1.27 (a) Semi-submersible foundation for offshore Fukushima (Japan); (b) details of Hywind wind turbine installation, which is spar buoy – floating system; and (c) example of TLP design. [Photo Courtsey: Fukushima Offshore Wind Consortium and GICON.]

(c)

Figure 1.27 *(Continued)*

Table 1.4 Definition of TRL.

TRL level as European Commission
TRL 1: Basic principles verified
TRL 2: Technology concept formulated
TRL 3: Experimental proof of concept
TRL 4: Technology validated in lab
TRL 5: Technology validated in relevant environment
TRL 6: Technology demonstrated in relevant environment
TRL 7: System prototype demonstration in operational environment
TRL 8: System complete and qualified
TRL 9: Actual system proven in operational environment

To reduce the cost of offshore wind power, the Carbon Trust initiated a competition called 'The Offshore Wind Accelerator'. The aim of this project was to reduce the cost of offshore wind by 10% by 2015. This cost reduction was to be achieved in a number of ways, including more cost-effective foundation arrangements. A number of novel foundation solutions were proposed, and after evaluation and investigation, these were narrowed down to four arrangements for further development, as shown in Figures 1.28 and 1.29. These solutions consist of:

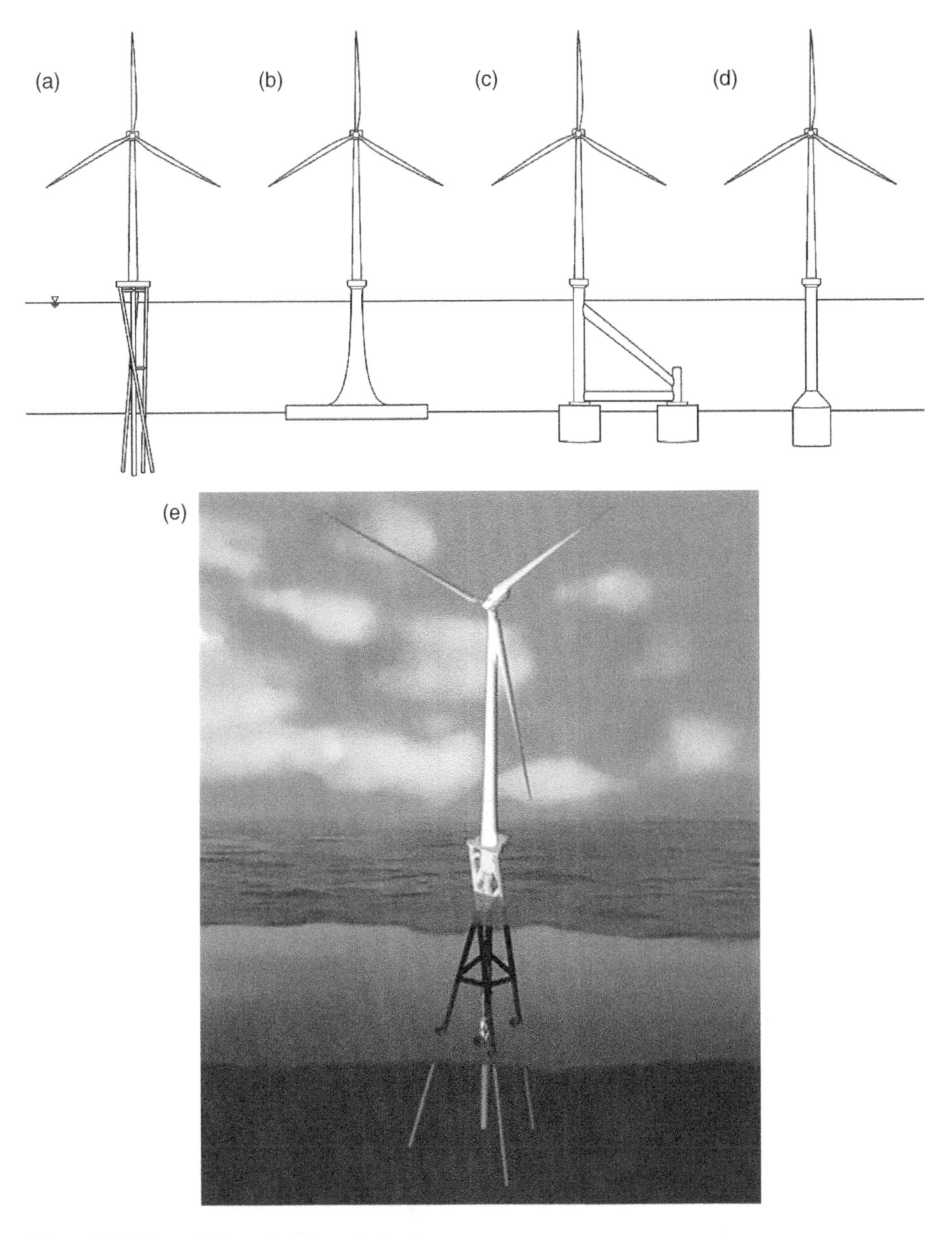

Figure 1.28 Foundations for future: (a to e).

1. Keystone innovative jacket, also known as twisted jacket or inward battered guided structure (IBGS), see Figure 1.28 (a) and 1.28 (e)
2. Gifford/BMT/Freyssinet gravity structure see 1.28 (b)
3. SPT Offshore & Wood Group tri-bucket (see Figure 1.30 for a schematic view) and 1.28 (c)
4. Universal foundations suction bucket monopile, see 1.28 (d)

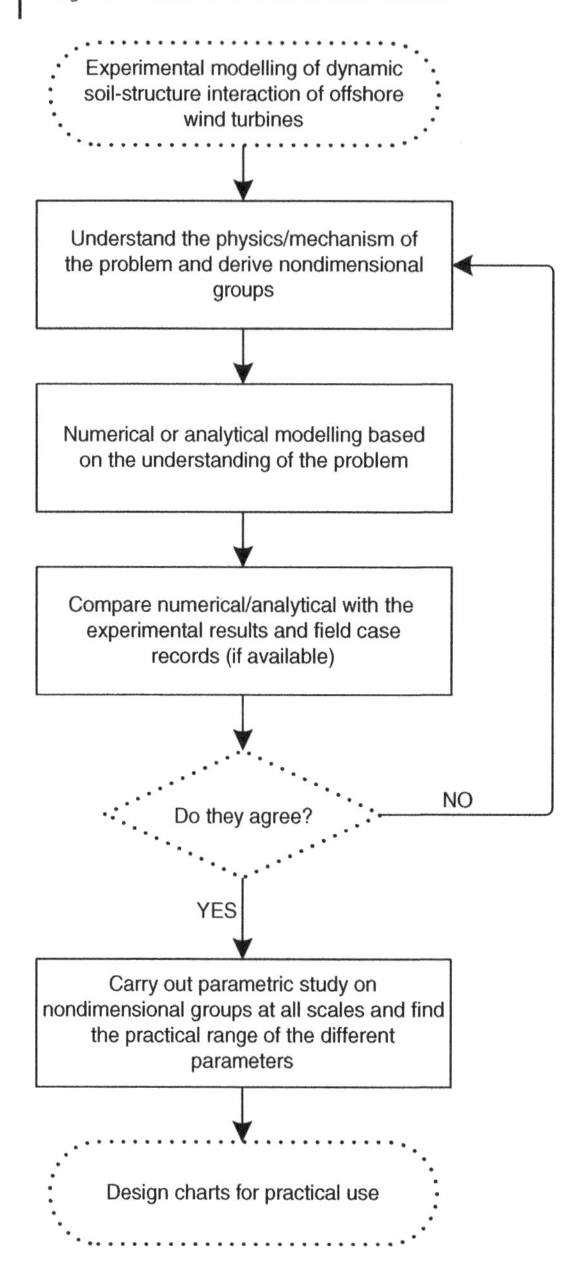

Figure 1.29 Flowchart showing the usefulness of scaled laboratory testing.

ASIDE

Current research activities are focussed on the use of screw piles and hybrid foundations (combination of shallow and deep foundations) as cost-effective solutions.

For TRL 4 level work, small-scale tests are carried out in a well-controlled laboratory. For offshore wind turbine foundations, tests must be carried out for the following purposes:

Figure 1.30 Tribucket also known as SIWT (self-installing wind turbine): (a) artistic impression; (b) Scaled model in a sand test bed; and (c) scaled model in a clay test bed.

1. Confirm and validate the mechanism of load transfer from superstructure to the ground through the foundation element. This is very important for a new concept of a foundation or a connection. For example, monopile type of foundation transfers load through overturning moments. On the other hand, jacket-type structures transfer loads through axial push–pull action. Load transfer are discussed in Chapter 2.
2. Find out the modes of vibration of the structures. These are carried out through free vibration/perturbation tests or white noise testing. In Chapters 2 and 3, it will be shown that the natural frequency of wind turbine structures is very close to the forcing frequencies due to wave and 1P (rotor frequency), 3P (blade passing frequency, discussed in Chapter 2) and as a result, these structures are sensitive to dynamics. Modes of vibration can strongly influence the foundation design, fatigue life, and wear and tear of the mechanical components in the RNA.
3. Offshore wind turbine foundations are subjected to hundreds of millions of cycles load, which can be cyclic or dynamic in nature. Scaled model tests can reveal the expected trends of behaviour of the foundations due to cyclic and dynamic soil-structure interaction. One of the uncertainties is the long-term nonrecoverable tilt of the foundation. Excessive tilt may lead to shutdown of the turbine.
4. The loads on a foundation are very complex and can be one-way cyclic or two-cyclic. The long-term effects of such one-way cyclic loading can be identified through scaled model tests.
5. Due to dynamic sensitivity, offshore wind turbines need damping. Trends and sources of damping can be identified through carefully designed scale model tests.
6. To identify any 'unknown-unknowns' for the problem under investigation through the tests. Experimental observations often unearth new design considerations.

1.6.1 Scaled Model tests

Behaviour of offshore wind turbines involves complex dynamic wind–wave–foundation–structure interaction and the control system (Section 1.4) adds further interaction. In wind tunnel tests, the aerodynamic effects are modelled efficiently and correctly (as far as practicable) and as a result the loads on the blade and towers can be simulated. On the other hand, in the wave tank the hydrodynamic loads on the substructure and scouring on the foundation can be modelled. In a geotechnical centrifuge, one can model the stress level in the soil, but the model package is spun at a high RPM, which will bring in unwanted vibrations in the small-scale model.

Ideally, a tiny wind tunnel, together with a tiny wave tank onboard a geotechnical centrifuge, may serve the purpose, but this is not viable and will add more uncertainty to the models than it tries to unearth. A model need not be more complex, however, and often simple experiments can unearth the governing laws. In every type of experiments, there will be cases where the scaling laws/similitude relationships will not be satisfied (rather violated), and these must be recognised while analysing the test results. Therefore, results of scaled-model tests for offshore wind turbine problems should not be extrapolated for prototype prediction through scaling factors. The tests must be carried out to identify trends and behaviours, and upscaling must be carried out through laws of physics, numerically or analytically. Figure 1.30 shows a suggested method for such purpose. It shows how small-scale tests can be used for developing design methods.

Derivation of scaling laws for model tests for monopiles and multipod-supporting wind turbines can be found in Bhattacharya et al. (2011b, 2013a,b). Discussion on the model testing and its applicability can be found in Bhattacharya et al. (2018).

1.6.2 Case Study of a Model Tests for Initial TRL Level (3–4)

Bhattacharya et al. (2013a,b) reported some aspects of TRL work for the foundation shown in Figures 1.28c and 1.30. Essentially, this is a seabed frame supported on three suction caissons. Three scaled models (1 : 100, 1 : 150, and 1 : 200) were constructed and tested in two types of ground (sand and clay). Free vibration tests were carried out to identify the modes of vibration and two closely spaced peaks were observed as shown in Figures 1.31a and 1.32. This was repetitive in all three models, which confirms a

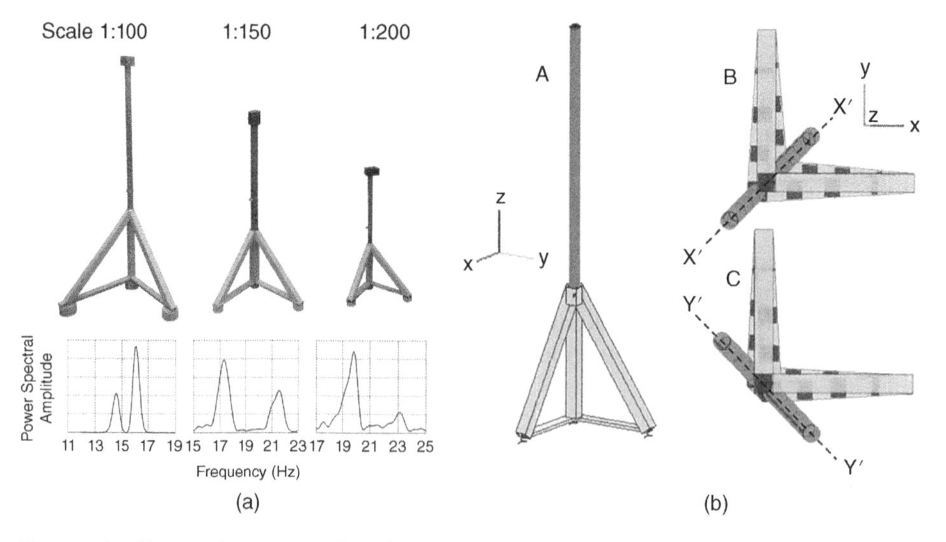

Figure 1.31 Observed experimental results and numerical modelling of the problem: (a) two closely spaced peaks suggesting rocking modes of vibration and (b) simulation of the problem numerically showing the modes of vibration.

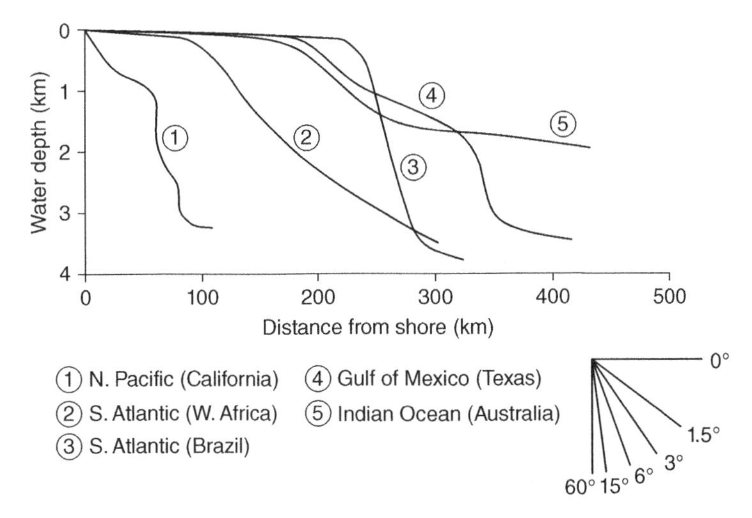

Figure 1.32 Water depth offshore. Courtesy: Randolph and Gourvenec (2011).

phenemenon later identified as rocking modes of vibration. This was later verified and validated through numerical modelling, which further showed that these foundation systems will vibrate about two principal axes due to the variability of the ground. For some foundation arrangement (symmetric), these two peaks will converge with cycles of loading due to ground-reaching steady-state behaviour. For asymmetric system, these peaks may not converge, and later chapters discuss how this will have negative outcome on the fatigue life. The readers are referred to Bhattacharya et al. (2013b) for further details on this TRL study.

1.7 On the Choice of Foundations for a Site

The choice of foundation will depend on the following: site condition, fabrication, installation, operation and maintenance, decommissioning, and economics. Bhattacharya (2017) defined an ideal foundation as follows:

1. A foundation that is capacity or 'rated power' (i.e. 5 or 8 MW rated power) specific but not turbine manufacturer specific. In other words, a foundation designed to support 5 MW turbine but can support turbines of any make. There are other advantages in the sense that turbines can be easily replaced.
2. Installation of foundation is not weather sensitive, i.e. not dependent of having a calm sea or a particular wind condition. The installation of the first offshore wind farm in the United States took more time due to the unavailability of a suitable weather window.
3. Low maintenance and operational costs, i.e. needs least amount of inspection. For example, a jacket-type foundation needs inspection at the weld joints.

It is economical for a wind farm to have a large number of turbines due to economies of scales, but this requires a large area. If the continental shelf is very steep, grounded (fixed) turbines are not economically viable; a floating system is desirable. Figure 1.32 shows water depth plotted against distance from the shore for some oceans.

Monopiles are currently preferred for water depths up to 30 m. The simple geometry allows automation of the manufacturing and fabrication process. The typical cost (2018 European Steel Price) of monopile steel is €2 kg^{-1} to manufacture. Welding can be carried out by robots and installation is simpler. However, if the diameter become larger (known as XL or XXL piles, which are over 8 m in diameter and weigh 1200 t), the transportation and installation becomes challenging and a limited number of installation contractors can carry out the work. Innovations are underway to install large-diameter piles using the vibro method, where a foundation is installed through vibration of the soil, effectively liquefying the soil around it.

An alternative to large-diameter monopiles is a three- or four-legged jacket on small diameter piles. Steel for jackets cost around €5 kg^{-1} to manufacture, which is more than double that of monopiles due to many tubular joints, and they are often welded manually.

Gravity-based foundations are cheaper to manufacture as compared to steel, but they require a large fabrication yard and storing area. As concrete foundations will be much heavier than equivalent steel foundations, large crane, and vessels are required to install. For an offshore site where the surface ground is rock, a gravity-based structure will be a preferred choice; see, for example, the French waters.

1.8 General Arrangement of a Wind Farm

Figure 1.33 shows the components of a typical wind farm. The turbines in a wind farm are connected by inter-turbine cables (electrical collection system) and are connected to the offshore substation. There are export cables from offshore to the shore. Figure 1.34a shows the photograph of a wind farm with many wind turbines and a substation. Figure 1.34b shows the details of the substructure of a monopile with J tubes for the electrical collection system. Figure 1.35 shows the photograph of a jacket-supported substation. Figure 1.36 shows the plan of Horns Rev wind farm.

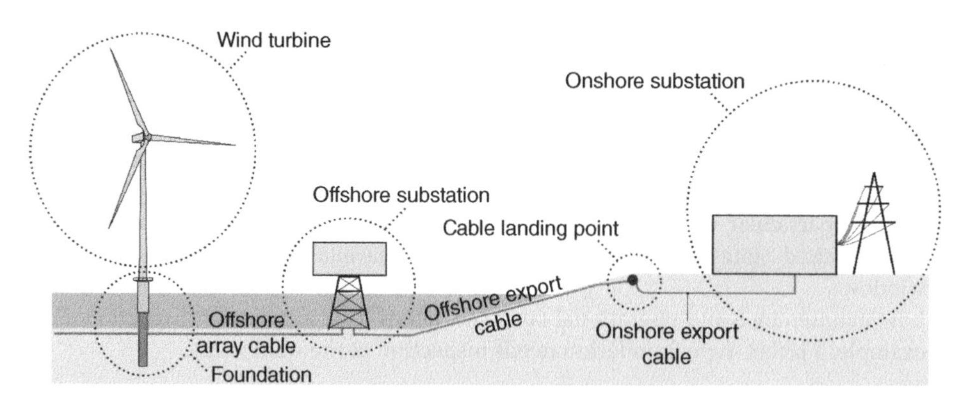

Figure 1.33 Overview of a wind farm.

(a) (b)

Figure 1.34 (a) Wind farm with offshore substation and (b) turbine J tubes for inter-turbine cables and electrical collection system.

Figure 1.35 Offshore substation.

1.8.1 Site Layout, Spacing of Turbines, and Geology of the Site

Wind turbines in a wind farm are spaced to maximise the amount of energy that can be generated without substantially increasing the CAPEX (Capital expenditure, i.e. upfront cost). If the farm is much spread out, i.e. large spacing of the turbines, the inter-array cable length will increase. This spacing is therefore an optimization problem between

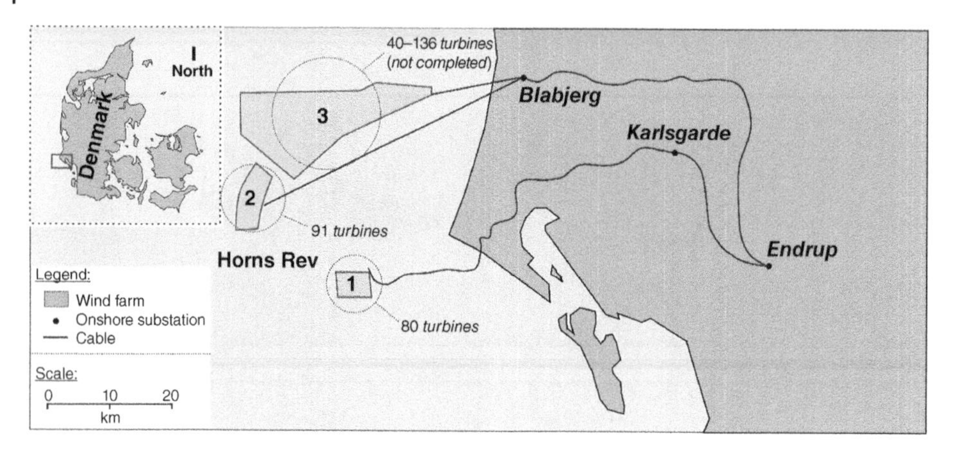

Figure 1.36 Aerial layout of a wind farm.

compactness of the wind farm (which minimises the CAPEX cost due to subsea cables) and the adequate separations between turbines so as to minimise the energy loss due to wind shadowing from upstream turbines.

Figure 1.37 shows the aerial photo of wake turbulence behind individual wind turbines that can be seen in the fog of the Horns Rev wind farm off the Western coast of Denmark (Photo credit Vattenfall Wind Power, Denmark).

Figure 1.37 Wake turbulence. [Photo Credit: Vattenfall; Photographer: Christian Steiness.]

The geometric layout of a wind farm can be a single line of array, or a square or a rectangle configuration. Due to advanced methods for optimisation having different constraints as well as site conditions, different layout patterns are increasingly being used. Typically, the spacing between turbines is equivalent to 3–10 times the rotor diameter and depends on the prevailing wind direction. The spacing should be larger than 3–4 times the rotor diameter perpendicular to the prevailing wind direction and 8–10 times the diameters for direction parallel to the wind direction. For example, for a prevailing southwesterly wind direction (which is typical of Northern Ireland), a possible site layout for a wind farm located in the area is shown in Figure 1.38. The spacing along the wind direction is kept at 6 times the rotor diameter (6D) but for across the wind, the spacing can be kept bit lower (4D). For Nordsee Ost wind farm, the layout has more turbines in the first row.

Due to the large spacing of the turbines (typically 800–1200 m apart), a small to medium size wind farm would extend over a substantial area. A typical size for a modern-day wind farm is 20 km × 6.5 km; see Figure 1.39 for the layout of the Sandbank wind farm of German North Sea. Due to the large coverage of area for a wind farm, there may be significant variation in the geological and subsurface conditions, as well as practical restraints. Examples include a sudden drop in the sea floor causing change in water depth, paleo channels, change in ground stratification, submarine slopes, presence of foreign objects such as shipwrecks, location of important utility lines (gas pipeline, fibre optic cables), etc. Detailed site investigation programmes consisting of geotechnical, geophysical are carried out to establish a 3D geological model, which often dictates the layout of the wind farm.

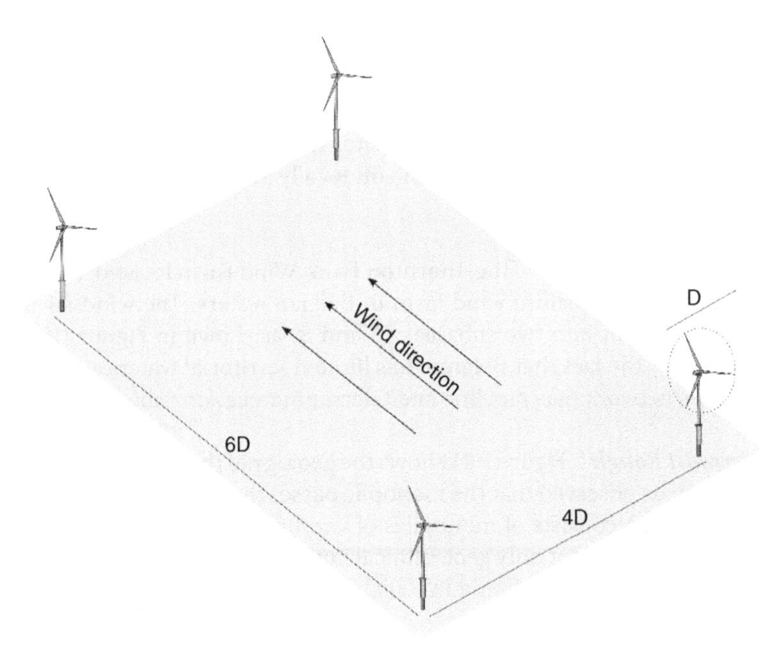

Figure 1.38 Spacing of turbines.

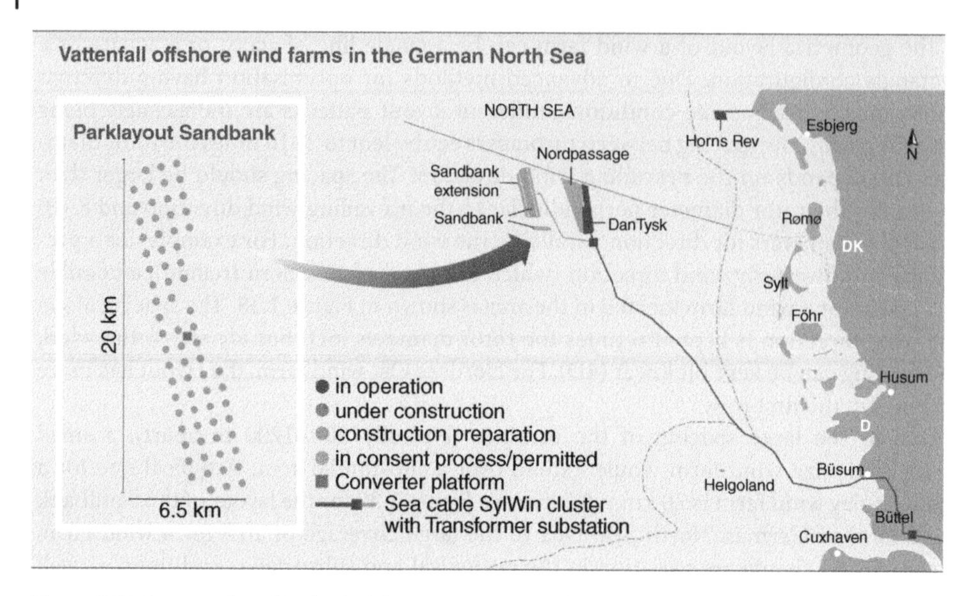

Figure 1.39 Layout of sandbank wind farm.

1.8.2 Economy of Scales for Foundation

Due to vast size of the wind farm, there will be varying seabed condition including water depth and distance from the shore. As a result, the loads on the foundations will change and ideally the best design will be to design each foundation individually, which will give rise to a customised foundation design for each turbine location. However, from an economic point of view, it is desirable to have few foundation types so that the overall economy is achieved and the process of fabrication and installation can be carried out efficiently using same installation vessel. Most North European developers prefer one type of foundation (either monopiles or jackets) in a site. This consideration often dictates the layout of the farm to avoid deeper water or soft locally available mud. Few case studies are discussed here.

Example 1.3 ***Thornton Bank Project*** The Thornton Bank Wind Farm located 30 km off the Belgian coast is the first offshore wind farm in Belgian waters. The wind farm will have 60 turbines and is split into two subareas (A and B) as shown in Figure 1.40. The site layout is dictated by the fact that Belgium has limited territorial waters and the location of major utilities network (gas pipeline and telecommunication cable).

Example 1.4 ***Westermost Rough*** Figure 1.41 shows the geology of the site at the westermost rough where it can be observed that the monopile passes through different geology. Therefore, the foundation consists of monopiles of varying length, wall thickness, and diameter. This also shows that not only geotechnical but also geological study needs to be carried out. Further details can be found in Kallehave et al. (2014).

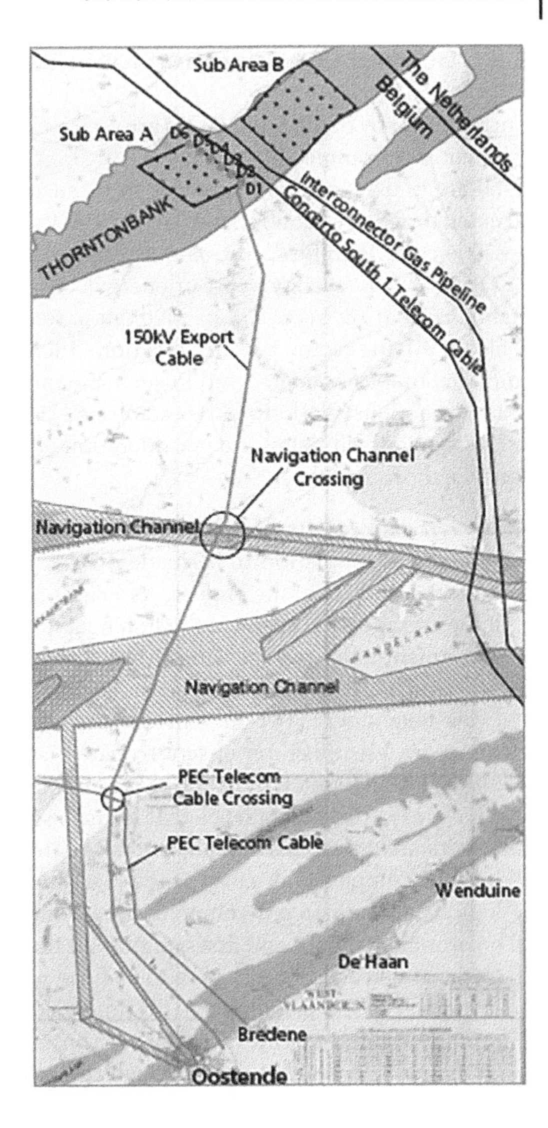

Figure 1.40 Thornton Bank wind farm. Courtesy: Thornton Bank Wind Farm developer.

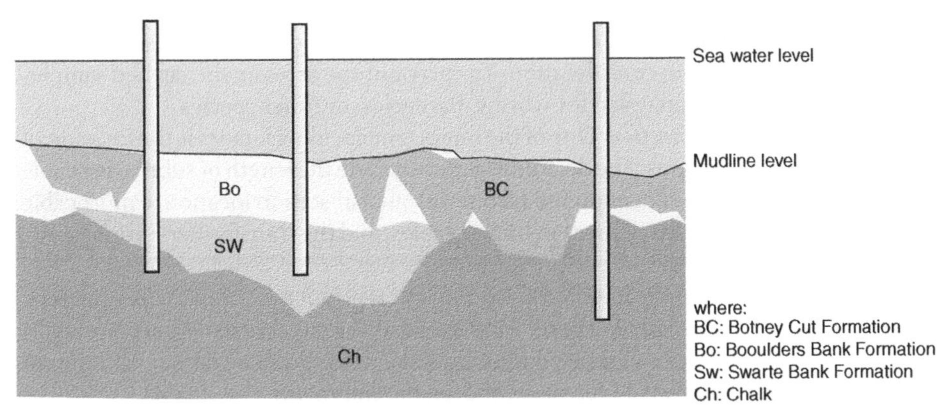

Figure 1.41 Monopile foundation in westermost rough.

1.9 General Consideration for Site Selection

Currently, many of the wind farms are operating from the subsidy provided by the government. For example, in the United Kingdom, schemes such as contract for difference (cfD) are in use. However, in order to be sustainable, large wind farms must be constructed to achieve economies of scales with the aim to produce electricity at the lowest possible cost. Therefore, cost of electricity from different sources are compared using LCOE or SCOE (society's cost of energy). As much of the installation, operation, and maintenance (O & M) will be carried out in rougher waters, time in construction (TIC) is also a driving factor for site selection. Therefore, every cost increasing part of the construction has to be lowered in such a manner that an optimal method for the construction and installation will be established.

This section will detail the considerations for choosing a particular site. The main considerations are:

1. *Wind resources.* A thorough knowledge of wind resources in an area is fundamental, as it allows estimation to be made on the wind farm productive and therefore the financial viability of the project. As a rule of thumb, a project is not financially viable if the average wind speed at the hub height is below 6 m s^{-1} and it is considered safe investment if the average wind speed at the hub is more than 8.5 m s^{-1}.
2. *Marine aspects.* Marine aspects would include water depth, wave spectrum at the site (wave height, wave period), current and tide data, exposure to waves and sediment transport, identification of scour-related issues, and if scour protection is needed. Often, installation of foundations creates obstacles in the local flow pattern of water, which may create turbulence that leads to scour.
3. *Environmental impact.* For all wind farm, an environmental impact assessment (EIA) must be completed as a part of the planning process and it covers the physical, biological, and human environment. This would involve collecting all types of existing environmental data and assessing for all the potential impacts that could arise due to the construction and operation of the wind farm. The impacts can range from favourable to less favourable to detrimental. Potential aspects on the biological environment include marine mammals, sea birds that use the area on a regular basis, birds from nearby areas that pass through the area during flight, fish, etc. Other aspects include effect of flora and fauna during the construction (e.g. noise due to piling or operating noise), and electro-magnetic field generated by subsea cable. The human environment includes change of landscape. Marine archaeology aspects such as shipwrecks are also taken in consideration. To carry out the assessment, seabed samples may be collected and analysed for worms, barnacles, or other species.
4. *Power export/grid connection.* One of the important deciding factors is the location of onshore grid connections. The deciding factors include the length of submarine cable required, which is dependent on the turbine layout, substation location, export cable routing (landfall), risk assessment of buried cables, and the transformer options – AC or DC.
5. *Economics.* Modelling of capital costs and LCOE is a function of many parameters: depth of water, distance from shore, wind speed at the site, port and harbour facilities near the site, socioeconomic conditions and access to skilled labour, location of national grid, and hinterland for the proposed development.

6. *Navigation*. This survey will investigate whether there is a need for exclusion zone due to fishing or navigation or military operations. Cables connecting the wind farms and the export cables are buried to depths of 2–3 m to avoid risk of entanglement with net. Navigation risk must be assessed.
7. *Consents and legislations*. Depending on the country, the consent requirements may change. For example, in the United Kingdom, any development more than 100 MW is classified as significant infrastructure project and requires development of a consent order from the Infrastructure Planning Commission (IPC). These rules are subject to amendment. Currently, the final decision rests with the Secretary of State for Energy and Climate Change.

Example 1.5 ***Burbo Wind Farm*** Vestas V80 make 3 MW wind turbines; see Figure 1.42 for the layout.

The location of Burbo wind farm is influenced by the following:

1. Average wind speed more than 7 m s^{-1}
2. Shallow water depth 0.5–8 m at Low tide
3. Good seabed condition for construction of foundation
4. Close to entrance of Mersey River
5. Proximity to Liverpool port
6. 6.4 km from Sefton coastline
7. Safe distance from navigation channel
8. Onshore export cable travelled 3.5 km underground to a substation to be fed in the grid

The following consents were taken for Burbo:

1. Consents are Section 36 of Electricity Act 1989 for wind turbine and cabling
2. Consent under Section 34 of Coastal Protection Act 1949 for Construction in navigable waters

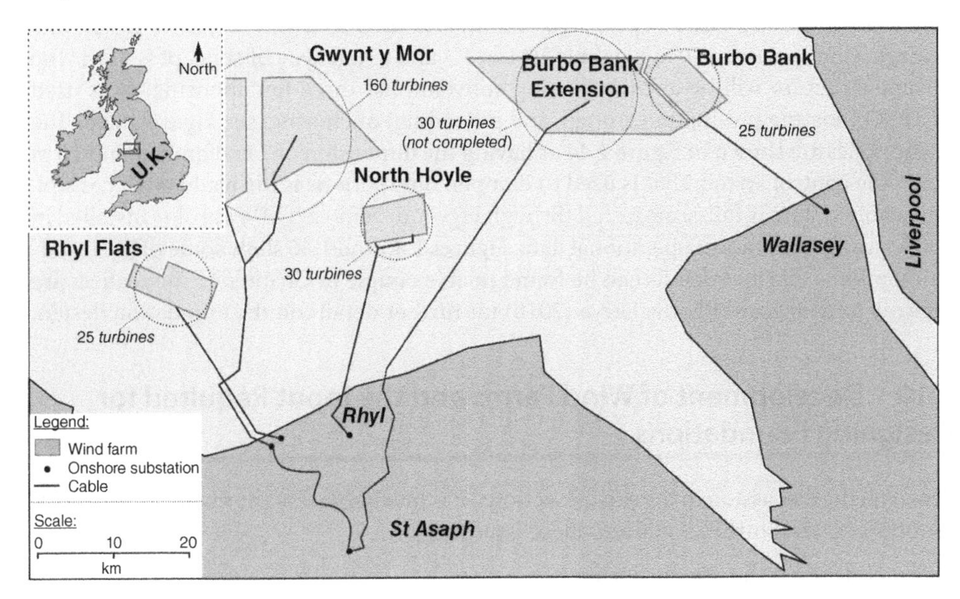

Figure 1.42 Location of the Burbo wind farm.

Figure 1.43 Case study Hywind wind park (location and schematic).

3. Permission from Port Authority
4. Permission under Section 57 of the Town and Country Planning Act 1990 for onshore cabling, interconnection facilities, and substation
5. License under Section 5 of Food and Environment Protection Act 1985 for the siting of wind turbines and deposit of scour protection material

Example 1.6 *Hywind Scotland Pilot Park* This is the first floating wind farm with rated capacity of 30 MW and having five turbines, located 30 km off the coast of Scotland; see Figure 1.43. In October 2017, Hywind started delivering power to the Scottish Electricity grid. The water depth is 95–120 m and the mean waves (Hs) is 1.8 m. The average wind speed at 100 m height is 10.1 m s^{-1} and a spar type of concept is used. The turbine structure will be anchored using conventional three-line mooring, as is used in FPSO (floating production storage and offloading) anchoring; see Figure 1.26c. The anchor piles are shown in Figure 1.44 as having the dimension of 7 m diameter and 16 m long. The control system that is used to dampen out motions is the blade pitch control. The whole system is full-scale tested through Hywind Demo 2.3 MW turbine installed in 2009, having five years of operational data. Figures 1.45 and 1.46 show some of the installation photos. Further details can be found in an example in Chapter 6. The readers are referred to Arany and Bhattacharya (2018) for further details on the foundation design.

1.10 Development of Wind Farms and the Input Required for Designing Foundations

Based on the discussion in the earlier sections, it is inevitable that the main design inputs for offshore wind turbines address these issues:

- Water depth at the specific location
- Turbine loads (dependent on size and weight)

Figure 1.44 Anchor piles used in the project (Source: Photo courtesy: Gustava Sanchez).

Figure 1.45 Installation of the turbine: (a) the substructure being pulled to the location; (b) lowering of the substructure; (c) anchored to the seabed with chains; and (d) the RNA assembly together with the tower is being transported for connecting to the substructure. Photo Courtesy: Statoil.

- Ground profile at the specific location
- Site specific loads due to waves, current, tide, and earthquakes
- Other considerations are construction and installation costs
- Time in construction (TIC)

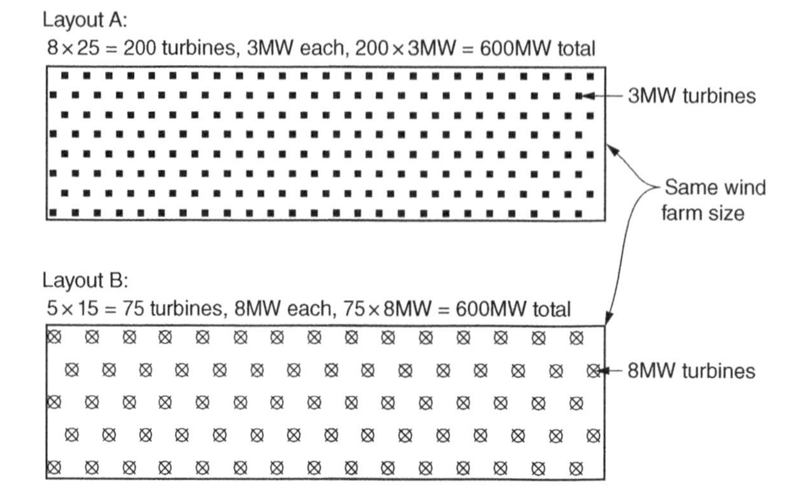

Figure 1.46 Two different layouts of an offshore wind farm occupying the same area A and of equal rated power of 600 MW. Layout A: 200 turbines of 3 MW in 8 × 25 layout. Layout B: 75 turbines (8 MW) in a 5 × 15 layout. [Figure Courtesy: Dr Arany.]

There can be conflicting requirements considering the site-specific conditions such as water depth, soil properties and site-specific loads and the requirements for a large number of wind turbines (economy of scales). Often, it is also necessary to avoid different support structures within one wind farm – so that one contractor and one installation vessel can be used. Therefore, optimization for the local site conditions through advanced layout for a large number of installations is necessary. Typically, one season of a year is suitable for offshore operations i.e. the installation of the wind turbines which asks for a fast and robust less risky installation method.

The key aspects for windfarm development are:

1. Turbine selection depending on the site condition and economic viability
2. Array design and the cable routing
3. Substructure and turbine design process
4. Selection of foundation
5. Electrical system
6. Cost modelling
7. Installation
8. Operation and maintenance (O & M)

1.11 Rochdale Envelope Approach to Foundation Design (United Kingdom Approach)

The design of offshore wind turbine foundations is an iterative process that requires many amendments and improvements as more data become available. It is therefore quite natural that in the consenting phase, the design is not finalised. This is because, detailed studies such as site investigation, EIA are not carried out until the business decision has been made. Also, this is a new industry and the pace of technological development is faster than the design cycle of offshore wind farms and as a result the appropriate turbine and foundation cannot be selected in advance. This results in the turbine, foundation type, foundation dimensions, appropriate transportation, installation technologies, etc. being not clearly defined at the consenting stage. For this reason, developers of almost all wind farms around the United Kingdom are using the Rochdale envelope approach.

Essentially, this methodology defines the turbine and foundation technology and design parameters very loosely so that necessary modifications throughout the design process can be made. This allows for changes in the following:

- Turbine type and size
- Number of turbines and rated power
- Wind farm layout
- Foundation type and size
- Seabed area usage together with sediment displacement volume

A simple example is introduced to explain the method. Assume that a certain area A of a fixed shape is available for the development of an offshore wind farm, with a given maximum water depth S, basic information about the seabed (e.g. bathymetry, soil types, and approximate layer thickness), and basic metocean data (e.g. 50-year significant wave height expected for the area, wind rose, etc.). For these particular conditions, a conservative estimate for the parameters of the planned wind farm needs to be provided in the framework of the Rochdale envelope approach. This may look as follows:

- *Turbine type and size*. The chosen turbine may vary between 3 and 8 MW. The rotor diameter will be in the range 90–164 m and the total height of the turbine from mean sea level to blade tip between 100 and 200 m.
- *Number of turbines and rated power*. The target-rated power output is 600 MW and this may be connected to an offshore substation with two transformers. Accordingly, the number of turbines may vary between 75 (75 × 8 MW) and 200 (200 × 3 MW). It is important to note here that larger turbines do not necessarily utilise the given offshore area better even if all other parameters like costs and installation times are constant. This is because the distance between turbines is a multiple of the turbine diameter, typically four to six times the rotor diameter. The power generated by a turbine is proportional to the square of the rotor diameter, i.e. $P \propto D^2$. The number of turbines that can be placed on a line of certain length is inversely proportional to the rotor diameter. Arany (2017) showed that the energy density (i.e. energy per unit area, which is a measure of effective use of seas and oceans) in a large wind farm made up of many rows is actually independent of the chosen turbine size. Therefore, the selection of the best turbine for a given site is not straightforward, and is a matter of engineering and financial judgement, which changes as the project progresses.
- *Wind farm layout*. The wind farm layout shall be a grid with shifted rows. The number of rows will vary between 5 and 8, the number of turbines in each row will vary between 15 and 25. The two layouts with the highest and lowest number of turbines are shown in Figure 1.46.

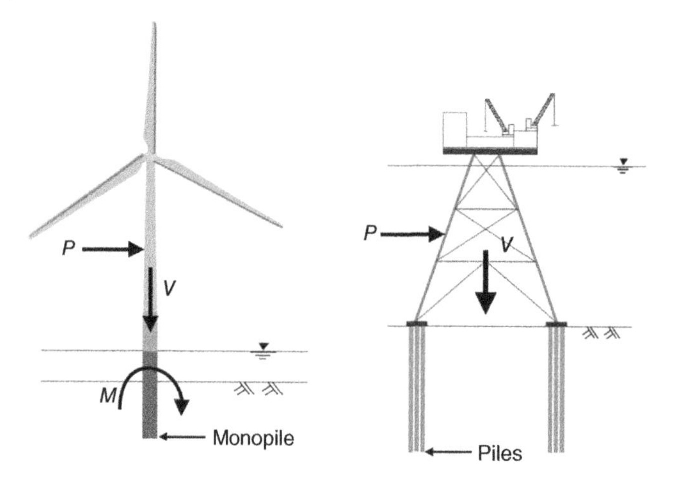

Figure 1.47 Offshore wind turbines and offshore oil and gas installations.

- *Foundation type, size, and numbers.* The foundations chosen will be monopiles, jackets, or tripods. The maximum diameter of a monopile foundation will be 9 m given the limitations of manufacturing and available installation vessels. Depending on the water depth and hub-height, the maximum leg spacing of a jacket will be 40 m and that of tripod maximum 50 m. The foundations may all be the same or different types combined throughout the wind farm. The number of foundations will vary between 75 and 200, based on the number of turbines.

The seabed area usage, sediment displacement volume, etc. can be determined similarly, aiming for a conservative estimate, as these are crucial for the consenting of the wind farm. Throughout the design cycle the foundations go through many design iterations, and the wide range of possibilities allowed in the Rochdale envelope. For this purpose, analysis have to be carried out on the expected sizes of support structures and foundations, as well as predicting the long-term behaviour. This helps to arrive at rational decision based on economics and engineering judgement.

1.12 Offshore Oil and Gas Fixed Platform and Offshore Wind Turbine Structure

Constructing stable platforms in deeper water and further offshore is not new, and considering offshore oil and gas, this is a very matured industry. Experiences regarding safe offshore operations such as installation, piling, mooring, and anchoring can be used. While the experience gained from offshore oil and gas operations can be used, it is considered important to highlight the significant differences between these two types of structures, which deserve special attention. As this book is related to foundation design, the differences will be limited from the point of view of foundation design. Figure 1.47 shows a typical monopile supported wind turbine and a pile-supported fixed offshore jacket structure. It is very clear that the ratio of horizontal load (P) to vertical load (V) is very high in offshore wind turbines when compared with fixed-jacket structures. As a result, monopile is a moment-resisting foundation.

Offshore wind turbine structures, due to their shape and form (i.e. a long slender column with a heavy mass as well as a rotating mass at the top) are dynamically sensitive because the natural frequencies of these slender structures are very close to the excitation frequencies imposed by the environmental and mechanical loads. For typical 3.6 MW turbines, the first natural frequency (eigen frequency) of the whole system is close to 0.3 Hz and for the corresponding 8 MW turbine is 0.22 Hz. The frequency of the rotor of the wind turbine is in the range of 0.2 Hz (see Table 1.1). Typical wind turbine blades weigh 30 t and as a result 90 t is rotating at the top of the tower. On the other hand, the natural frequencies of offshore oil and gas platforms is more than 0.6 Hz and the most important cyclic/dynamic loading is the wave having frequencies 0.1 Hz (typical North Sea value). As the forcing frequencies are not very close to the natural frequencies making oil and gas platforms less sensitive to dynamics.

ASIDE

Often to explain the problem to school children, an analogy may be used. It is effectively a large washing machine on the top of a flagpole, see Figure 1.48. The flagpole is supported on a fencepost. The whole challenge is how to design the fencepost.

Figure 1.48 Analogy of a wind turbine structure.

There are, however, obvious differences between those two types of foundations:

1. Offshore oil and gas platforms are supported on many small diameter piles. Piles for offshore platform structures are typically 60–110 m long and 1.8–2.7 m in diameter and monopiles for offshore wind turbines are commonly 30–40 m long and 3.5–6 m in diameter.
2. The fixity or the boundary condition of oil and gas platform piles are very different from that of the monopiles. The oil and gas platform piles can under lateral loads translate laterally but cannot rotate. Therefore, degradation in the upper soil layers resulting from cyclic loading is less severe for offshore piles, which are significantly restrained from pile head rotation, whereas monopiles are free-headed. Free-headed piles allow more deformation and, as a result, high strain levels in the soil.
3. Beam on nonlinear Winkler springs (known as 'p-y' method in American Petroleum Institute, API code or Det Norske Veritas, DNV code) is used to obtain pile-head deflection under cyclic loading for offshore oil and gas piles, but its use is limited for wind turbines application for two reasons:
   1. The widely used API model is calibrated against response to a small number of cycles (maximum 200 cycles) for offshore fixed-platform applications. In contrast, for real offshore wind turbines, 10^7–10^8 cycles of loading are expected over a lifetime of 20–25 years.
   2. Under cyclic loading, the API or DNV model always predicts degradation of foundation stiffness in sandy soil. However, recent work suggests that the foundation stiffness for a monopile in sandy soil will actually increase as a result of densification of the soil next to the pile.

1.13 Chapter Summary and Learning Points

The key lessons are:

1. From the turbine manufacturer, one needs the operating range of the turbine (1P range). This is necessary for setting the target natural frequency of the whole system. The theory for target frequency is discussed in Section 2.2 (Chapter 2). Practical examples to compute target frequency for the chosen turbine at a given location are shown in Examples 6.1 in Chapter 6.
2. For foundation design, one needs to know the mass of the RNA assembly and tower dimensions and mass. This will be provided by the turbine manufacturer.

2

Loads on the Foundations

Learning Objectives

Offshore wind turbine installation is a unique type of structure due to their geometry (i.e. mass and stiffness distribution along the height) and the loads acting on it. The aim of this chapter is to list the loads and show that the environmental loads are a mixture of cyclic and dynamic components. The loads depend on the location of the wind farm (wave period, fetch, wind turbulence) together with the size and type of the turbine. The main purpose of a foundation is to transfer these loads safely (without excessive deformation and yielding) to the surrounding soil. Therefore, the first step in any foundation design is the estimation of the loads that the foundation must carry over the lifetime. There can be additional regional specific loads on the foundations. For example, for offshore locations in India, Taiwan, and China, seismic considerations may dominate the design.

If a structure is subjected to dynamic effects, the loads that the foundation must resist are amplified and the well-known 'Dynamic Amplification Factor' concept may be used to estimate them. The load that needs to be resisted by the foundation also depends on the foundation system and effectively load transfer mechanism. This is often known as load flow path, i.e. how the superstructure loads are flowing to the ultimate 'sink', which is the ground.

The aims of this chapter are as follows:

(a) Discuss the dynamic sensitivity of wind turbine structure together with the modes of vibration of these structures. In other words, show that the global natural frequencies of the structures are very close to forcing frequencies. This is necessary to appreciate the dynamic amplification of loads.
(b) Load transfer from superstructure to the supporting ground for different types of foundations. This will be shown for both grounded systems as well as floating systems.
(c) List the loads acting on the wind turbine and different methods to estimate them.

2.1 Dynamic Sensitivity of Offshore Wind Turbine Structures

Offshore wind turbines (OWTs), due to their shape and form (slender column with a heavy mass as well as a rotating mass at the top), are dynamically sensitive. The natural

Design of Foundations for Offshore Wind Turbines, First Edition. Subhamoy Bhattacharya.

Companion website: www.wiley.com/go/bhattacharya/offshorewindturbines

frequency of these slender structures is very close to the excitation frequencies imposed by the environmental and mechanical loads. The aim of the foundation is to transfer the loads of the substructure and superstructure safely to the ground. This chapter reviews the loads acting on the wind turbine structure. Apart from the self-weight of the whole system, there are four main lateral loads acting on an OWT structure: wind, wave, 1P (rotor frequency), and 2P/3P (blade-passing frequency) loads.

The main cyclic/dynamic loads acting on the wind turbine are as follows:

(a) *Wind.* The load produced by the thrust of the wind on the blades and tower. The cyclic component of the load depends on the turbulence of the wind at that location (wind speed variation over a period) and turbine operating characteristics.
(b) *Wave.* The load caused by waves crashing against the substructure (i.e. the part of the structure exposed to the wave), the magnitude of which depends on the wave height and wave period;
(c) *1P load.* Load caused by the vibration at the hub level due to the mass and aerodynamic *imbalances* of the rotor. This load has a frequency equal to the rotational frequency of the rotor (referred to as 1P loading). Since most of the industrial wind turbines are variable-speed machines, 1P is not a single frequency but a frequency band between the frequencies associated with the lowest and the highest rpm (revolutions per minute). Ranges of typical 1P frequencies for different wind turbines are given in Chapter 1 (Table 1.1) and are noted by cut-in (RPM) and 'rated (RPM)'.
(d) *Blade-passing load (2P/3P).* Loads in the tower due to the vibrations caused by blade shadowing effects (referred to as 2P/3P). As the blades of the turbine pass through the front of the tower, the *temporary* shadowing effect reduces the thrust on the tower. Figure 2.1 shows the shadowing effect due to two configurations of the blades, and the differential load on the tower may be noted. This is a dynamic load having frequency equal to three times the rotational frequency of the turbine (3P) for three-bladed wind turbines. For a two-bladed turbine, this will be two times (2P) the rotational frequency of the turbine (1P). The 2P/3P loading is also a frequency band like 1P and is simply obtained by multiplying the limits of the 1P band by the number of the turbine blades.

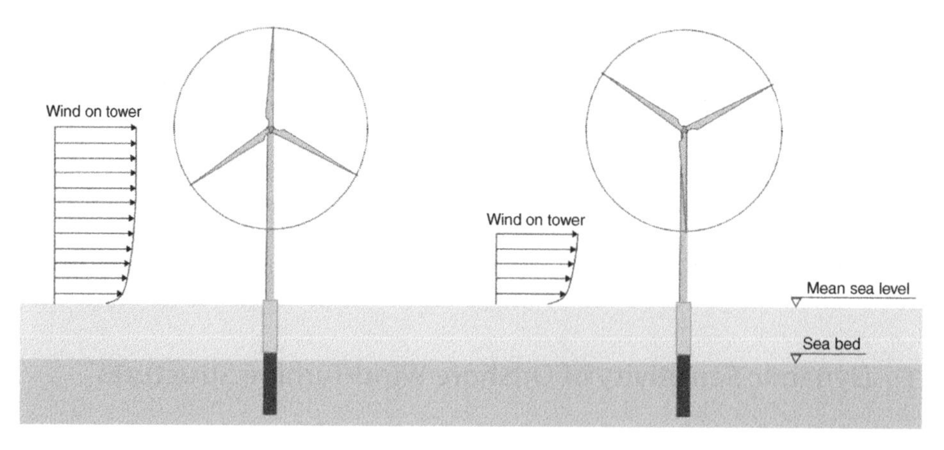

Figure 2.1 Explanation of 3P loading.

Example 2.1 ***Cyclic and Dynamic Loading*** Figure 2.2a shows a schematic representation of the time history (wave form) of the main loads on a monopile supported wind turbine. Each of these loads has unique characteristics in terms of magnitude, frequency, and number of cycles applied to the foundation. The loads imposed by the wind and the wave are random in both space (spatial) and time (temporal), and therefore they are better described statistically. Apart from the random nature, these two loads may also act in two different directions (often termed as *wind-wave misalignment*) guided by the control system to have a steady power output. 1P loading is caused by mass and aerodynamic imbalances of the rotor and the forcing frequency equals the rotational frequency of the rotor. On the other hand, 2P/3P loading is caused by the blade shadowing effect, wind shear (i.e. the change in wind speed with height above the ground), and rotational sampling of turbulence. Figure 2.2b shows a simplified version of the different loads.

As a rule of thumb, if the natural frequency of the structure is more than five times the forcing frequency, the loading can be considered cyclic and inertia of the system may be ignored. For example, for a 3 MW wind turbine having a natural frequency of 0.3 Hz, any load having frequency more than 0.06 Hz is dynamic. Therefore, wave loading of 0.1 Hz is dynamic.

Example 2.2 ***Dynamic Amplification Factor (DAF)*** If the acting loads are dynamic, there are additional considerations to take into effect the inertia of the system. If the frequency of loading approaches that of the natural frequency of the structure, the displacements experienced will be higher. This can be estimated using the concept of a dynamic amplification factor (DAF).

The DAF can be estimated by the following formula:

$$DAF = \frac{1}{\sqrt{(1-\beta^2)^2 + (2\xi\beta)^2}}$$

$$\beta = \frac{f}{f_0}\left(= \frac{\text{Excitation frequency}}{\text{Natural frequency}}\right) \tag{2.1}$$

$$\xi = \text{Damping ratio}$$

The readers are referred to standard textbooks on dynamics. A wind turbine structure passes through air, water, and soil, and therefore estimation of damping is very complex. Discussion on damping is provided in Chapter 3 (Section 3.3.2.).

2.2 Target Natural Frequency of a Wind Turbine Structure

To ascertain if the loads are dynamic and to estimate them, one needs to find the frequency of the load as well as that of the structure. One of the first steps in a design is to fix the *target frequency of the whole wind turbine structure*. Figure 2.3 shows a mechanical model of a monopile-supported wind turbine structure that can be used to estimate the natural frequency of the structure. In the model, the foundation is replaced by a set of springs and is often known as macro-element model. Based on a pile geometry and ground profile (i.e. ground stiffness along the depth of the ground), one can obtain the initial stiffness of the foundation(i.e. K_L, K_R, and K_{LR}). Methods to estimate these

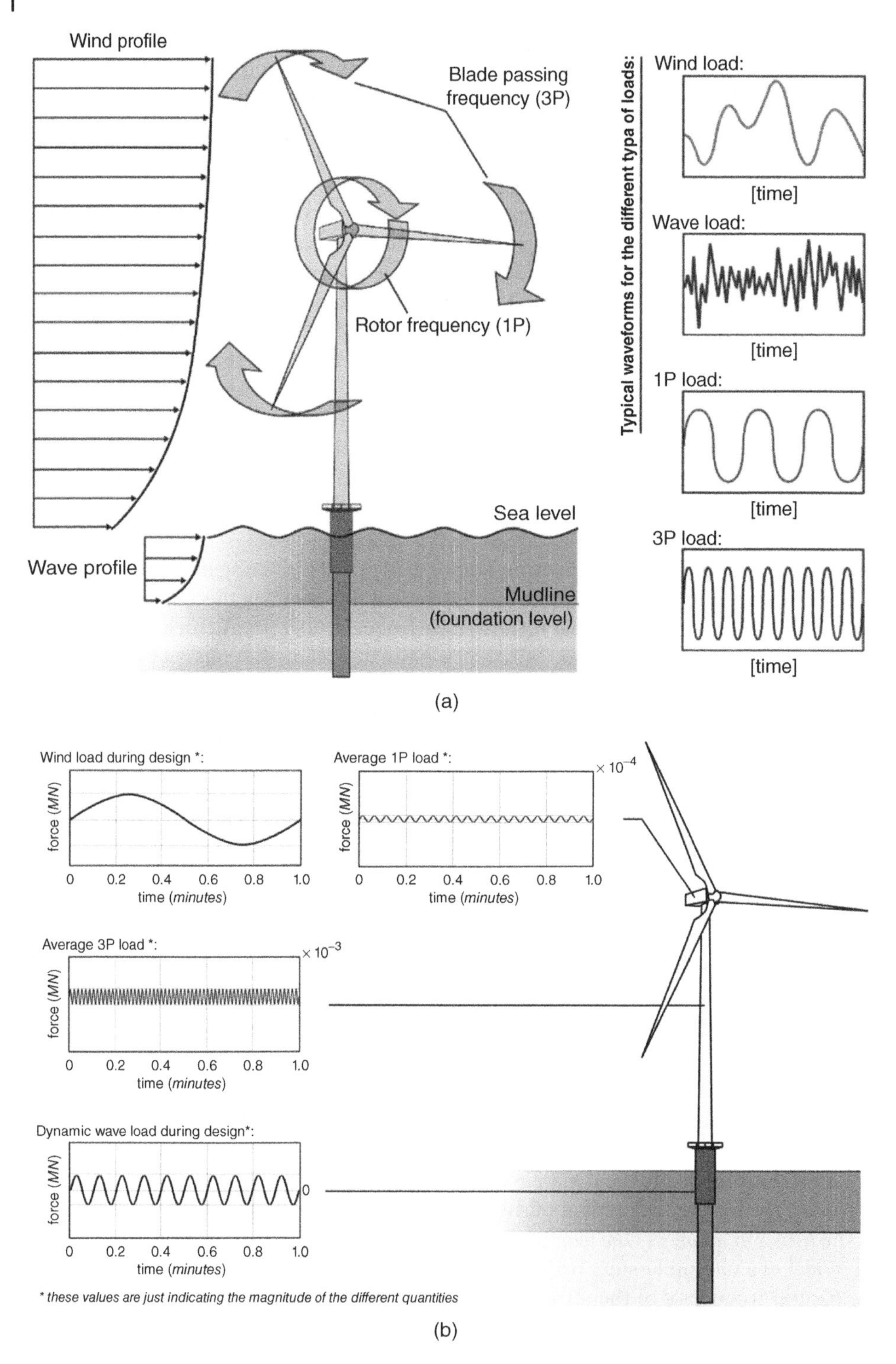

Figure 2.2 (a) Realistic distribution of main four types of loads. (b) Schematic diagram showing the loads on a wind turbine structure.

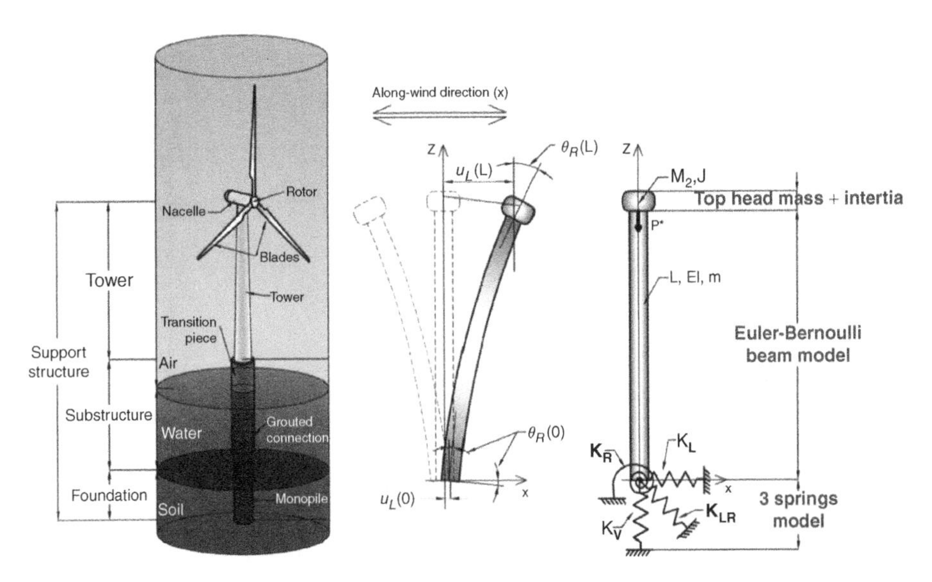

Figure 2.3 Simplified diagram of a wind turbine structure.

are explained in Chapter 5 and worked examples are provided in Chapter 6. K_L represents lateral stiffness i.e. force required for unit lateral displacement of the pile head (unit of MN m^{-1}), whereas K_R represents moment required for unit rotation of the pile head (unit of GNm rad^{-1}). K_{LR} is the cross-coupling spring and will be explained in subsequent chapters. Once K_L, K_R, and K_{LR} are known, one can predict the first natural frequency of the whole system for sway-bending modes. K_V is the vertical stiffness of the foundation and is necessary for axial vibration analysis.

The target frequency of the whole system depends on the location and the turbine used: wind turbulence information, wave period, and spectrum as well as operating range of the turbine (1P range). The turbulent wind velocity and the wave height on sea are both variables and are best treated statistically using power spectral density (PSD) functions. In other words, instead of time domain analysis the produced loads are more effectively analysed in the frequency domain, whereby the contribution of each frequency to the total power in wind turbulence and in ocean waves is described. Representative wave and wind (turbulence) spectra can be constructed by a Discrete Fourier Transform (DFT) from site-specific data. However, in the absence of such data, theoretical spectra can also be used. Some codes of practice specify Kaimal spectrum for wind and the JONSWAP (Joint North Sea Wave Project) spectrum for waves. Details on the construction of wind and wave spectrum are provided in the next few sections.

Figure 2.4 plots a schematic PSD function for typical wind and waves. The figure also shows the 1P and 3P frequency bands, which depend on the range of 1P frequency. It is obvious to avoid the excitation frequencies from resonance point of view. Det Norske Veritas (DNV) code suggests that first natural frequency should not be within 10% of the 1P and 3P ranges and is indicated as '*safety margin*' in Figure 2.4. It is therefore clear that the first natural frequency of the wind turbine needs to be fitted in a very narrow band. For some cases, 1P and 3P ranges may even coincide, leaving no gap, and in such cases, the control system is designed to leapfrog some frequencies in 1P range.

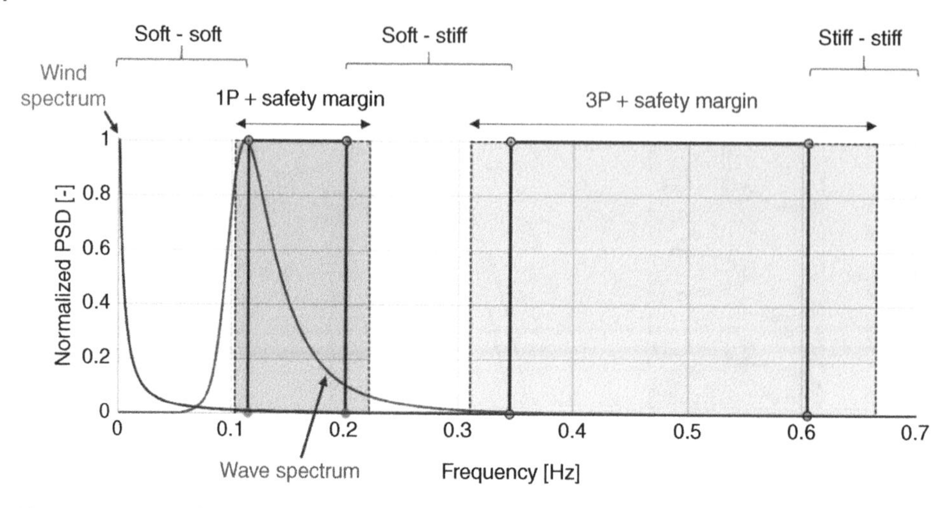

Figure 2.4 Range of frequencies acting on a wind turbine structure.

Design is therefore a collaborative effort between turbine manufacturer and substructure designer. In the current practice, the turbine manufacturer will provide the detailed design of the tower and the RNA. The substructure designer will carry out the design of transition piece and the foundation.

It is clear from the frequency content of the applied loads that the designer has to select a system frequency i.e. global frequency of the overall wind turbine including the foundation, which lies outside these frequencies to avoid resonance and ultimately increased fatigue damage. From the point of view of first natural frequency f_0 of the structure, three types of designs are possible:

(1) *Soft-soft* design where f_0 is placed below the 1P frequency range ($f_0 < f_{1P,\,min}$), which is a very flexible structure and almost impossible to design for a grounded system.
(2) *Soft-stiff* design where f_0 is between 1P and 3P frequency ranges ($f_{1P,\,max} < f_0 < f_{3P,\,min}$) and this is the most common in the current offshore development.
(3) *Stiff-stiff* designs where f_0 have a higher natural frequency than the upper limit of the 3P band ($f_0 > f_{3P,\,max}$) and will need a very stiff support structure.

It is important to note that from the point of view of dynamics, OWT designs are only conservative if the prediction of the first natural frequency are sufficiently accurate. Unlike in the case of some other offshore structures (such as the ones used in the oil and gas industry), under-prediction of f_0 is unconservative. The safest solution would seem to be placing the natural frequency of the wind turbine well above the 3P range. However, stiffer designs with higher natural frequency require massive support structures and foundations involving higher costs of materials, transportation, and installation. Thus, from an economic point of view, softer structures are desirable, and it is not surprising that almost all of the installed wind turbines are 'soft-stiff' designs, and this type is expected to be used in the future as well. It is clear from the above discussion that designing soft-stiff wind turbine systems demands the consideration of dynamic amplification and also any potential change in system frequency due to the effects of cyclic/dynamic loading on the system i.e. dynamic-structure-foundation-soil-interaction.

Example 2.3 ***Target Frequency for NREL Turbine*** Figure 2.5 shows the main frequencies for a three-bladed National Renewable Energy Laboratory (NREL) standard 5 MW wind turbine with an operational interval of 6.9–12.1 rpm. The RPM can be converted to Hz and therefore the rotor frequency (often termed 1P) lies in the range 0.115–0.2 Hz. The corresponding 'blade passing frequency' for a three-bladed turbine can be obtained by multiplying 1P range by 3 and therefore lies in the range 0.345–0.6 Hz. The figure also shows typical frequency distributions for wind and wave loading. The spectrum for wind is Froya based on Andersen and Løvseth (1992) and for the wave is Pierson-Moskowitz (1964) is used. Details on the construction of wind and wave spectrum is provided in the next section. For wave, the peak frequency is shown as 0.1, which is typical of North Sea. The dotted line on both sides of 1P and 3P frequency range reflects the DNV guidelines of avoiding target frequency within 10%.

Soil-structure interaction (SSI) may alter the foundation stiffness i.e. K_L, K_R, and K_{LR} values in Figure 2.3 and the global natural frequency of the wind turbine system will alter. If the foundation gets stiffer due to increase in the ground stiffness, the natural frequency will increase and the maximum upper bound value is fixed-base frequency. On the other hand, if the ground stiffness decreases due to SSI, the natural frequency of the whole system will reduce. For a soft-stiff system, any increase or decrease will bring the whole system in the excitation frequency range.

Extensive studies by Arany et al. (2016) showed that amongst the three foundation stiffness terms (K_L, K_R, and K_{LR}), rocking stiffness (K_R) dominates the natural frequency calculations for monopile-supported OWT. Figure 2.6 shows the natural frequency of 12 operating wind turbines where the normalised natural frequency (f/f_{FB}) is plotted against the normalised rotational stiffness $\eta_R = \frac{K_R L}{EI}$ where EI and L are the average stiffness and length of the tower, respectively. The study clearly shows that the fundamental

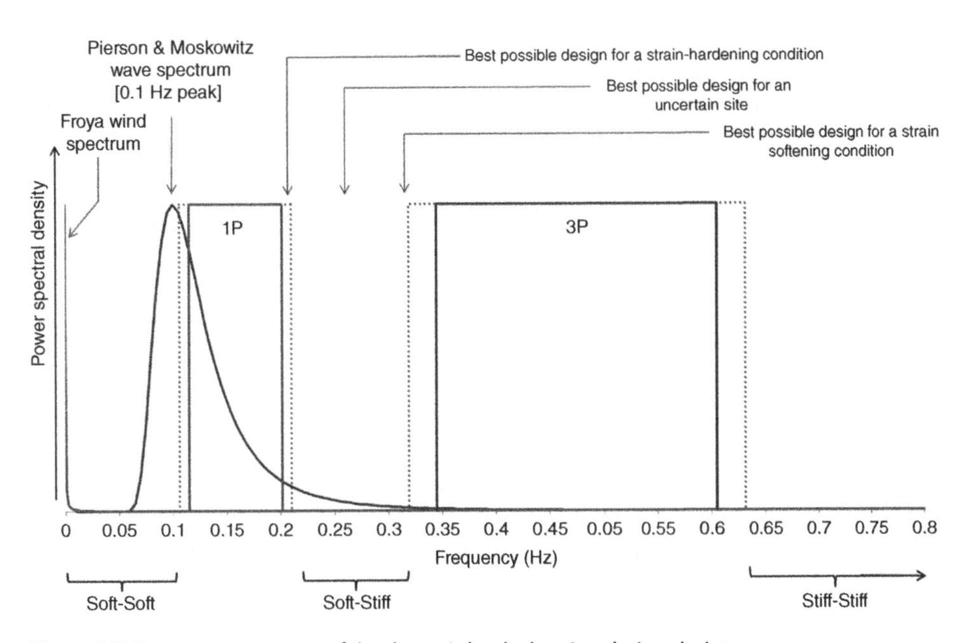

Figure 2.5 Frequency spectrum of the dynamic loads showing design choices.

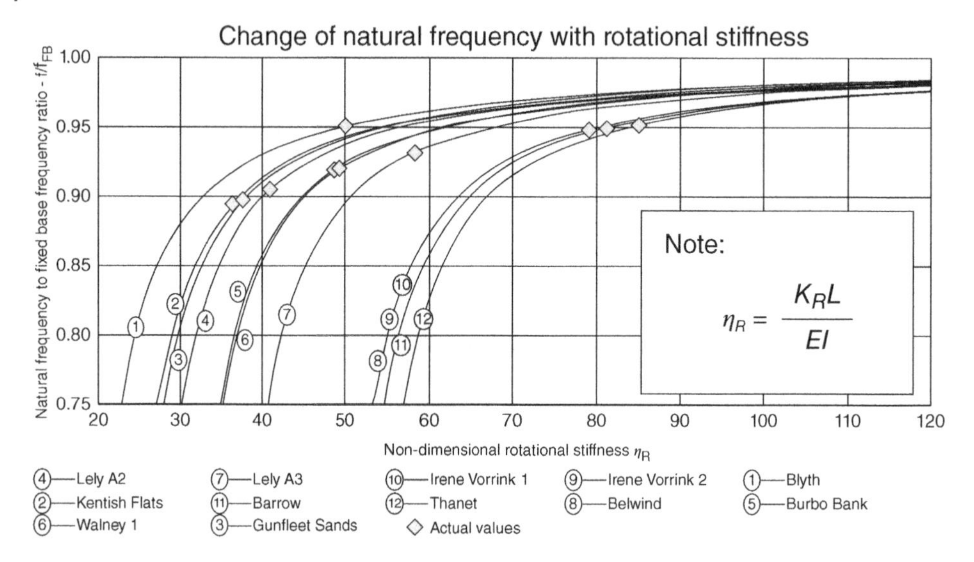

Figure 2.6 Change in natural frequency with nondimensional rotational stiffness (K_R). Explanation of the notations are given in Figure 2.3.

natural frequency of these installed turbines is about 90–95% of the fixed base frequency. A few additional points may be noted:

(a) K_R is the foundation stiffness defined in Figure 2.3, dependent on the soil stiffness. Following the curves shown in Figure 2.6, it may be observed that any change in soil stiffness therefore will alter the natural frequency of the whole system, affecting dynamic behaviour as well as fatigue.
(b) This behaviour is nonlinear and for soft-stiff design, increase or decrease in natural frequency can impinge in forcing frequencies; see Figures 2.4 and 2.5.
(c) The above discussion shows the importance of understanding the change in soil stiffness over time. It will be shown in Chapter 4 that advanced soil testing can be used for such purposes.

Theoretically, any ground can be classified as strain-hardening or strain-softening, depending on the change of ground stiffness under repeated strains. Based on the evidences so far, pure sandy sites will exhibit strain-hardening under a low to moderate strain level of loading. On the other hand, a clay site under large strain will exhibit strain-softening due to progressive accumulation of pore water pressure. Loose to medium-dense sand may liquefy momentarily under large strain levels as in earthquakes and can be classified as strain-softening. Figure 2.5 suggests best-possible target frequencies for two types of sites described in this section.

2.3 Construction of Wind Spectrum

Turbulence of wind is usually estimated as a fluctuating wind speed component (u) superimposed on the mean wind speed $\overline{U}$, and therefore the total wind speed can be written as $U = \overline{U} + u$. The degree of turbulence is usually characterised by the turbulence intensity I, given by Eq. (2.2).

$$I = \sigma_U / \overline{U} \tag{2.2}$$

where σ_U is the standard deviation of wind speed around the mean $\overline{U}$ (which is usually taken over 10 minutes). The turbulence intensity varies with mean wind speed, with site location, and with surface roughness, and is also modified by the turbine itself.

Taylor (1938) states in his frozen turbulence hypothesis that the characteristics of eddies can be considered constant (frozen) in time, and vortices travel with the mean horizontal wind speed as shown in Figure 2.7. This assumption is found to be acceptable for wind turbine applications, and the turbulence is usually analysed in the frequency domain by a PSD function, which describes the contribution of different frequencies to the total variance of the wind speed. The frequency of turbulence is connected to the size of eddies. A larger eddy means low-frequency variation in wind speed, while smaller vortices induce short, high-frequency wind-speed variations. If characteristic size of an eddy is $d[m]$ and it travels with $\overline{U}[m/s]$ speed, the travel time through the rotor in $\tau = d/\overline{U}$ time. The frequency connected to this time period is $f = 1/\tau[Hz]$.

Through this, the length scale and time scale of turbulence can be connected. The typical length scales of high energy-containing large turbulent eddies are in the range of several kilometres. The large eddies tend to decay to smaller and smaller eddies with higher frequencies as turbulent energy dissipates to heat. Kolmogorov's law describing this process states that the asymptotic limit of the spectrum is $f^{-5/3}$ at high frequency.

There are two families of spectra commonly used in wind energy applications: the von Kármán and the Kaimal spectra. The main difference is that the Kaimal spectrum is somewhat less peaked and the energy is contained in a bit wider frequency range. Kaimal spectrum is more suitable for modelling the atmospheric boundary layer, and the von Kármán spectrum is better for wind tunnel modelling. International Electrotechnical Commission (IEC) code suggests using the Mann spectrum (modified von Kármán type spectrum) and the Kaimal spectrum. On the other hand, DNV suggests the Kaimal spectrum, and it is used throughout this book. There are also other spectra, such as Froya, shown in Figure 2.5.

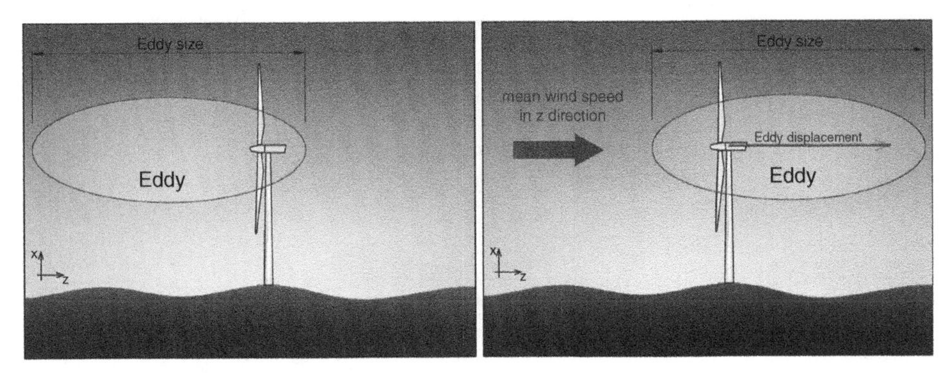

Figure 2.7 Taylor's frozen turbulence hypothesis: an eddy travels with the mean wind speed while its size and characteristic parameters remain constant.

2.3.1 Kaimal Spectrum

Some relevant details of Kaimal spectrum connected to foundation design are given in this section:

1. The power spectrum of turbulence can be modified by the landscape. If the inhomogeneity of the surface is high, the turbulence intensity increases. Another important aspect is whether the stratification is stable, unstable, or neutral. Neutral conditions rarely occur, but near-neutral conditions are typical for medium and high wind speeds and are necessary for fatigue damage calculation.
2. The theoretical Kaimal spectrum for a fixed reference point in space in neutral stratification of the atmosphere $S_{uu}(f)$ as suggested by DNV can be written as:

$$S_{uu}(f) = \frac{\sigma_U^2\left(\frac{4L_k}{\overline{U}}\right)}{\left(1+\frac{6fL_k}{\overline{U}}\right)^{\frac{5}{3}}} \tag{2.3}$$

   where L_k is the integral length scale (formula available in the DNV code), f is frequency, $\overline{U}$ is the mean wind speed (from site measurements), and σ_U is the standard deviation of wind speed. These can be estimated from measurements or calculated using Eq. (2.2), and turbulence intensity values may be obtained from standards IEC 61400-1 and IEC 61400-3.

Based on the DNV code, $L_k = 5.67z$ for $z < 60m$ and, $L_k = 340.2m$ for $z \geq 60m$ where z denotes the height above sea level.

Figure 2.8 shows a comprehensive wind spectrum following van der Hoven where both long-term variations of wind speed and turbulence effects are plotted. van der Hoven observed a significant four-day synoptic peak and in some cases a small diurnal peak in the spectrum of the horizontal wind speed. The Kaimal spectrum describes wind

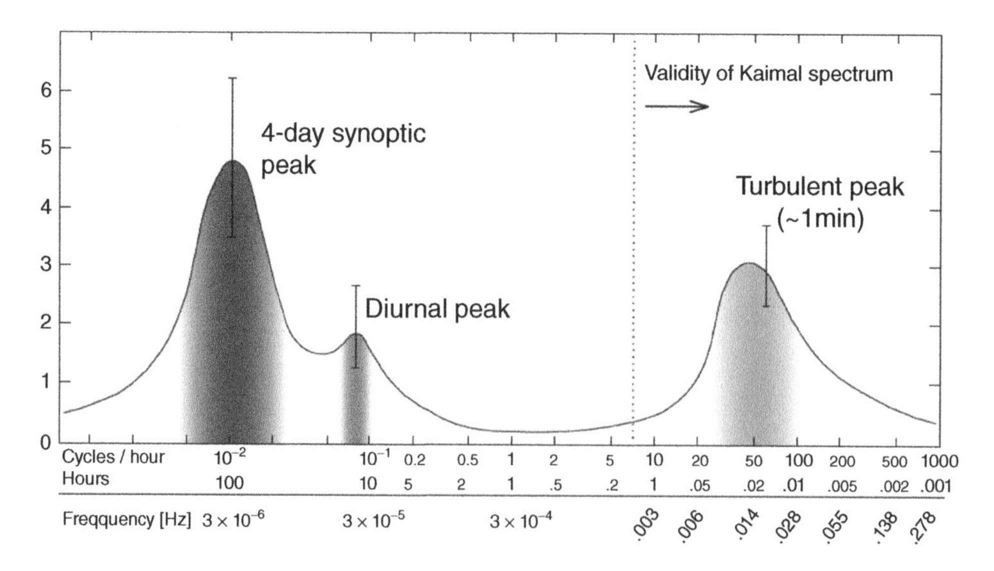

Figure 2.8 Van der Hoven spectrum and the validity of the Kaimal spectrum.

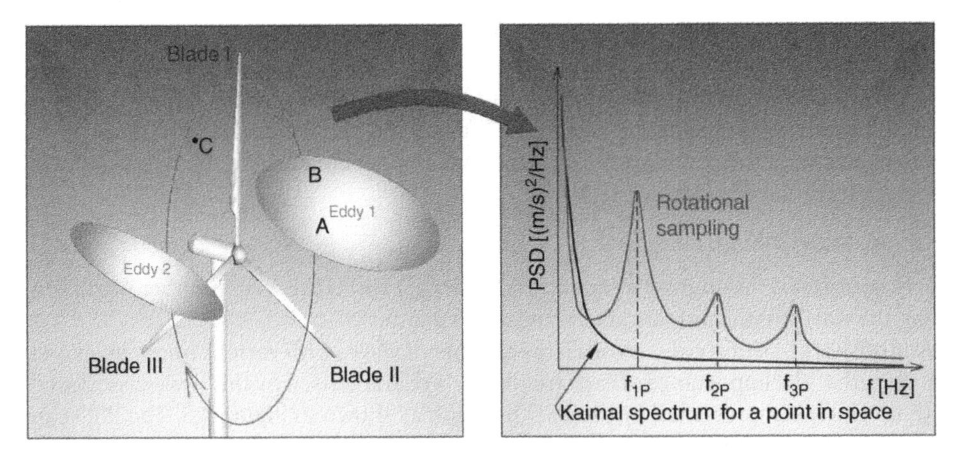

Figure 2.9 Coherence: Coherence between points A and B is high, while between C and either A or B the coherence is small. Rotational sampling: the PSD of the wind speed 'seen' by Blade 1 while rotating differs from the Kaimal spectrum for a point in space. The graph is a schematic representation of the PSD of rotational sampling.

turbulence on a short time scale (shown by a dotted line in Figure 2.8). This represents only the high-frequency end of the spectrum, omitting the diurnal and synoptic peaks. Therefore, even though the Kaimal spectrum can be calculated for arbitrarily low frequency, its validity is limited to the high-frequency variations. The lowest frequency that is typically considered corresponds to the time interval of 10 minutes, because the mean is taken over 10 minutes and therefore ($T = 600\,[s] \rightarrow f = 1/600[Hz] \approx 0.0017[Hz]$).

A small local gust (say of 20 m diameter) passing through the rotor has little or no effect on distant parts of the rotor area. Referring to Figure 2.9, the wind speeds of points A and B are closely related, and they show a high coherence, while a distant point, such as C in the figure, has a low coherence with either A or B. In other words, low-frequency variations affect a larger area of the rotor than high-frequency variations. It should also be noted that through their rotary motion the blades experience a different spectrum from the Kaimal spectrum for a point in space. The blade will pass through any given eddy once in every revolution. In Figure 2.9, Blade I, for example, passes through Eddy 1 and Eddy 2 once in each revolution. Therefore, when rotational sampling is considered, that is, the wind speed 'seen' by the rotating blade is sampled, the PSD will show peaks at the rotational frequency f_{1P} and at higher harmonics ($f_{2P} = 2f_{1P}, f_{3P} = 3f_{1P}$). This effect is more important for blade load analysis; usually, only higher harmonics are transferred to the hub, as typically all blades pass through the same eddies.

Other available design wind spectra are Davenport (1961), Simiu and Leigh (1984), and Ochi and Shin (1988). For example, Ochi and Shin (1988) empirically derived a spectrum describing the wind over a sea based on offshore observations.

2.4 Construction of Wave Spectrum

The wind blowing over the sea generates wind waves because of the increased pressure on the free surface of water. First, small waves are produced with high frequency and low wave height, and the energy is gradually transferred towards the higher amplitude waves with lower frequency and longer wavelength. The developing sea state depends on many factors, including but not limited to the water depth, the shape of the sea bottom, the mean wind speed, and the fetch. The latter is the typical leeward distance to shore considering the prevailing wind direction. The dependence on the water depth is apparent from the dispersion relation:

$$\omega^2 = gk\tanh(kS) \tag{2.4}$$

where $\omega[rad/s]$ is the angular frequency, $k = 2\pi/\lambda[1/m]$ is the wave number with $\lambda[m]$ being the wavelength, and $S[m]$ is the mean sea depth.

Any offshore location consists of a large number of waves with various frequencies and wavelengths. The importance of each frequency is characterised by the power associated with it, which is represented by the PSD function as shown in Figure 2.5. The PSD can be produced from site measurements of the wave height using DFT, or alternatively the JONSWAP spectrum $S_{ww}(f)$ suggested by many codes such as DNV:

$$\begin{aligned} S_{ww}(f) &= \frac{\alpha g^2}{(2\pi)^4 f^5} e^{-\frac{5}{4}\left(\frac{f}{f_p}\right)^4} \gamma^r \quad \gamma = 3.3 \\ r &= e^{-\frac{(f-f_p)^2}{2\sigma^2 f_p^2}} \quad \alpha = 0.076\left(\frac{\overline{U}_{10}^2}{Fg}\right)^{0.22} \quad f_p = \frac{22}{2\pi}\left(\frac{g^2}{\overline{U}_{10}F}\right)^{\frac{1}{3}} \\ \sigma &= \begin{cases} 0.07 & f \le f_p \\ 0.09 & f > f_p \end{cases} \end{aligned} \tag{2.5}$$

where

f is frequency,
α is the intensity of the spectrum,
F is the fetch,
f_p is the peak frequency,
γ is the peak enhancement factor,
g is the gravitational constant,
$\overline{U}_{10}$ is the mean wind speed at 10 m height above sea level.

Both JONSWAP spectrum as well as Pierson-Moskowitz spectrum represents the frequency content of a sea state developed in a constant wind speed condition after a sufficiently long time. The Pierson-Moskowitz spectrum assumes a fully developed sea, and that the process of transferring energy from high to low-frequency waves and the wave-wave interaction have reached a steady state. Furthermore, the waves are in equilibrium with the wind. This assumption requires a sufficiently long fetch (i.e. about 5000 wavelengths), and that the constant wind velocity has maintained for sufficiently long time (i.e. about 10000 wave periods). JONSWAP spectrum takes fetch into account and thus considers a developing sea and as a result the peak frequency of the spectrum depends on the mean wind speed and the fetch. The longer the fetch, the more developed the sea is and the more energy is in the low-frequency waves. The fetch can greatly differ for offshore wind farms located in different coasts. Figure 2.7 shows a relatively sheltered sea location where the sea is typically not fully developed. A schematic wind rose is placed at the location of the turbine.

2.4.1 Method to Estimate Fetch

A method to estimate the fetch is to take the average of the distances to leeward shores (e.g. F_S, F_{SSW}, F_{SW}, etc.), adding weights to the distances based on the significance of the direction. For example, in Figure 2.10, the prevailing wind is blowing from south-southwest (SSW) and southwest (SW); therefore, the weights of the distances F_{SSW}, F_{SW} will be the highest in the weighted average of the distances F_i.

2.4.2 Sea Characteristics for Walney Site

Figure 2.11 shows the significant wave height (Figure 2.11a) and the peak wave period (Figure 2.11b), as functions of the mean wind speed for Walney 1 wind farm location. Curves (1)–(5) represent the JONSWAP spectrum for increasing mean wind speed and for different fetch. On the other hand, curve (6) shows the Pierson-Moskowitz spectrum. Curves (1)–(5) in Figure 2.12a show the JONSWAP spectrum for increasing mean wind speeds keeping the fetch constant at 60[km], while curves (1)–(5) in Figure 2.12b presents the JONSWAP spectrum for increasing fetch, keeping the mean wind speed constant at 10 [m s^{-1}].

2.4.3 Walney 1 Wind Farm Example

Figure 2.13 shows the main frequencies from a three-bladed Siemens 3.6 MW wind turbine having a rotational operating interval of 5–13 RPM (lowest is 5RPM i.e. 0.083 Hz and rated is 13RPM i.e. 0.216 Hz). The site is characterised by a mean wind speed of 9 m s^{-1} and has a peak wave frequency of 0.197 Hz. This example considers Kaimal spectrum for wind and JONSWAP spectra for wave. Details of the calculations are provided later in the chapter and in Chapter 6.

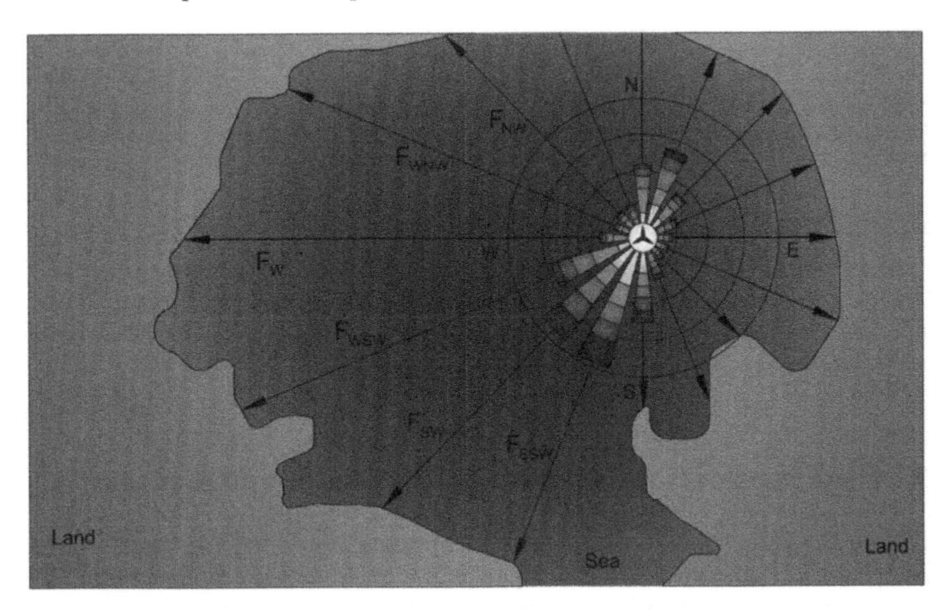

Figure 2.10 Estimation of the fetch F_i where i represents the directions of the 16-point compass rose (e.g. S for south, SW for southwest, SSW for south-southwest, etc.).

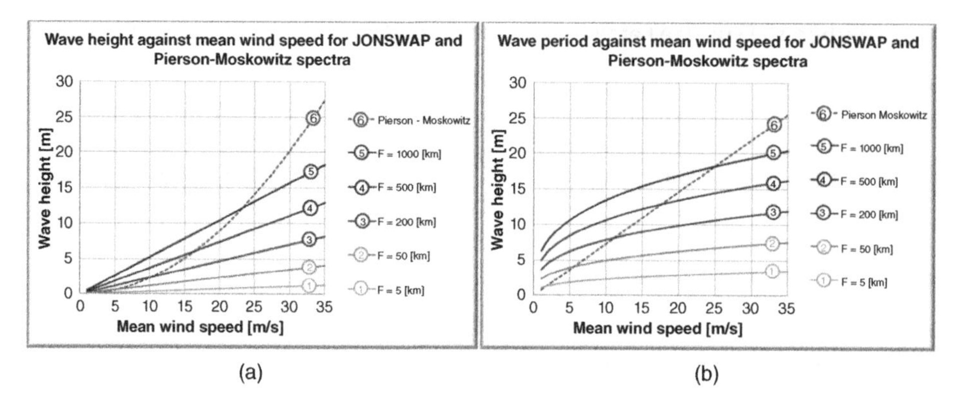

Figure 2.11 Wave height and wave period as a function of mean wind speed based for Walney 1 site. The water depth is taken as 21.5 m.

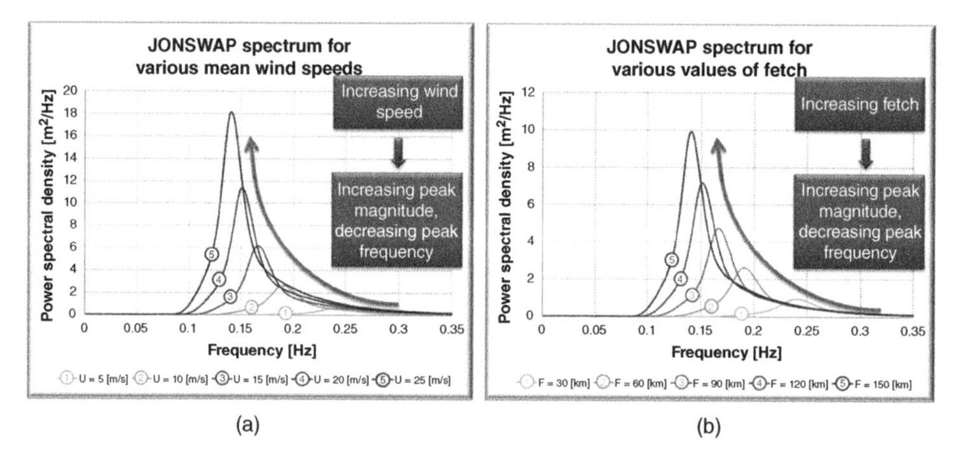

Figure 2.12 (a) JONSWAP spectrum for several values of the mean wind speed with a constant fetch of 60 km. (b) JONSWAP spectrum for several values of fetch F keeping the wind speed constant at 10 m s^{-1}.

ASIDE: Choosing Target Frequencies

In this example, choosing target frequency is very difficult due to overlapping of 1P and 3P frequencies. In such cases, the RNA control design is programmed to leapfrog certain frequencies and is done in consultation with the sub-structure designer. This shows integrated nature and complexity in the design process.

2.5 Load Transfer from Superstructure to the Foundation

The loads on the foundation depend on the foundation system. In this chapter, three distinct cases are presented: (i) single foundation for a grounded system, for example monopile, mono-caisson or a Gravity Base; (ii) multiple foundation for a grounded system; (iii) foundation for a particular floating system. The difference between the load transfer processes of single foundations and multiple foundations is explained through

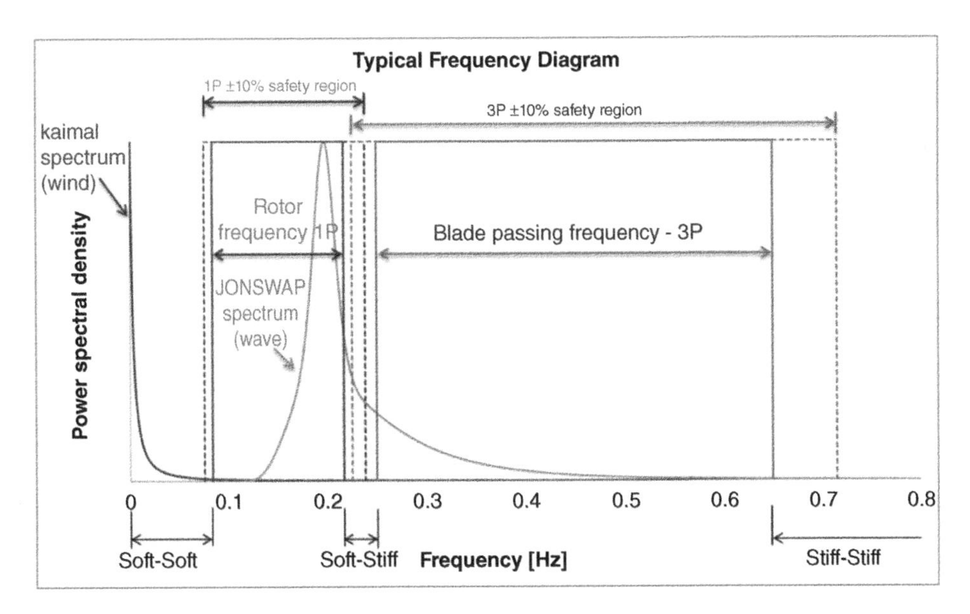

Figure 2.13 Frequency diagram for Siemens SWT-107-3.6 offshore wind turbine with operating rotational wind speed range between 5 and 13 rpm (using the mean wind speed U = 9 m/s and fetch F = 60 km. The frequency diagram includes wind spectrum, wave spectrum, and 1P and 3P frequency bands. The amplitudes are normalised to unity to focus on the frequency content of the loading.

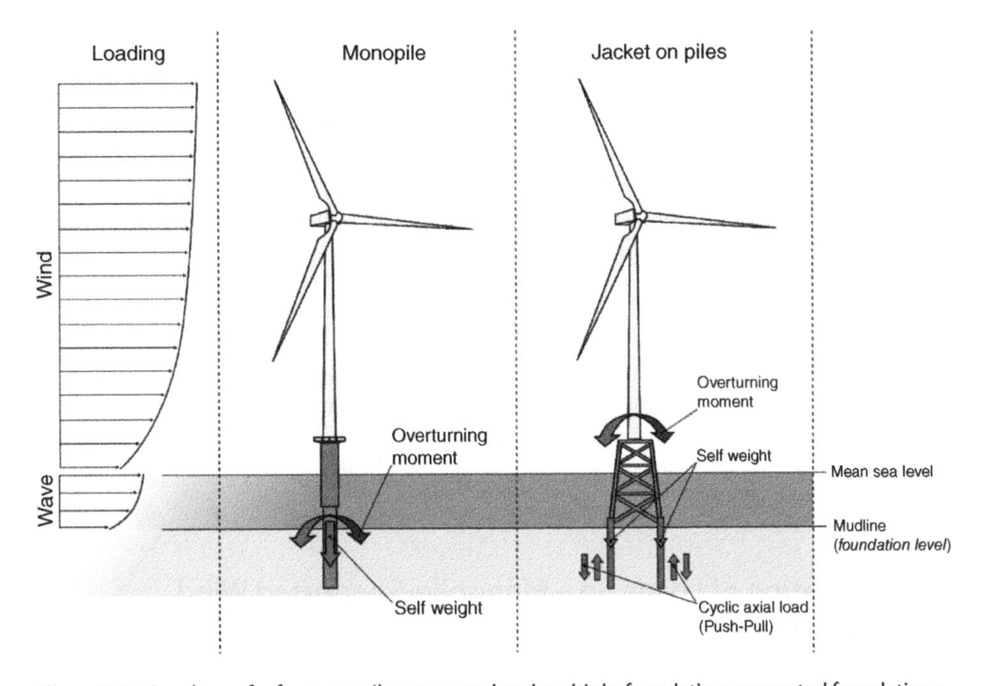

Figure 2.14 Load transfer for monopile-supported and multiple-foundation-supported foundations.

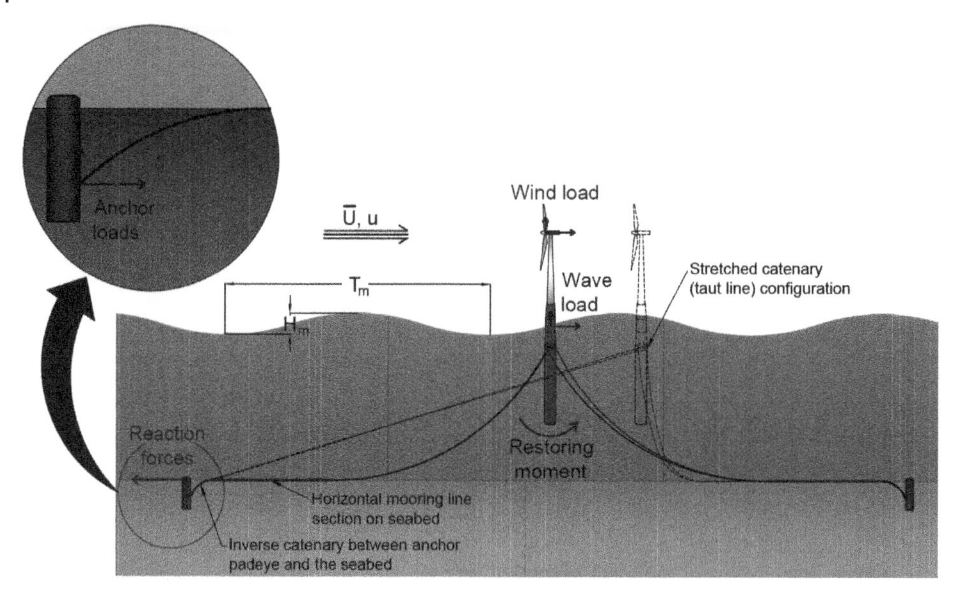

Figure 2.15 Load transfer for floating offshore wind turbine system.

Figure 2.14 by taking the example of the single large diameter monopile and multiple piles supporting a jacket. In the case of monopile-supported wind turbine structures or, for that matter, any single foundation, the load transfer is mainly through overturning moments where the monopile/foundation transfers loads to the surrounding soil and therefore there is lateral foundation–soil interaction. On the other hand, for multiple support structure, the load transfer is mainly through push-pull action – i.e. axial load, as illustrated in the figure.

Figure 2.15 shows a particular type of floating system known as spar-supported floating OWTs with catenary mooring and suction caisson anchors. This is effectively the example of the Hywind concept, the first floating offshore wind farm. For foundation design, it is necessary to estimate an upper bound for the ultimate load on the anchor. This can be obtained by taking the configuration where the mooring line is completely stretched and there is no part of it lying on the seabed. This is very similar to the configuration of a single taut mooring line. In this case the load is transferred directly to the anchor without the effect of soil friction on a horizontal section of the mooring line. Furthermore, in this configuration the angle of the mooring line at the seabed is also maximal, which impacts the inverse catenary shape at the anchor. This configuration is also shown in Figure 2.15. This is very similar to the *broken line configuration* scenario for the case of FPSO (floating production storage and offloading) anchor design; see Bhattacharya et al. (2006).

2.6 Estimation of Loads on a Monopile-Supported Wind Turbine Structure

The loads acting on the wind turbine rotor and substructure are ultimately transferred to the foundation and can be classified into two types: static or dead load due to the self-weight of the components and the cyclic/dynamic loads arising from the wind, wave, and 1P and 3P loads. However, the challenging part is the dynamic loads acting on the wind turbine and the salient points are as follows:

1. The rotating blades apply a cyclic/dynamic lateral load at the hub level (top of the tower), and this load is determined by the turbulence intensity in the wind speed. The magnitude of the dynamic load component depends on the turbulent wind speed component.
2. The waves crashing against the substructure apply a lateral load close to the foundation. The magnitude of this load depends on the wave height and wave period, as well as the water depth.
3. The mass imbalance of the rotor and hub and the aerodynamic imbalances of the blades generate vibration at the hub level and apply lateral load and overturning moment. This load has a frequency equal to the rotational frequency of the rotor (referred to as 1P loading in the literature). Since most industrial wind turbines are variable speed machines, 1P is not a single frequency but a frequency band between the frequencies associated with the lowest and the highest rpm (revolutions per minute).
4. The blade-shadowing effects (referred to as 2P/3P in the literature) also apply loads on the tower. This is a dynamic load having frequency equal to three times the rotational frequency of the turbine (3P) for three-bladed wind turbines and two times (2P) the rotational frequency of the turbine for two-bladed turbines. Rotational sampling of the turbulence by the blades and wind shear may also produce 2P/3P loads at the foundation. The 2P/3P excitations also act in a frequency band like 1P and are simply obtained by multiplying the limits of the 1P band by the number of the turbine blades.

This section presents a calculation procedure that can be easily carried out in a spreadsheet program. The output of such a calculation will be relative wind and the wave loads. For simplified design purposes, it is assumed that the wind and wave are perfectly aligned. This is a fair assumption for deeper water further offshore projects where the fetch distance is high.

The peak frequency of the wind turbulence can be obtained theoretically from the Kaimal spectrum. In the absence of site-specific data, and for foundation design purposes, the wind load can be assumed to act at the hub level with a time period for wind given by $T = 4L_k/\overline{U}$ (where $L_k$is the integral length scale and $\overline{U}$ is the wind speed) and typical values are about 100 seconds.

2.6.1 Load Cases for Foundation Design

IEC codes (IEC 2005; IEC 2009a; IEC 2009b) as well as the DNV code (DNV 2014) describe hundreds of load cases that need to be analysed to ensure the safe operation of wind turbines throughout their lifetime of 20–30 years. However, in terms of foundation design, not all these cases are significant or relevant. The main design requirements for foundation design as will be explained in Chapter 3 are ULS (Ultimate Limit State), FLS (Fatigue Limit State), and SLS (Serviceability Limit State). Five load cases important for simplified foundation design are identified and described in Table 2.1.

Table 2.1 Representative design environmental scenarios as load cases chosen for foundation design.

#	Name and description	Wind model	Wave model	Alignment
E-1	Normal operational conditions *Wind and wave act in the same direction (no misalignment).*	NTM at U_R (U-1)	1-yr ESS (W-1)	Collinear
E-2	Extreme wave load scenario *Wind and wave act in the same direction (no misalignment).*	ETM at U_R (U-2)	50-yr EWH (W-4)	Collinear
E-3	Extreme wind load scenario *Wind and wave act in the same direction (no misalignment).*	EOG at U_R (U-3)	1-yr EWH (W-2)	Collinear
E-4	Cut-out wind speed and extreme operating gust scenario *Wind and wave act in the same direction (no misalignment).*	EOG at U_{out} (U-4)	50-yr EWH (W-4)	Collinear
E-5	Wind-wave misalignment scenario *Same as E-2, except the wind and wave are misaligned at an angle of $\phi = 90°$. The dynamic amplification is higher in the cross-wind direction due to low aerodynamic damping.*	ETM at U_R (U-2)	50-yr EWH (W-4)	Misaligned at $\phi = 90°$

All design load cases are built as a combination of four wind and four sea states. The wind conditions are:

(U-1) *Normal turbulence scenario.* The mean wind speed is the rated wind speed (U_R) where the highest thrust force (Th) is expected, and the wind turbulence is modelled by the Normal Turbulence Model (NTM). The NTM standard deviation of wind speed is defined in IEC (2005).

(U-2) *Extreme turbulence scenario.* The mean wind speed is the rated wind speed (U_R), and the wind turbulence is very high. The extreme turbulence model (ETM) is used. The ETM standard deviation of wind speed is defined in IEC (2005).

(U-3) *Extreme gust at rated wind speed scenario.* The mean wind speed is the rated wind speed (U_R) and the 50-year extreme operating gust (EOG) calculated at U_R hits the rotor. The EOG is a sudden change in the wind speed and is assumed to be so fast that the pitch control of the wind turbine has no time to alleviate the loading. This assumption is very conservative and is suggested to be used for simplified foundation design. The EOG speed is defined in IEC (2005).

(U-4) *Extreme gust at cut-out scenario.* The mean wind speed is slightly below the cut-out wind speed of the turbine (U_{out}) and the 50-year EOG hits the rotor. Due to the sudden change in wind speed the turbine cannot shut down. Note that the EOG speed calculated at the cut-out wind speed is different from that evaluated at the rated wind speed (IEC 2005).

The wave conditions are:

(W-1) *One-year extreme sea state (ESS).* A wave with height equal to the 1-year significant wave height $H_{S,1}$ acts on the substructure.
(W-2) *One-year extreme wave height (EWH).* A wave with height equal to the 1-year maximum wave height $H_{m,1}$ acts on the substructure.
(W-3) *50-year ESS.* A wave with height equal to the 50-year significant wave height $H_{S,50}$ acts on the substructure.
(W-4) *50-year EWH.* A wave with height equal to the 50-year maximum wave height $H_{m,50}$ acts on the substructure.

The one-year ESS and EWH are used as a conservative overestimation of the normal wave height (NWH) prescribed in IEC (2009a). It is important to note here in relation to the ESS that the significant wave height and the maximum wave height have different meanings. The significant wave height H_S is the average of the highest one-third of all waves in the three-hour sea state, while the maximum wave height H_m is the single highest wave in the same three-hour sea state.)

According to the relevant standards (IEC 2005; IEC 2009a; DNV 2014), in the probability envelope of environmental states the most severe states with a 50-year return period has to be considered, and not 50-year return conditions for wind and wave unconditionally (separately). Indeed, extreme waves and high wind speeds tend to occur at the same time; however, the highest load due to wind is not expected when the highest wind speeds occur. This is partly because the pitch control alleviates the loading above the rated wind speed, but also because turbines shut down at high wind speeds for safety reasons. Idle or shut-down turbines, as well as turbines operating close to the cut-out wind speed have a significantly reduced thrust force acting on them compared to the thrust force at the rated wind speed due to the reduced thrust coefficient, as shown in Figures 2.17 and 2.18.

The highest wind load is expected to be caused by scenario U-3 and the highest wave load is due to scenario W-4. In practice, the 50-year extreme wind load and the 50-year extreme wave load have a negligible probability to occur at the same time, and the DNV (2014) code also doesn't require these extreme load cases to be evaluated together. The designer has to find the most severe event with a 50-year return period based on the joint probability of wind and wave loading. Therefore, for the ULS analysis (for details see Chapter 3), two combinations are suggested by Arany et al. (2017):

(1) The ETM wind load at rated wind speed combined with the 50-year EWH – the combination of wind scenario U-2 and wave scenario W-4. This will provide higher loads in deeper water with higher waves.
(2) The 50-year EOG wind load combined with the one-year maximum wave height. This will provide higher loads in shallow water in sheltered locations where wind load dominates.

These scenarios are somewhat more conservative than those required by standards, and can be adopted for simplified analysis. From the point of view of SLS and FLS, the single largest loading on the foundation is not representative, because the structure is expected to experience this level of loading only once throughout the lifetime.

The load cases in Table 2.1 are considered to be representative of typical foundation loads in a conservative manner and may serve as the basis for conceptual design of foundations. However, detailed analysis for design optimization and the final design may require addressing other load cases as well. These analyses require detailed data about the site (wind, wave, current, geological, geotechnical, bathymetry data, etc.) and also the turbine (blade profiles, twist and chord distributions, lift and drag coefficient distributions, control parameters and algorithms, drive train characteristics, generator characteristics, tower geometry, etc.).

2.6.2 Wind Load

The thrust force (Th) on a wind turbine rotor due to wind can be estimated in a simplified manner as

$$Th = \frac{1}{2}\rho_a A_R C_T U^2 \tag{2.6}$$

where ρ_a is the density of air, A_R is the rotor swept area, C_T is the thrust coefficient, and U is the wind speed. The wind speed can range from cut-in to cut-out, with the appropriate thrust coefficient. The thrust coefficient can be approximated in the operational range of the turbine using three different sections (as shown in Figures 2.16 and 2.17):

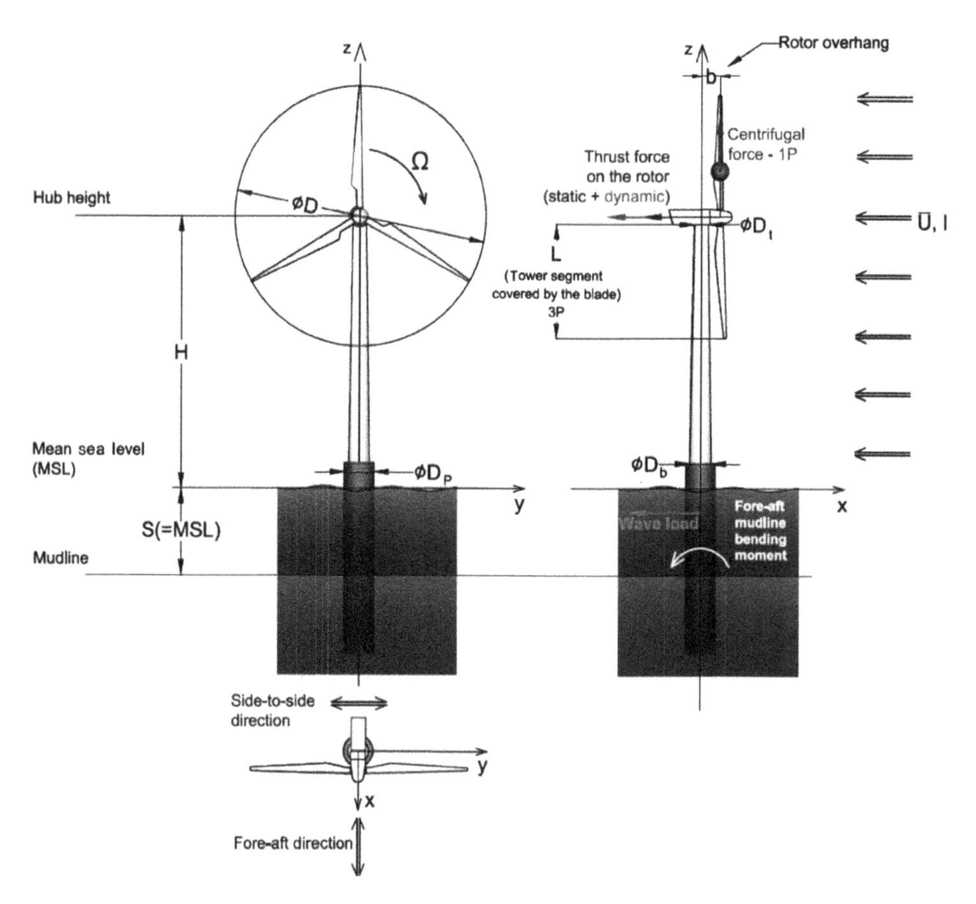

Figure 2.16 Definition of the geometry, axes, loads, and directions of the offshore wind turbine structure.

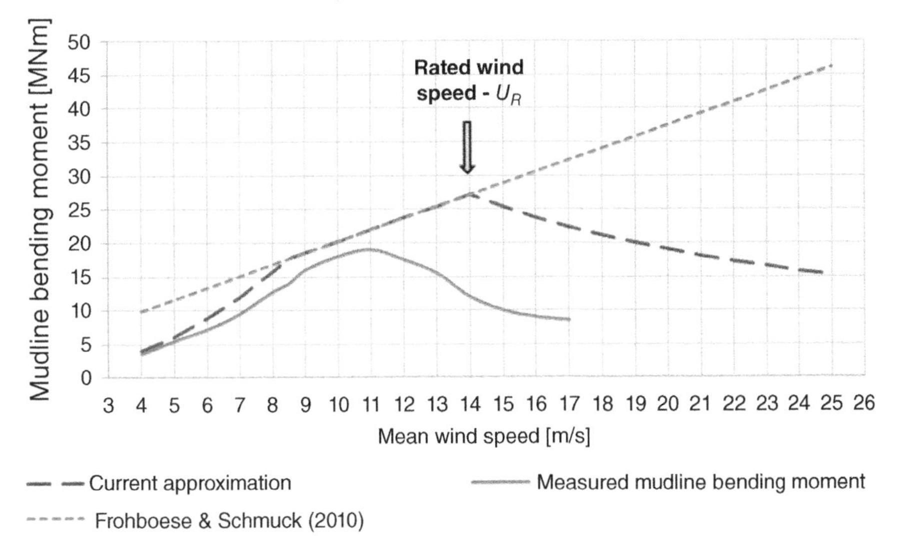

Figure 2.17 Measured and approximated mudline bending moments of Horns Rev offshore wind turbine. The measured data are obtained from Hald et al. (2009).

(1) Between cut-in (U_{in}) and rated wind speed (U_R), the method of Frohboese and Schmuck (2010) is recommended following Arany et al. (2017)

$$C_T = \frac{3.5\left[\frac{m}{s}\right]\left(2U_R + 3.5\left[\frac{m}{s}\right]\right)}{U_R^2} \approx \frac{7\left[\frac{m}{s}\right]}{U_R} \tag{2.7}$$

(2) After rated wind speed, when the pitch control is active, the power is assumed to be kept constant, and thus the thrust coefficient is expressed as

$$C_T = 3.5\left[\frac{m}{s}\right] U_R\left(2U_R + 3.5\left[\frac{m}{s}\right]\right)\cdot\frac{1}{U^3} \approx 7\left[\frac{m}{s}\right]\cdot\frac{U_R^2}{U^3} \tag{2.8}$$

(3) The thrust coefficient is assumed not to exceed 1; therefore, in the low-wind-speed regime where the formula of Frohboese and Schmuck (2010) overestimates the thrust coefficient, the value is capped at 1.

When the wind speed is changing slowly, the thrust force follows the mean thrust curve as the pitch control follows the change in wind speed. However, when a sudden gust hits the rotor, the pitch control's time constant might be too high to follow the sudden change. If this is the case, then the thrust coefficient is 'locked' at its previous value while the wind speed in Eq. (2.11) changes to the increased wind speed due to the gust.

Assuming a quasi-static load calculation method, the wind speed can be divided into two parts, a mean wind speed $\overline{U}$, and a turbulent wind speed u component. For each load case, the mean wind speed $\overline{U}$ and the turbulent wind speed component u are defined separately, and the total wind speed is expressed as

$$U = \overline{U} + u \tag{2.9}$$

Using this assumption, the wind load can be separated into a mean thrust force (or static force) and a turbulent thrust force (or dynamic force).

$$Th = Th_{mean} + Th_{turb} = \frac{1}{2}\rho_a A_R C_T \overline{U}^2 + \frac{1}{2}\rho_a A_R C_T (2\overline{U}u + u^2) \tag{2.10}$$

In this equation, the thrust coefficient C_T is calculated as shown in Eq. (2.11).

2.6.2.1 Comparisons with Measured Data

Figure 2.17 plots the measured mean mudline-bending moment for a Vestas V80 wind turbine at the Horns Rev offshore wind farm obtained from Hald et al. (2009) and compares it with the approximation based on Eq. (2.10). This is based on the measured thrust force acting at the hub level, and the corresponding thrust coefficient is shown in Figure 2.18.

The maximum thrust force occurs around the rated wind speed (or, more specifically, before the pitch control activates). In reality, the pitch control activates somewhat below the rated wind speed (as can be seen in Figures 2.17 and 2.18). This is to ensure a smooth transition between the section where the power is proportional to the cube of the wind speed and the pitch controlled region where the power is kept constant to avoid repeated switching around the rated wind speed (Burton et al. 2001).

Therefore, in the measured scenario the maximum of the thrust force (and thus the bending moment) occurs at around 11[m s^{-1}], as opposed to the theoretical approximation, in which the maximum is at $U_R = 14[m/s]$. Taking the maximum to be at U_R is, however, conservative due to the formula used for the approximation. If the value of

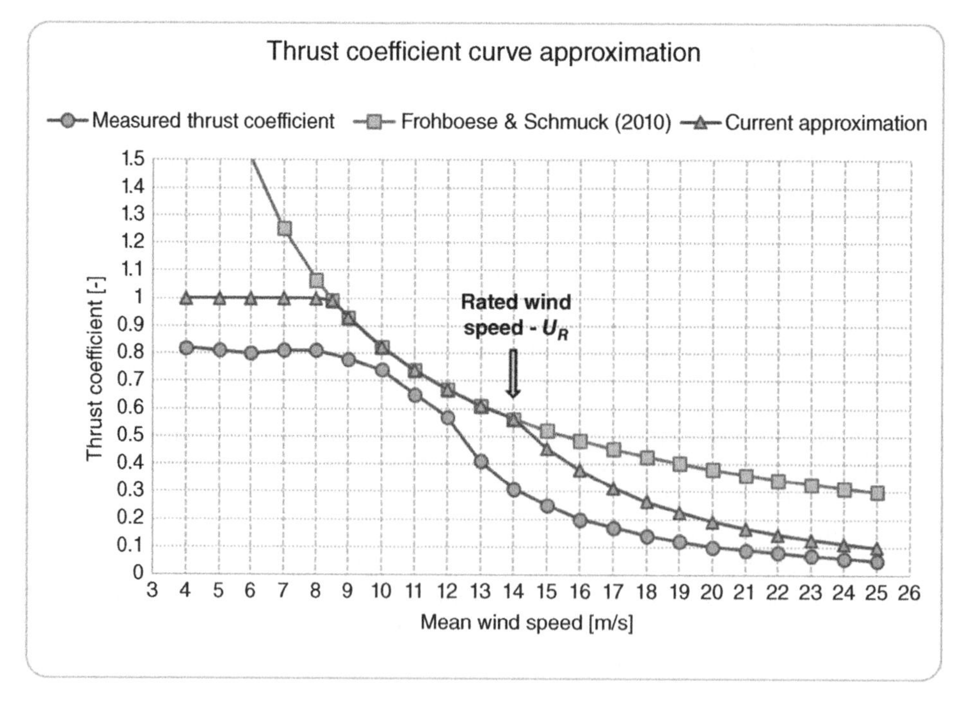

Figure 2.18 Measured (Hald et al. 2009) and approximate thrust coefficients of the Vestas V80 turbine at Horns Rev.

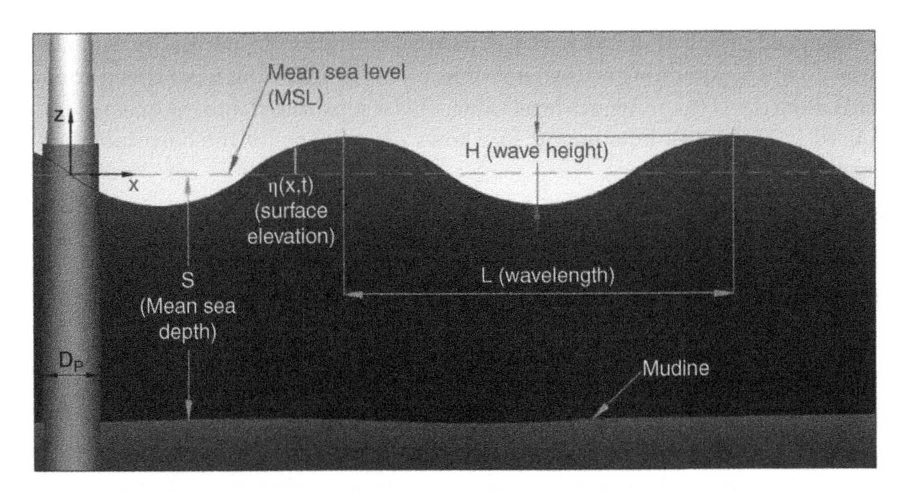

Figure 2.19 Definition of wave terminology.

11[m s^{-1}] is substituted into the formula in Eq. (2.10), one still arrives at a conservative approximation for the mean thrust force, and therefore this value (at which pitch control activates) may also be used. It should be noted, however, that typically the nature of the pitch control mechanism and the thrust coefficient curve of the turbine are not available in the early design phases. Consequently, the use of U_R is suggested, which is typically available and produces a conservative approximation.

Wind scenario U-1: Normal Turbulence (NTM) at Rated Wind Speed (U_R) This scenario is typical for normal operation of the turbine. The standard deviation of wind speed in normal turbulence following IEC (2005) can be written as

$$\sigma_{U,NTM} = I_{ref}(0.75U + b) \quad \text{with} \quad b = 5.6\,[m/s] \tag{2.11}$$

where I_{ref} is the reference turbulence intensity (expected value at U = 15 [m/s]).

For the calculation of the maximum turbulent wind speed component u_{NTM}, the time constant of the pitch control is assumed to be the same as the time period of the rotation of the rotor. In other words, it is assumed that the pitch control can follow changes in the wind speed that occur at a lower frequency than the rotational speed of the turbine. Then u_{NTM} may be determined by calculating the contribution of variations in the wind speed with a higher frequency than $f_{1P,\,max}$ to the total standard deviation of wind speed. From the Kaimal spectrum used for the wind turbulence process, this can be calculated using Eq. (2.16).

$$\sigma_{U,NTM,f>f_{1P}} = \sqrt{\int_{f_{1P,max}}^{\infty} S_{uu}(f)df} = \sigma_{U,NTM}\sqrt{\int_{f_{1P,max}}^{\infty} \frac{\frac{4L_k}{U_R}}{\left(1+\frac{6L_k}{U_R}f\right)^{\frac{5}{3}}}df}$$
$$= \sigma_{U,NTM}\sqrt{\frac{1}{\left(\frac{6L_k}{U_R}f_{1P,max}+1\right)^{\frac{2}{3}}}} \tag{2.12}$$

The turbulent wind speed encountered in normal operation in normal turbulence conditions is found by assuming normal distribution of the turbulent wind speed component and taking the 90% confidence level value. This is substituted into the quasi-static equation used in Eq. (2.14). Equation (2.15) shows expressions for the corresponding mudline moment.

$$u_{NTM} = 1.28\sigma_{U,NTM,f>f_{1P}} \tag{2.13}$$

$$F_{wind,NTM} = \frac{1}{2}\rho_a A_R C_T (U_R + u_{NTM})^2 \tag{2.14}$$

$$M_{wind,NTM} = F_{wind,NTM}(S + z_{hub}) \tag{2.15}$$

Wind scenario U-2: Extreme Turbulence (ETM) at Rated Wind Speed (U_R) The ETM is used to calculate the standard deviation of wind speed at the rated wind speed, and from that the maximum wind load under normal operation in extreme turbulence conditions. The standard deviation of wind speed in ETM is given in IEC (2005) as

$$\sigma_{U,ETM} = cI_{ref}\left[0.072\left(\frac{U_{avg}}{c} + 3\right)\left(\frac{U_R}{c} - 4\right) + 10\right] \quad \text{with} \quad c = 2\,[\text{m/s}] \tag{2.16}$$

where U_{avg} is the long-term average wind speed at the site. The maximum turbulent wind speed component u_{ETM} is determined similarly to the previous case.

$$\sigma_{U,ETM,f>f_{1P}} = \sqrt{\int_{f_{1P,max}}^{\infty} S_{uu}(f)df} = \sigma_{U,ETM}\sqrt{\int_{f_{1P,max}}^{\infty} \frac{\frac{4L_k}{U_R}}{\left(1 + \frac{6L_k}{U_R}f\right)^{\frac{5}{3}}}df}$$
$$= \sigma_{U,ETM}\sqrt{\frac{1}{\left(\frac{6L_k}{U_R}f_{1P,max} + 1\right)^{\frac{2}{3}}}} \tag{2.17}$$

The turbulent wind speed encountered in normal operation in extreme turbulence conditions, which is used for cyclic/dynamic load analysis, is found by assuming normal distribution of the turbulent wind speed component. As opposed to the normal turbulence situations, the 95% confidence level value is taken. This is substituted into the quasi-static equation used in Eq. (2.14). Equation (2.20) shows the expression for mudline moments.

$$u_{ETM} = 2\sigma_{U,ETM,f>f_{1P}} \tag{2.18}$$

$$F_{wind,ETM} = \frac{1}{2}\rho_a A_R C_T (U_R + u_{ETM})^2 \tag{2.19}$$

$$M_{wind,ETM} = F_{wind,ETM}(S + z_{hub}) \tag{2.20}$$

Wind Scenario U-3: Extreme Operating Gust (EOG) at Rated Wind Speed (U_R) The maximum force is assumed to occur when the maximum mean thrust force acts and the 50-year EOG hits the rotor. Due to this sudden gust, the wind speed is assumed to change so fast that the pitch control doesn't have time to adjust the blade pitch angles. This assumption is very conservative as the pitch control in reality has a time constant that would allow for some adjustment of the blade pitch.

The methodology for the calculation of the magnitude of the 50-year extreme gust is described in DNV (2014). This methodology builds on the long-term distribution of 10-minutes mean wind speeds at the site, which is typically represented by a Weibull distribution. The cumulative distribution function (CDF) can be written in the following form:

$$\Phi_{U10}(K,s) = 1 - e^{-\left(\frac{u}{K}\right)^{s}} \tag{2.21}$$

where K and s are the Weibull scale and shape parameters, respectively. From this the CDF of 1-year wind speeds can be obtained using

$$\Phi_{U10,1-year}(K,s) = \Phi_{U10}(K,s)^{52596} \tag{2.22}$$

where the number 52596 represents the number of 10-minutes intervals in a year ($52596 = 365.25[\text{days/year}] \cdot 24[\text{hours/day}] \cdot 6[10\text{ min intervals/hour}]$).

From this, the 50-year extreme wind speed, which is typically used in wind turbine design for extreme wind conditions, can be determined by the wind speed at which the CDF is 0.98 (i.e. 1-year 10-minutes mean wind speed that has 2% probability).

$$U_{10,50-year} = K\left[-ln\left(1 - 0.98^{\frac{1}{52596}}\right)\right]^{\frac{1}{s}} \tag{2.23}$$

The extreme gust speed is then calculated at the rated wind speed from

$$u_{EOG} = min\left\{1.35(U_{10,1-year} - U_R); \frac{3.3\sigma_{U,c}}{1 + \frac{0.1D}{\Lambda_1}}\right\} \tag{2.24}$$

where D is the rotor diameter, $\Lambda_1 = L_k/8$ with L_k being the integral length scale, $\sigma_{U,c} = 0.11U_{10,1-year}$ is the characteristic standard deviation of wind speed, $U_{10,1-year} = 0.8U_{10,50-year}$. Using this, the total wind load is estimated as

$$F_{wind,EOG} = Th_{EOG} = \frac{1}{2}\rho_a A_R C_T (U_R + u_{EOG})^2 \tag{2.25}$$

and using the water depth S and the hub height above sea level z_{hub}, the mudline bending moment (without the load factor γ_L) is given as

$$M_{wind,EOG} = F_{wind,EOG}(S + z_{hub}) \tag{2.26}$$

Wind Scenario U-4: Extreme Operating Gust (EOG) at the Cut-Out Wind Speed (U_{out}) This load case is examined here because intuitively it may seem natural to expect the highest loads when the turbine is operating at the highest operational wind speed, however, this is not the case. Wind load caused by the EOG at the highest operational wind speed (the cut-out wind speed U_{out}) is calculated taking into consideration that the thrust coefficient expression of Frohboese and Schmuck (2010) is no longer valid. The thrust coefficient is determined from the assumption that the pitch control keeps the power constant. This means that the thrust force is inversely proportional to the wind speed above rated wind speed U_R and the thrust coefficient is inversely proportional to the cube of the wind speed.

$$C_T = \frac{7\left[\frac{m}{s}\right] U_R^2}{U^3} \tag{2.27}$$

The EOG speed at cut-out wind speed $u_{EOG,U_{out}}$ is determined as given (note that this differs for different mean wind speeds, i.e. the value is not the same at U_R and at U_{out}). The thrust force and moment are then given by:

$$T_{wind,U_{out}} = \frac{1}{2}\rho_a A_R C_T(U_{out})(U_{out} + u_{EOG,U_{out}})^2 \tag{2.28}$$

$$M_{wind,U_{out}} = (S + z_{hub})T_{wind,U_{out}} \tag{2.29}$$

2.6.2.2 Spectral Density of Mudline Bending Moment

The spectral density of the turbulent thrust force on the rotor $S_{FF,\,wind}(f)$ can be written as:

$$S_{FF,wind}(f) = \rho_a^2 \frac{D^4\pi^2}{16} C_T^2 \overline{U}^2 \sigma_U^2 \widetilde{S}_{uu}(f) = \rho_a^2 \frac{D^4\pi^2}{16} C_T^2 \overline{U}^4 I^2 \widetilde{S}_{uu}(f) \quad \widetilde{S}_{uu}(f) = \frac{S_{uu}(f)}{\sigma_U^2} \tag{2.30}$$

where D is the diameter of the rotor, $S_{uu}(f)$ is the Kaimal spectrum, $\widetilde{S}_{uu}(f)$ is the normalised Kaimal spectrum, ρ_a is the density of air, C_T is the thrust coefficient, I is the turbulence intensity, and σ_U is the standard deviation of wind speed. The fore-aft bending moment at the mudline is simply given by:

$$M_{wind} = Th(H + S) \tag{2.31}$$

where H is the hub height above mean sea level and S is the mean sea depth (see Figure 2.6).

Similarly, the load can be reduced to any other cross section, such as the transition piece (TP). The mudline moment spectrum associated with wind speed fluctuations can be expressed as:

$$S_{MM,wind}(f) = \rho_a^2 \frac{D^4\pi^2}{16}[C_T(\overline{U})]^2 \overline{U}^4 I^2 (H + S)^2 \widetilde{S}_{uu}(f) \tag{2.32}$$

2.6.3 Wave Load

In simplified load calculation methodologies, to determine the wave loading, simple linear waves are assumed. Higher-order theories like Stokes waves or Dean's stream function theory would provide better estimates, especially in shallow waters. However, the linear theory allows for simpler load calculation and its application is justified for foundation design loads.

A simplified approach to wave load estimation is Morison's (or MOJS) equation (Morison et al. 1950). In these equations, the diameter of the substructure is taken as $D_S = D_P + 2t_{TP} + 2t_G[m]$ to account for the transition piece (TP) and the grout (t_G) thickness. The circular substructure area A_S is also calculated from this diameter. The methodology in this paper builds on linear (Airy) wave theory, which gives the surface elevation η, horizontal particle velocity w, and the horizontal particle acceleration $\dot{w}$ as

$$\eta(x,t) = \frac{H_m}{2}\cos\left(\frac{2\pi t}{T_S} - kx\right) \tag{2.33}$$

$$w(x,z,t) = \frac{\pi H_m \cosh(k(S+z))}{T_S \sinh(kS)}\cos\left(\frac{2\pi t}{T_S} - kx\right) \tag{2.34}$$

$$\dot{w}(x,z,t) = \frac{-2\pi^2 H_m \cosh(k(S+z))}{T_S^2 \sinh(kS)}\sin\left(\frac{2\pi t}{T_S} - kx\right) \tag{2.35}$$

where x is the horizontal coordinate in the along-wind direction ($x = 0$ at the turbine, see Figure 2.19) and the wave number k is obtained from the dispersion relation

$$\omega^2 = gk\tanh(kS) \quad \text{with} \quad \omega = \frac{2\pi}{T_S} \tag{2.36}$$

The force on a unit length strip of the substructure is the sum of the drag force F_D and the inertia force F_I:

$$dF_{wave}(z,t) = dF_D(z,t) + dF_I(z,t) = \frac{1}{2}\rho_w D_S C_D w(z,t)|w(z,t)| + C_m \rho_w A_S \dot{w}(z,t) \tag{2.37}$$

where C_D is the drag coefficient, C_m is the inertia coefficient, and ρ_w is the density of seawater. The total horizontal force and bending moment at the mudline is then given by integration as

$$F_{wave}(t) = \int_{-S}^{\eta} dF_D dz + \int_{-S}^{\eta} dF_I dz \tag{2.38}$$

$$M_{wave}(t) = \int_{-S}^{\eta} dF_D(S + z_{hub})dz + \int_{-S}^{\eta} dF_I(S + z_{hub})dz \tag{2.39}$$

The peak load of the drag and inertia loads occur at different time instants, and therefore the maxima are evaluated separately. The maximum of the inertia load occurs at the time instant $t = 0$ when $\eta = 0$ and the maximum of the drag load occurs when $t = T_S/4$ and $\eta = H_m/2$. The maximum load is then obtained by carrying out the integrations:

$$F_{D,max} = \frac{1}{2}\rho_w D_S C_D \frac{\pi^2 H_S^2}{T_S^2 \sinh^2(kS)} P_D(k,S,\eta) \tag{2.40}$$

$$M_{D,max} = \frac{1}{2}\rho_w D_S C_D \frac{\pi^2 H_S^2}{T_S^2 \sinh(kS)} Q_D(k,S,\eta) \tag{2.41}$$

$$P_D(k,S,\eta) = \frac{e^{2k(S+\eta)} - e^{-2k(S+\eta)}}{8k} + \frac{S+\eta}{2} \tag{2.42}$$

$$Q_D(k,S,\eta) = \left(\frac{S+\eta}{8k} - \frac{1}{16k^2}\right)e^{2k(S+\eta)} - \left(\frac{S+\eta}{8k} + \frac{1}{16k^2}\right)e^{-2k(S+\eta)} + \left(\frac{S+\eta}{2}\right)^2 + \frac{1}{8k^2} \tag{2.43}$$

$$F_{I,max} = \frac{1}{2}\rho_w C_m D_S^2 \frac{\pi^3 H_S}{T_S^2 \sinh(kS)} P_I(k, S, \eta) \tag{2.44}$$

$$M_{I,max} = \frac{1}{2}\rho_w C_m D_S^2 \frac{\pi^3 H_S}{T_S^2 \sinh(kS)} Q_I(k, S, \eta) \tag{2.45}$$

$$P_I(k, S, \eta) = \frac{\sinh(k(S+\eta))}{k} \tag{2.46}$$

$$Q_I(k, S, \eta) = \left(\frac{S+n}{2k} - \frac{1}{2k^2}\right) e^{k(S+\eta)} - \left(\frac{S+\eta}{2k} - \frac{1}{2k^2}\right) e^{-k(S+\eta)} + \frac{1}{k^2} \tag{2.47}$$

In the simplified method for obtaining foundation loads, it can be conservatively assumed that the sum of the maxima of drag and inertia loads is the design wave load. This assumption is conservative, because the maxima of the drag load and inertia load occur at different time instants. All wave scenarios (W-1)–(W-4) are evaluated with the same procedure, using different values of wave height H and wave period T.

Example 2.4 ***PSD of Mudline Bending Moment*** Establishing a mudline moment spectrum for a given site with a given wind turbine and support structure and using practical approximations, the inertia coefficient, the water density, the diameter of the pile, the water depth (if the tidal and surge variations are neglected), and the fetch can be considered constant and the following equation may be used:

$$S_{MM,waves} = S_{MM,waves}(f, k, \overline{U}) \tag{2.48}$$

In sufficiently deep water ($S > \lambda/2$) or sufficiently shallow water $\left(S < \frac{\lambda}{10}\right)$, the following approximations may be used for the dispersion relation:

$$\text{Deep water}: \omega^2 = gk \rightarrow k = 4\pi^2 f^2/g \cdot S > \lambda/2$$

$$\text{Shallow water}: \omega^2 = gk^2 S \rightarrow k = 2\pi f/(gS) \cdot S < \lambda/10$$

Expressing the wave number with the frequency, Eq. (2.24) can be further reduced to:

$$S_{MM,waves}(f) = S_{MM,waves}(f, \overline{U}) \tag{2.49}$$

The two main parameters in the representation of the JONSWAP spectrum as expressed in Eq. (2.5) are the mean wind speed $\overline{U}_{10}$ and the fetch F. It should be noted that such direct relationship between wind speed and wave height cannot always be established, perhaps because of the presence of swell waves (i.e. waves generated by storms far away from the wind turbine). Therefore, it is often more practical to use the significant wave height $H_{1/3}$ and the time period of waves T_p as the main parameters. They are also more practical in some cases because they are observable and can be measured. Such representation is shown in the DNV-RP-205 code for environmental conditions and environmental loads.

$$S_J(f) = 2\pi \cdot A_\gamma H_{1/3}^2 T_p^{-4} f^{-5} e^{-1.25\, T_p^4 f^{-4}} \cdot \gamma^r$$

$$r = e^{-\frac{T_P^2}{2\sigma}(f-f_p)^2} \quad \sigma = \begin{cases} 0.07 & f \leq 1/T_p \\ 0.09 & f > 1/T_P \end{cases} \quad A_\gamma = 1 - 0.287\ln(\gamma) \tag{2.50}$$

where $S_J(f)$ is the JONSWAP spectrum, f is the frequency of waves, A_γ is a normalising factor, $H_{1/3}$ is the significant wave height, T_p is the peak wave period, and γ is the peak enhancement factor.

Arany et al. (2015a) developed expressions for PSD of the mudline bending moment, which can be written as (Eq. (2.22)).

$$S_{MM,waves}(f) = C_m^2 \rho_w^2 \frac{D_P^4 \pi^6}{4} \frac{f^4}{\sinh^2(kS)} \times \left[\left(\frac{S}{2k} - \frac{1}{2k^2} \right) e^{kS} - \left(\frac{S}{2k} + \frac{1}{2k^2} \right) e^{-kS} + \frac{1}{k^2} \right]^2 S_{ww}(f) \tag{2.51}$$

2.6.4 1P Loading

A wind turbine is subjected to cyclic loading with the rotational frequency, and the source of this load is mainly the rotor mass imbalance and aerodynamic imbalance (due to differences in the pitch of individual blades). The amplitude of this forcing depends on the extent of the imbalances, and typical values can be used based on the literature. A simple method is shown to estimate the fore-aft bending moment at the mudline caused by the mass imbalance; however, the calculation of the effect of blade pitch misalignment requires more input information and more sophisticated methods. The mass imbalance can be modelled as an added lumped mass on the rotor at θ azimuthal angle from Blade I. and at R distance from the centre of the hub, as shown in Figure 2.20. Here the imbalance is assumed to be on Blade I ($\theta = 0$).

$$I_m = mR \tag{2.52}$$

where I_m is the mass imbalance with units of [kg · m], m is a lumped mass, and R is the radial distance from the centre of the hub along Blade I. The centrifugal force at any time

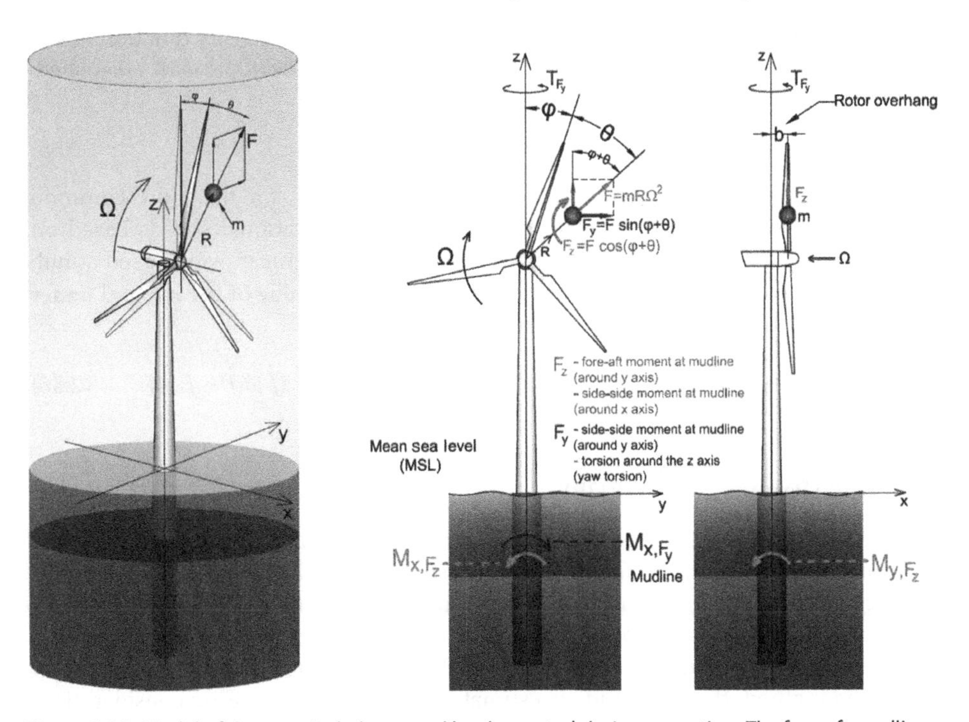

Figure 2.20 Model of the mass imbalance and loads exerted during operation. The fore-aft mudline bending moment caused by the wind load and waves (assumed collinear) is M_{y,F_z}.

can be calculated from the centrifugal acceleration $a = R\Omega^2$ with Ω being the angular frequency and $f = \Omega/(2\pi)$ the frequency of rotation. The lever arm of the centrifugal force F_{cf} is called the rotor overhang b (see Figure 2.20), with which the bending moment is expressed:

$$F_{cf} = ma = mR\Omega^2 = I_m\Omega^2 = 4\pi^2 I_m f^2 \quad M_{1P} = 4\pi^2 b I_m f^2 \tag{2.53}$$

It is to be noted here that the centrifugal force also produces torsion in the tower (around the x-axis), as well as moments in side-to-side direction (around the z-axis), as shown in Figure 2.20. The bending moment in the side–side direction caused by the centrifugal force is much higher than the fore–aft moment because the arm of the force is $H + MSL$ instead of b. The effect of the gravity force acting on the mass imbalance can be considered negligible.

Example 2.5 ***Spectral Analysis of 1P Load*** The spectral analysis of a load that is essentially a sinusoidal function of a certain frequency gives a Dirac-delta function. This is a function defined as $\delta(f - f_{1P}) = 0$ for all values of f except for $f = f_{1P}$ where $\delta(f - f_{1P}) \neq 0$ (with f_{1P} being the frequency of the wind turbine's rotation). The magnitude of this function at $f = f_{1P}$ is strictly undefined. However, the conditions in Eq. (2.54) apply.

$$\begin{aligned} &\int_{-\infty}^{\infty} \delta(f - f_{1P})df = 1 \\ &\int_{-\infty}^{\infty} M_{1P}^2(f)\delta(f - f_{1P})df = M_{1P}^2(f_{1P}) \end{aligned} \tag{2.54}$$

This means that the condition that the integral of the spectral density function has to give the variance of the signal, that is, the square of the amplitude of the load, is satisfied. Thus, the spectrum of the 1P loading is given in Eq. (2.55).

$$S_{MM,1P} = M_{1P}^2(f)\delta(f - f_{1P}) \tag{2.55}$$

For constant-speed wind turbines, f_r is constant; however, for the more common variable-speed wind turbines, which operate at different rotational speeds based on the wind speed, the frequency of the rotation depends on the mean wind speed at hub height $f_{1P} = f_{1P}(\overline{U})$. A higher frequency also means a higher value of the integral under the Dirac-delta curve, that is

$$\overline{U}_1 < \overline{U}_2 \rightarrow f_1 < f_2 \rightarrow \int_{-\infty}^{\infty} M_{1P}^2(f)\delta(f - f_1)df \leq \int_{-\infty}^{\infty} M_{1P}^2(f)\delta(f - f_2)df \tag{2.56}$$

2.6.5 Blade Passage Loads (2P/3P)

The wind produces drag force on the tower, which can be considered constant at a given mean wind speed, ignoring buffeting and vortex shedding on the tower and also without the effect of the rotating. When a blade passes in front of the tower, it disturbs the flow downwind and decreases the load on the tower. The frequency of this load loss is three times the rotational frequency of the turbine 3P (2P in case of two-bladed designs). In a simplified method, the magnitude is estimated by a simple geometric consideration: the upper part of the face area of the tower is partly covered by the blade when the blade is in a downward pointing position ($\varphi = \pi$). The drag force on the covered part of the tower is taken to be zero and the blade causes a load loss on the tower. When the blade is in the downward direction, it covers the tower from $z = H - L$ to $z = H$. The total moment of drag force on this upper section without the effect of blade passage:

$$M_{drag} = \int_{H-L}^{H} \frac{1}{2}\rho_a C_D D_T(z) U^2(z)(z+S)dx \quad U(z) = \overline{U}\left(\frac{z}{H}\right)^{\beta} \tag{2.57}$$

where H is the hub height from mean sea level, L is the length of the blades, ρ_a is the density of air, C_D is the drag coefficient, $D(z)$ is the diameter of the tower at z (assuming the diameter linearly decreases between the bottom and top diameters), z is the vertical coordinate (zero at mean sea level), S is the mean sea depth, and $U(z)$ is the power law velocity profile using the exponential wind profile with $\beta = 1/7 \approx 0.143$. If the ratio of the face area of the blade and the area of the top part of the tower (see Figure 2.16) is R_A, then the 3P moment amplitude can be written:

$$M_{3P} = R_A M_{drag} \tag{2.58}$$

Example 2.6 ***Spectral Analysis of 3P Load*** Similarly, to the 1P loading, the frequency of this loading is constant at a given rotational speed of the turbine; therefore, its power spectrum is a Dirac-delta function. The integral under the curve equals to the square of the 3P moment, with the amplitude of the Dirac-delta undefined. The integral is not directly dependent on the frequency of rotation; however, it depends on the mean wind speed, and the mean wind speed and the rotational speed of the turbine are connected through turbine characteristics.

$$\begin{aligned} &\int_{-\infty}^{\infty} \delta(f - f_{3P})df = 1 \\ &\int_{-\infty}^{\infty} M_{3P}^2(f)\delta(f - f_{3P})df = M_{3P}^2(f_{3P}) \\ &S_{MM,3P} = M_{1P}^2(f)\delta(f - f_{1P}) \end{aligned} \tag{2.59}$$

2.6.6 Vertical (Deadweight) Load

The total vertical load on the foundation is calculated as

$$V = mg \tag{2.60}$$

where m is the total mass of the structure

$$m = m_{RNA} + m_T + m_{TP} + m_P \tag{2.61}$$

where m_{RNA} is the total mass of the rotor-nacelle assembly, $m_T = \rho_T D_T \pi t_T L_T$ is the total weight of the tower, $m_{TP} = \rho_{TP}(D_P + 2t_G + t_{TP})\pi t_{TP} L_{TP}$ is the mass of the transition piece, and $m_P = \rho_P D_P \pi t_P (L_P + L_S)$ is the mass of the pile.

2.7 Order of Magnitude Calculations of Loads

The following sections discuss calculations for 1P loading, 3P loading, and typical values for wind and wave loading.

2.7.1 Application of Estimations of 1P Loading

To determine the 1P moment spectrum, one needs a typical value of the mass imbalance of the rotor. Some values are available for a somewhat smaller wind turbine studied in the literature. For the 2 MW Vestas V80 turbine, mass imbalance values of about 350–500 [kgm] were applied in these studies. The imbalance value is estimated for the 3.6 MW Siemens wind turbine by assuming that the imbalance is proportional to the mass of the rotor and also to the diameter of the turbine. The mass ratio and the diameter ratio of the rotors are calculated and the imbalance is scaled up by both ratios.

The mass of the rotor of the Vestas turbine is 37.5 tonnes, while that of the Siemens turbine is 95 tonnes. The diameter of the rotor of Vestas turbine is 80 m and that of the Siemens is 107 m. Therefore, the estimated imbalance value for the Siemens SWT-107-3.6 turbine is estimated as follows:

$$I_{m,SWT107} = I_{m,V80} \cdot \frac{M_{SWT107}}{M_{V80}} \cdot \frac{D_{SWT107}}{D_{V80}} = 500 \cdot \frac{95}{37.5} \cdot \frac{107}{80} \approx 1694\,[kg \cdot m] \quad (2.62)$$

The original imbalance value of V80 is a value typical for an average operational wind turbine. As an upper bound estimate, one may consider $I_m = 2000\,[kgm]$. The distance between the axis of the tower and the centre of the hub (i.e. the rotor overhang) is estimated as $b = 4\,[m]$. (For clarity, see Figure 2.20 for rotor overhang.) This way the maximum of the fore–aft bending moment caused by the imbalance can be written as:

$$M_{1P} = 4 \cdot 4\pi^2 \cdot 2000 \cdot f^2 = 3.1583 \cdot 10^5 \cdot f^2 \quad (2.63)$$

The maximum bending moment occurs at the highest rotational speed of $\Omega = 13\,[rpm]$, that is $f = 0.2167\,[Hz]$, its value is $M_{1P} = 0.015\,[MNm]$.

Table 2.2 shows an example for this case study based on Arany et al. (2015a) and the readers are referred to this publication for further details.

The readers are also referred to Example 6.7 of Chapter 6 and Arany et al. (2015a) for further details.

2.7.2 Calculation for 3P Loading

The 3P moment can be determined by estimating the total drag moment on the top part of the tower, which is covered by the downward-pointing blade, and then reducing this

Table 2.2 Moments due to 1P loads on the foundation.

Mean wind speed at hub height $\bar{U}[m/s]$	Rotational speed $\Omega\,[rpm]$	Fore–aft DAF [–]	1P fore–aft mudline bending moment $M_{1P}[MNm]$ /(with DAF)	Side-to-side DAF [–]	1P side-to-side mudline bending moment $M_{1P,\,side-to-side}[MNm]$ /(with DAF)
5	5.8	1.09	0.002/(0.002)	1.09	0.077/(0.084)
9	9	1.25	0.007/(0.009)	1.25	0.187/(0.234)
15	13	1.71	0.015/(0.025)	1.72	0.389/(0.669)
20	13	1.71	0.015/(0.025)	1.72	0.389/(0.669)

Table 2.3 3P loading and drag load on the tower.

Wind speed $\overline{U}[m/s]$	Total drag force on the tower $[MN]$	Total drag moment on the tower $[MNm]$	3P freq. [Hz]	DAF [–]	3P force $F_{3P}[MN]$ /(with DAF)	3P moment $M_{3P}[MNm]$ /(with DAF)
5	0.0019	0.176	0.29	3.77	0.001/(0.003)	0.069/(0.262)
9	0.0062	0.570	0.45	1.23	0.003/(0.004	0.225/(0.275)
15	0.0173	1.584	0.65	0.36	0.008/(0.003)	0.625/(0.225)
20	0.0308	2.816	0.65	0.36	0.014/(0.005)	1.111/(0.401)
6.125 (resonance)	0.0289	0.189	0.335	10	0.001/(0.013)	0.104/(1.042)

Table 2.4 Typical wind and wave loads for various turbine sizes for a water depth of 30 m.

		Turbine rated power				
Parameter	Unit	3.6 MW	3.6 MW	5.0 MW	6–7 MW	8 MW
Rotor diameter	m	107	120	126	154	164
Rated wind speed	m s^{-1}	13	13	11.4	13	13
Hub height	m	75	80	85	100	110
Mean thrust at hub (from wind)	MN	0.50	0.60	0.60	1.00	1.20
Max thrust at hub (from wind)	MN	1.00	1.20	1.20	2.00	2.30
Mean mudline moment from wave M_{mean}	MNm	53	69	70	135	165
Max mudline moment from wave M_{max}	MNm	103	136	137	265	323
Water depth	m	30	30	30	30	30
Maximum wave height	m	12	12	12	12	12
Typical monopile diameter	m	5.5	6	6.5	7	7.5
Horizontal wave force	MN	3.67	4.2	4.8	5.43	6.1
Mudline moment from waves	MNm	104	120	137	155	175
Unfactored design moment (wind+wave)	MNm	207	256	274	420	498

moment by the ratio of the face area of the blade and the face area of the top part of the tower. The drag moment is estimated by the method presented in Section 2.3. The density of air is $\rho_a = 1.225\ [kg/m^3]$, the drag coefficient of the tubular tower at high Reynolds number is $C_D \approx 0.5\ [-]$, and the linearly decreasing diameter of the tower can be written as:

$$D(z) = D_b - (D_b - D_t) \cdot \frac{z}{H} = 5 - 2 \cdot \frac{z}{83.5} \tag{2.64}$$

with the z coordinate running from mean sea level along the length of the tower, D_b and D_t are the bottom and top diameters of the tower, respectively. Using the exponential wind profile and the water depth $S = 21.5\ [m]$, the moment can be written as:

$$M_{towertop} = 0.306\overline{U}^2 \int_{31.5}^{83.5} \left(5 - 2 \cdot \frac{z}{83.5}\right) z^{2/7}(z + 21.5)dx = 4019 \cdot \overline{U}^2\ [Nm] \tag{2.65}$$

which gives $M_{drag} \approx 0.326[MNm]$ for $\overline{U} = 9\ [m/s]$. Both the area of the blade and the area of the top part of the tower are approximated as trapezoids, the areas are calculated as:

$$A_{\text{blade}} = \frac{(d_{\text{root}} + d_{\text{tip}})}{2} L = \frac{4+1}{2} \bullet 52 = 130\ [\text{m}^2]$$
$$A_{\text{towertop}} = \frac{D_{\text{lower}} + D_{\text{top}}}{2} L = \frac{4.25+3}{2} \bullet 52 = 188.2[\text{m}^2] \tag{2.66}$$

The magnitude of the load loss is then approximated as written in Eq. (2.45).

$$M_{3P} = M_{\text{towertop}} \frac{A_{\text{blade}}}{A_{\text{towertop}}} = 0.326 \cdot \frac{130}{188.2}[MNm] = 0.225[MNm] \tag{2.67}$$

The drag load on the tower is calculated by integrating along the whole tower:

$$M_{\text{drag}} = \int_0^H \frac{1}{2}\rho_a C_D D(z) U(z)^2 (z+S) dz \quad F_{drag} = \int_0^H \frac{1}{2}\rho_a C_D D(z) U(z)^2 dz \tag{2.68}$$

The 3P forces and moments are estimated in Table 2.3 for several values of the mean wind speed. Note that in the vicinity of 6.125 m s^{-1} (~6.7 rpm rotational speed), the DAF gets very high and the 3P moment is an order of magnitude higher than without DAF.

2.7.3 Typical Moment on a Monopile Foundation for Different-Rated Power Turbines

Based on the method described in the chapter and developed in Arany et al. (2015a,b, 2017), Table 2.4 shows typical values of thrust due to the wind load acting at the hub level for five turbines ranging from 3.6 to 8 MW. The thrust load depends on the rotor diameter, wind speed, controlling mechanism, and turbulence at the site. The mean and maximum mudline bending moment on a monopile is also listed. Wave loads strongly depend on the pile diameter and the water depth and it is therefore difficult to provide a general value. Table 2.4 contains a relatively severe case of 30 m water depth and a maximum wave height of 12 m.

Typical values of wave loading ranges between 2 and 10 MN acting at about three-fourths of the water depth above the mudline, which must be added to the wind thrust. Typical peak wave periods are around 10 seconds. The pattern of overturning moment on the monopile is schematically visualised in Figure 2.21. In the figure, a typical value of the peak period of wind turbulence is taken and can be obtained from wind spectrum data.

Note. It may be noted that only wind and wave moments are considered in the Figure 2.21. 1P and 3P moments are ignored, as they are about three orders of magnitude lower.

2.8 Target Natural Frequency for Heavier and Higher-Rated Turbines

Figure 2.22 shows a 1P range frequency plots for different turbines ranging from 2 to 8 MW. It is clear that as the turbine size/rated power increases, the target frequency in

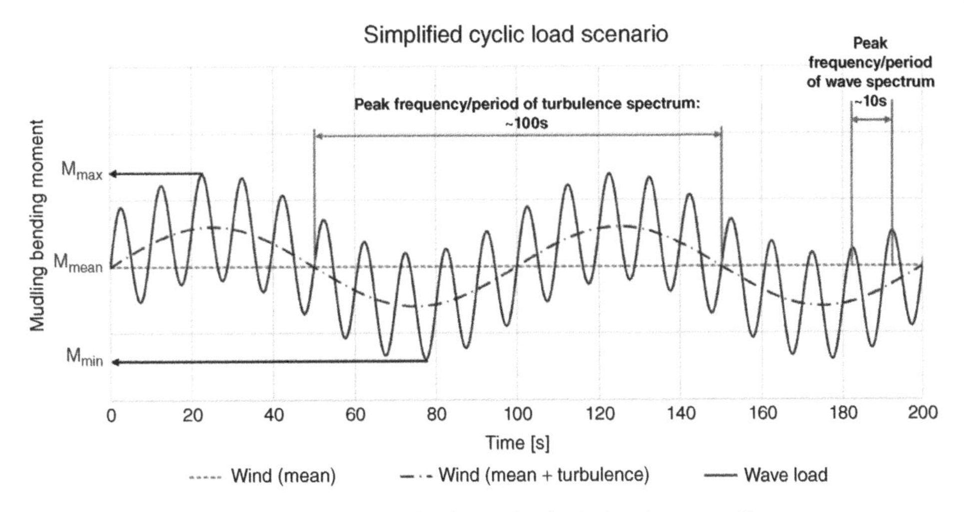

Figure 2.21 Mudline moment due to wind and wave load, wind and waves collinear.

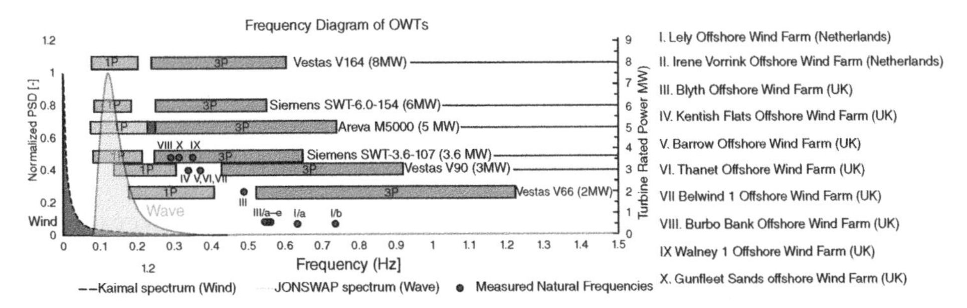

Figure 2.22 Importance of dynamics with deeper offshore and larger turbines.

a *soft-stiff* design is moving towards the left of the spectrum. For example, the target frequency of a 3 MW turbine is in the range of 0.35 Hz. In contrast, the target frequency for an 8 MW turbine is 0.22–0.24 Hz, which can be very close to wave frequencies. This can also be explained through a Campbell diagram plotted in Figure 2.23, which shows the narrow band of the target frequency for an 8 MW turbine.

ASIDE:

(1) For a soft-stiff 3 MW wind turbine generator (WTG) system, 1P and 3P loading can be considered as dynamic (i.e. ratio of the loading frequency to the system frequency very close to 1). Most of the energy in wind turbulence is in lower frequency variations (typically around 100 seconds peak period), which can be considered as cyclic. On the other hand, 1P and 3P dynamic loads change quickly in comparison to the natural frequency of the WTG system and therefore the ability of the WTG to respond depends on the characteristics, and dynamic analysis is therefore required.

(2) It is easily inferred that for large turbines (8 MW) sited in deeper waters, the wave loads will be highly dynamic (target frequency of the WTG system is 0.22 Hz and the most waves are in the frequency range of 0.05–0.2 Hz) and may control the design.

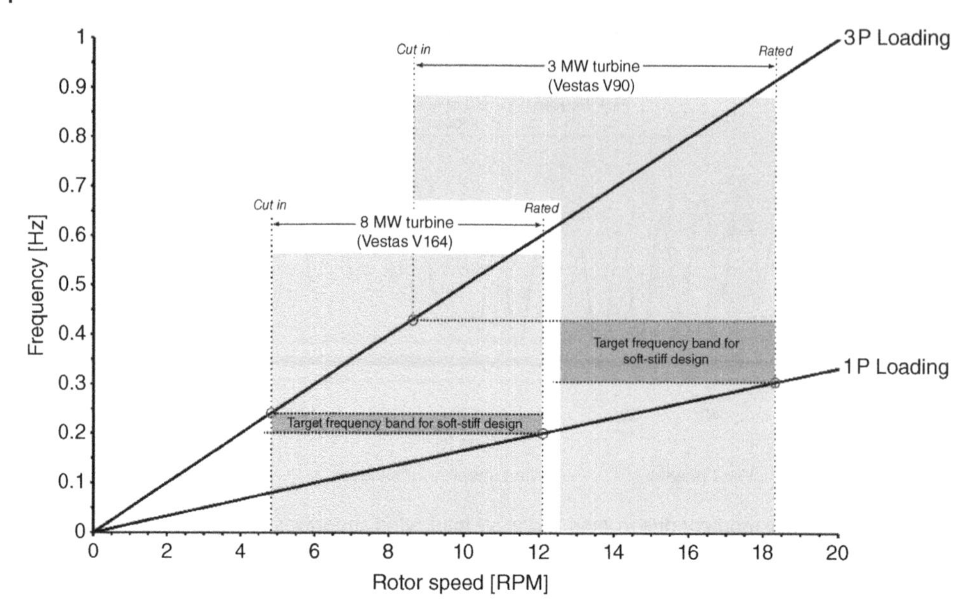

Figure 2.23 Campbell diagram for a 3 and 8 MW turbine.

2.9 Current Loads

Current loads act on the substructure due to the horizontal movement of water particles causing drag. Several effects may be the source of ocean currents and the most important from the point of view of OWT foundations are wind-induced currents and tidal currents. An approach to model currents is to choose a velocity profile, along the water depth and apply it to the substructure. Generally speaking, the velocity of current in the upper regions are driven by wind and in deeper down by tidal and other effects. The velocity typically reduces to zero at the seabed.

Since current speeds are subjected to high levels of uncertainty, a conservative approach to current load modelling is to assume a constant current velocity along the depth. The velocity may be assumed to be the maximum value expected in 50 years and may be taken as 1% of the 50-year extreme mean wind speed. Using the constant profile, the load due to current can be estimated similar to the drag load acting on the substructure.

2.10 Other Loads

There are other loads that need to considered while designing a wind farm and are listed below:

(1) Ice loads on the blades
(2) Floating ice sheet impact loads on the substructure
(3) Ship impact loads on the substructure
(4) Loads due to installation errors and residual stresses on the foundation

(5) Start-up loads and emergency shut-down loads
(6) Earthquake and tsunami loads
(7) Emergency fault case loads

Specialist literature and codes are recommended for further reading. Only earthquake and tsunami loads are discussed here.

2.11 Earthquake Loads

Many of the potential offshore wind farm locations are seismically active and therefore require proper analysis to safeguard investments. Wind energy production from offshore wind farms is a reality nowadays around the world. Figure 2.24 shows the main countries that are developing and investing in offshore wind power according to the Global Wind Energy Council following De Risi et al. (2018). In the same figure is shown a global seismic hazard map in terms of peak ground acceleration (PGA) with probability of exceedance of 10% in 50 years based on GSHAP model. Several countries are in high seismic regions, including the United States, China, India, and South-East Asia, and are also adjacent to subduction zones shown by blue lines. In these zones magnitude M9-class megathrust earthquakes can potentially occur.

Figure 2.25 shows normalised spectra for typical wind, sea wave, and ground shaking and on the same figure range of periods for the first vibration modes and higher modes are also superimposed. It is possible to observe that the first modes usually fall in the range of periods where the spectral content of earthquake ground motions are in decay but are still relatively high, and this depends on the characteristics of earthquake scenarios and local site conditions. Therefore, it is essential to understand which seismic events and soil conditions may excite such periods and if higher modes play significant role in the seismic behaviour.

Design of OWT structures in seismic regions involves following steps:

(1) A detailed seismic hazard analysis (SHA) must be conducted. One of the important outcomes will be PGA expected at the site during its design life in deterministic and probabilistic format.

Figure 2.24 Map of countries investing in offshore wind farms (red boundaries), subduction trenches (blue lines), and global seismic hazard map, see De Risi et al. (2018).

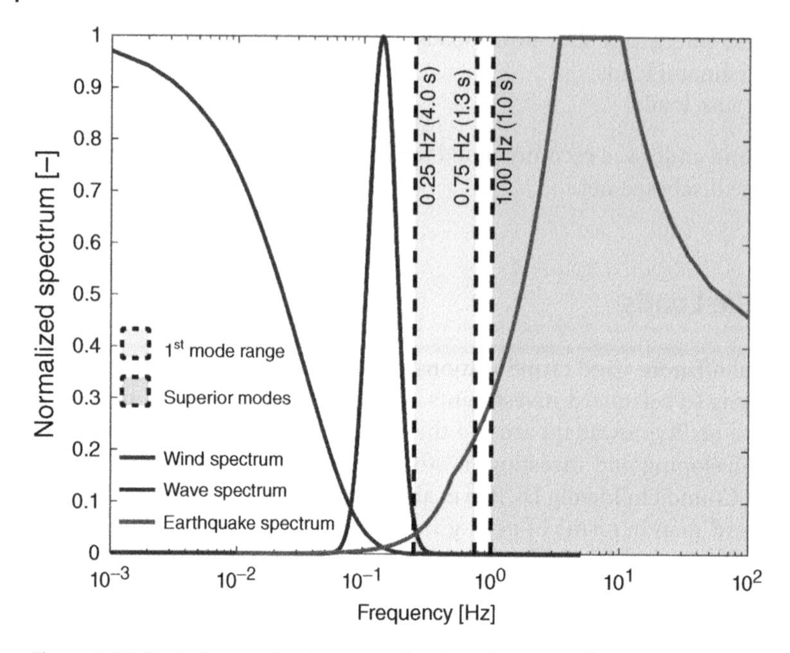

Figure 2.25 Typical normalised spectra of actions due to wind, sea waves, and earthquake ground motions. The yellow and orange bands represent the ranges of vibration periods for conventional wind turbines corresponding to main and higher modes, respectively, in De Risi et al. (2018).

(2) Identification of potential seismic hazards at the site. This must also include cascading events. Examples are:
   (a) Effect of large fault movements.
   (b) Shaking with no-liquefaction of the subsurface. This includes inertial effects on the structure and inertial bending moment on the foundation piles.
   (c) Shaking + Liquefaction of the subsurface. Liquefaction may lead to a large, unsupported length of the pile and elongate the natural vibration period of the whole structure. The ground may liquefy very quickly or may take time and is a function of ground profile and type of input motion. In such scenarios, the transient effects of liquefaction need to be considered, as it will affect the bending moment in the piles.
   (d) Shaking + Liquefaction + Tsunami.
   (e) Shaking + Landslide.
   (f) Earthquake sequence such as: Foreshock + Mainshock + Aftershock.
(3) Generation of input motion for the site depends on the seismo-tectonics of the area. This includes faulting pattern, distance of the site from earthquake source, wave travelling path, geology of the area, etc. This can be either synthetic (artificially generated) or recorded ground motion from previous earthquakes.
(4) Site response analysis is how the ground will behave under the action of the input motion.
(5) Dynamic SSI analysis incorporates the knowledge of the site response into the calculations.

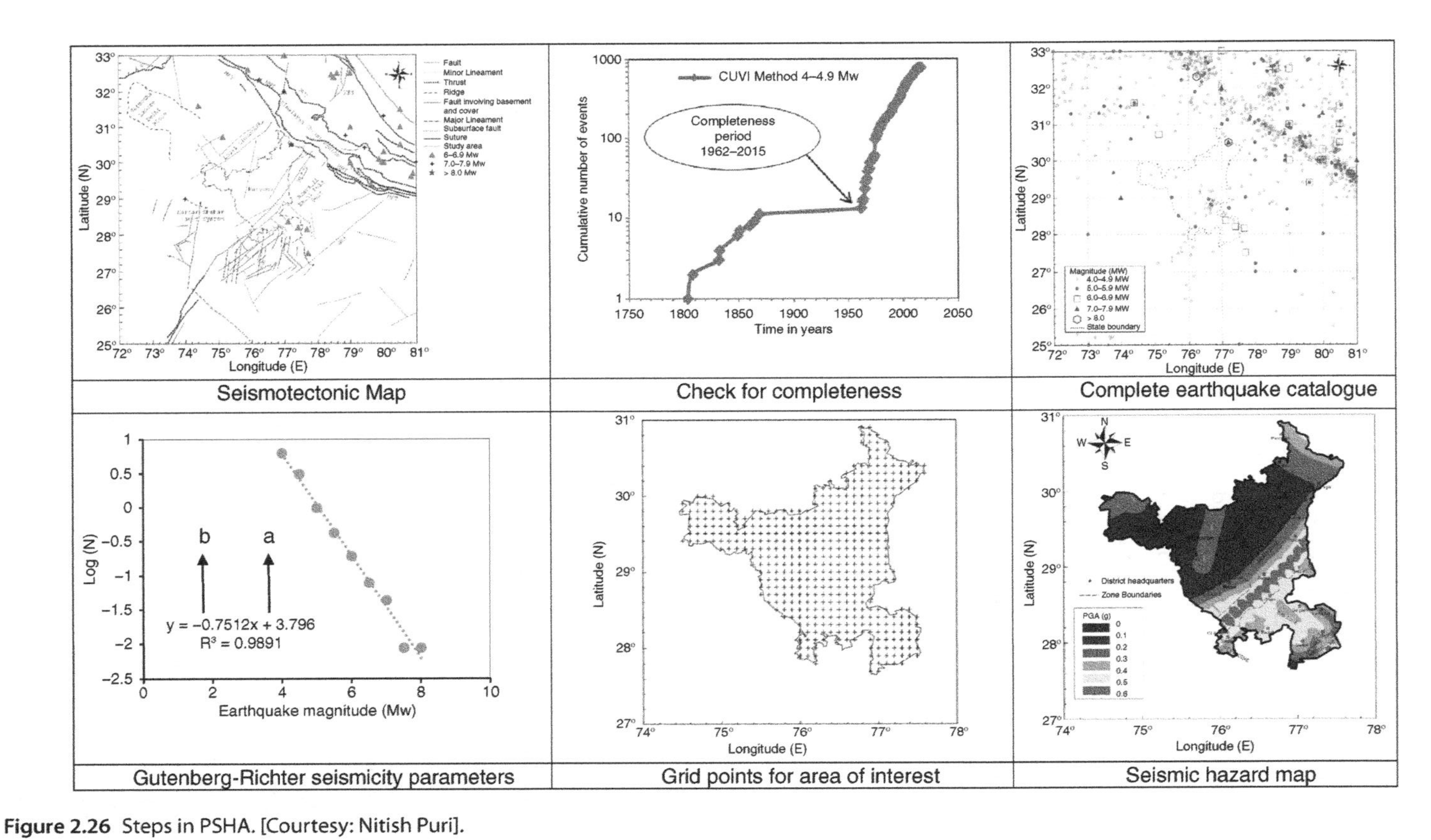

Figure 2.26 Steps in PSHA. [Courtesy: Nitish Puri].

2.11.1 Seismic Hazard Analysis (SHA)

A detailed SSA can be divided into three parts: (i) Estimation of seismic hazard by carrying out deterministic seismic hazard analysis (DSHA) and probabilistic seismic hazard analysis (PSHA); (ii) evaluation of local site effects by carrying out nonlinear ground response analysis; and (iii) preparation of hazards maps and response spectra.

Two approaches, probabilistic (PSHA) and deterministic (DSHA), are commonly adopted for seismic hazard assessment. In DSHA, a particular earthquake scenario is assumed based on earthquake data and tectonic setup of the study area. The hazard is estimated on the basis of attenuation characteristics of the region. For a worst-case scenario, DSHA is more useful. Hazard is evaluated considering the closest distance between the source and the site of interest as well as the maximum magnitude for the fault.

To evaluate seismic hazard deterministically for a particular site or region, all possible sources of seismic activity are identified and their potential for generating strong ground motion is evaluated. It is used widely for nuclear power plants, large dams, large bridges, hazardous waste containment facilities, and also as a 'cap' for PSHA. On the contrary, in probabilistic method (PSHA), the uncertainties in earthquake size, location, and time of occurrence are explicitly considered, see Figure 2.26 for the steps.

The procedure followed in carrying out the SHA is as follows:

1. Consider an area covering 300–500 km radius around point of interest as seismic study area.
2. Collect all the tectonic information such as location, magnitude, and depth of earthquakes occurred, and satellite imageries of the fault lines for the seismic study area.
3. All this information has to be merged to prepare a single map (seismotectonic map), which represents tectonic setting of the seismic study area.
4. Prepare an earthquake catalogue for the study region, which includes all recordings from pre-instrumental and instrumental period. Try to collect data from all the possible sources. Check the catalogue for duplicity of events, clustering of events, and completeness.
5. Determine Gutenberg-Richter seismicity parameters a and b for the estimation of rate of occurrence of different magnitudes of earthquake and return periods. For DSHA, this step can be avoided.
6. Identify the tectonic features likely to generate significant ground motions in the study area. Also assign maximum observed magnitude (Mobs) to each source.
7. Calculate the maximum magnitude potential (M_{max}) for all the seismogenic sources present in the study area. For PSHA, it is important to consider uncertainty in size of earthquake.
8. Consider a suitable grid on the area of interest for e.g. $0.2° \times 0.2°$ for a country, $0.1° \times 0.1°$ for a state, and $0.005° \times 0.005°$ for a city.
9. Compute shortest distance of grid points to each seismogenic source. For PSHA, it is important to consider uncertainty in the location of earthquake.
10. Select a suitable ground motion prediction equation (GMPE), which accurately represents the tectonic setup of the study area. In case of nonavailability of regional GMPEs, GMPEs developed for other regions can also be used, provided their source and site characteristics are similar to the study area. For PSHA, it is necessary to handle epistemic uncertainty in the GMPEs. This can be handled by adopting logic trcc approach.

11. For the computation of hazard at a grid point, compute ground motion parameters with respect to various seismogenic sources and identify the source causing maximum ground motion at the point of interest. For that grid point, consider the maximum magnitude potential of that seismogenic source as the controlling earthquake.
12. Site amplification factors can be estimated by carrying out linear, equivalent linear, or nonlinear ground response analysis.

2.11.2 Criteria for Selection of Earthquake Records

Ground motion records to be used in analysis should represent the potential earthquake hazards at the site i.e. consideration of magnitude, distance, site conditions, source mechanism, directivity, and other effects obtained from SSA. In this regard, the appendix of ISO 19901-2 states that a minimum of four records should be used for analysis, and that each record should be selected according to the following criteria:

> Given the magnitude and distance of events dominating extreme level earthquake (ELE) ground motions, the earthquake records for time history analysis can be selected from a catalogue of historical events. Each earthquake record consists of three sets of tri-axial time histories representing two orthogonal horizontal components and one vertical component of motion. In selecting earthquake records, the tectonic setting (e.g. faulting style) and the site conditions (e.g. hardness of underlying rock) of the historical records should be matched with those of the structure's site.

Three approaches are often used, and all these methods have their advantages and disadvantages and are discussed later in this section through an example.

2.11.2.1 Method 1: Direct Use of Strong Motion Record

In this method, a strong motion is chosen to match the seismicity of the place, either in terms of moment magnitude or expected peak bed rock acceleration (PBRA) also known as PGA. It may be difficult to find a recorded motion satisfying both the requirements. For example, a site may expect a magnitude 6.0 earthquake and a PBRA of 0.2 g. Therefore, a designer has to look for the recorded data matching these characteristics. This method is rarely used in practice, as it is quite unlikely that the shape of the spectra will match the expected bedrock spectra at a particular site. This will be illustrated later using an example.

2.11.2.2 Method 2: Scaling of Strong Motion Record to Expected Peak Bedrock Acceleration

In this method, an earthquake record is chosen from a database to match the moment magnitude and then the maximum amplitude is scaled to match the expected PBRA.

2.11.2.3 Method 3: Intelligent Scaling or Code Specified Spectrum Compatible Motion

Traditionally, seismic hazard at a site for design purposes has been represented by design spectra. Thus, all seismic design codes and guidelines require scaling of selected ground

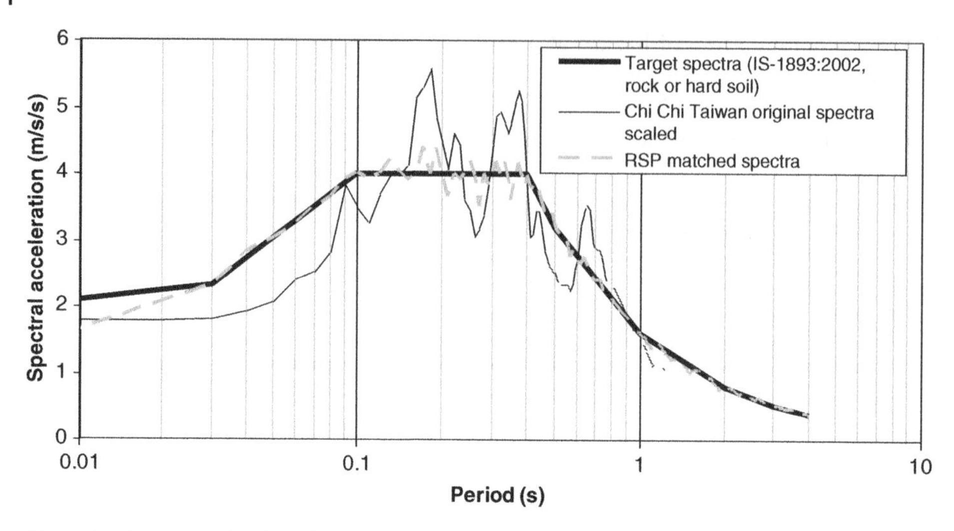

Figure 2.27 An example of earthquake spectra of selected motion matched to code-specified spectra.

motion time histories so that they match or exceed the controlling design spectrum within a period range of interest. Figure 2.27 shows the response spectra of IS: 1893 (2002) for hard soil or rock together with a Chi-Chi (Taiwan) earthquake motion and spectrally matched motion. Several methods of scaling time histories are available that involve rigorous mathematics.

Essentially, in all these methods, an input motion is selected from a strong motion database, and the motion is manipulated in such a way so as to obtain a motion that matches the target design spectra. The manipulation can be carried out in frequency-domain where the time-frequency content of the recorded ground motions is manipulated in order to obtain a good match. Mathematically, this can be expressed as follows, following Alexander et al. (2014): a process that can transform a known signal or input motion $x(t)$ into a similar input motion $y(t)$ that matches the spectrum. The additional aim is that transformed signal $y(t)$ should maintain very similar non-stationary characteristics such as general envelope, time location of large pulses, and variation of frequency content with time. In other words, the aim signal $y(t)$ should look, for all intents and purposes, like a real earthquake.

Figure 2.28 shows the response spectra for different types of earthquake that can also be used to estimate the inertia forces on the structure. For example, for a three-second period wind turbine structure, the spectral acceleration will be function of the location and can be read off the ordinate (y-axis).

ASIDE: Attraction of Scaling Input Motion in the Way as Described Above

One of the main sources of uncertainty in earthquake engineering is the earthquake themselves, i.e. their spatiotemporal incidence and time-series. Bounding this uncertainty in the predicated ground motion time-series is a complex question. As a result, design codes around the world, concentrate on satisfying a response spectrum of some kind. However, for major or critical infrastructure projects (such as nuclear reactor or

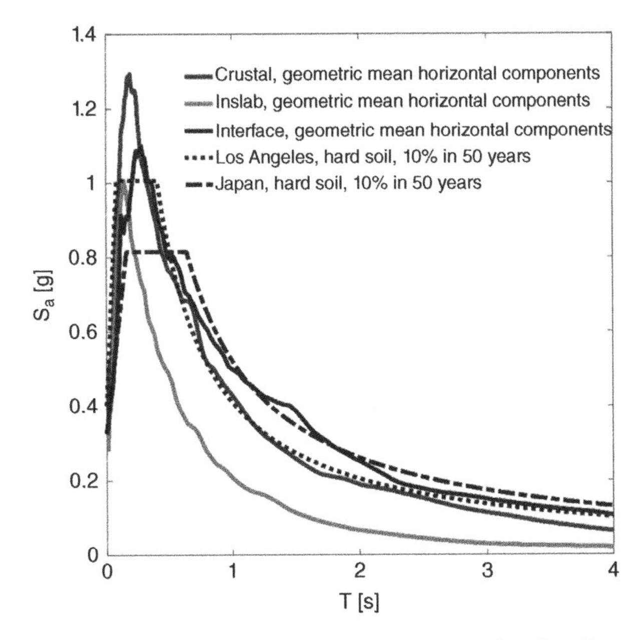

Figure 2.28 Response spectra for different types of earthquakes.

important lifeline structures such as bridges or hospitals) ground motion time-series are needed for time-history analyses. Now the question is what input motion to choose and how many of them are required? The brute-force method is to randomly sample all accelerograms in order to obtain a reasonable good estimate of the population statistics. This, of course, would require thousands or even hundreds of thousands of records to be effective. Therefore, a selective or stratified sampling of a small set of real accelerograms can be selected that satisfies certain geophysical and structural criteria. Spectrally matched records are therefore often used in practice. While this approach can be easily criticised as it modifies the recorded time series by adjusting its time-frequency content. It is also considered overly conservative as it matches a flat and smooth broadband spectrum unlike most real earthquakes. The perceived attractiveness of spectrally matched records is that fewer accelerograms need be used in time-history analyses.

2.11.3 Site Response Analysis (SRA)

Once the bedrock motion is obtained, the vertically propagating shear waves are convoluted through the soil depth using a SRA to obtain the site effects. Figure 2.29 shows a schematic diagram of the methodology. Various software exist in performing this particular function such as SHAKE 91, DEEPSOIL, DMOD, EERA, Cyclic 1D. For liquefiable deposits, Cyclic 1D can be used to perform the SRA, as it uses an advanced constitutive model for the liquefied soil developed by Parra (1996) and Yang (1999) in its analysis. Literature can be found on the use of various site response analysis programs, which capture various nonlinear aspects of the soil in Hashash et al. (2010).

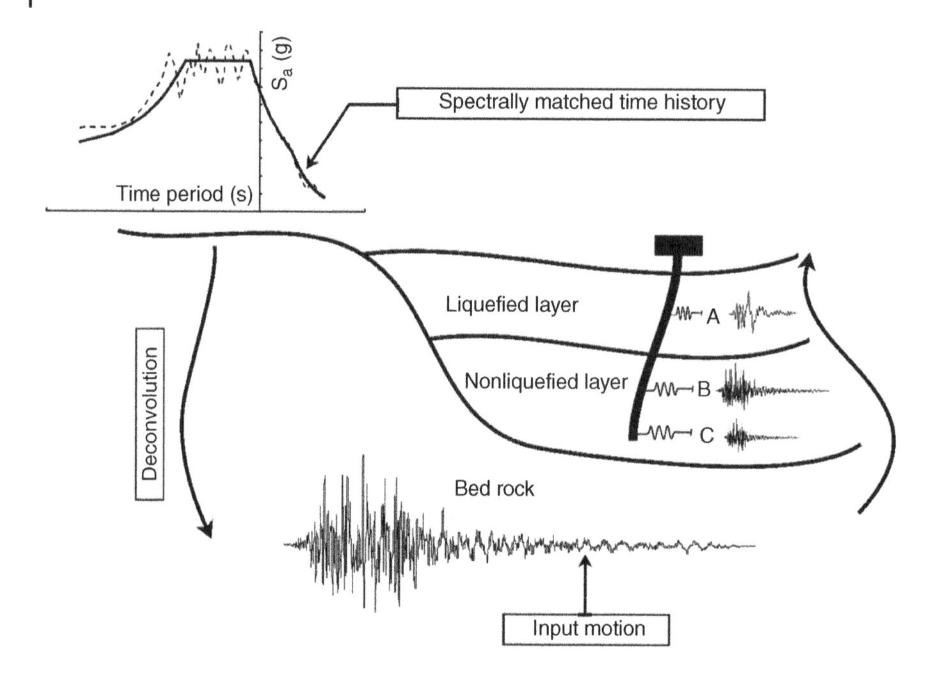

Figure 2.29 Schematic of ground response analysis.

2.11.4 Liquefaction

Apart from site response analysis, it is necessary to predict if the subsurface will liquefy. The part of the foundation in liquefiable soil needs to be considered as laterally unsupported. Therefore, the length of the tower will effectively increase leading to elongation of natural period. From the point of analysis, liquefaction may be assumed to progress along the depth of the soil layer in top-down fashion. A modal analysis can be carried out to study the variation of depth of liquefaction (DL) with the first natural period of the soil-pile system. The results are plotted in Figure 2.30, which is a graphical representation of normalised elongation of period for various depths of liquefaction. In Figure 2.30, DL represents depth of liquefaction, L is the embedded length of the pile, T is the time period, and $T_{initial}$ is the initial time period of the structure. However, this is momentary, and as the earthquake stops, the ground resolidifies and most of the ground stiffness is regained. Three cases are considered in the study:

1. Post liquefaction, the American Petroleum Institute (API) prescribed soil spring stiffness is degraded 92% using a p-multiplier as per Brandenberg (2005).
2. Zero initial stiffness and strength are provided to the p-y spring for the liquefied soil as per RTRI (1999).
3. Hyperelastic p-y spring for liquefiable soils (Lombardi et al. (2017).

As the eigenvalue analysis is linear, only the initial stiffness of the p-y curves (reduced API, hyperelasticity, zero strength API) is used in the model to compute the natural frequencies of the soil-pile system.

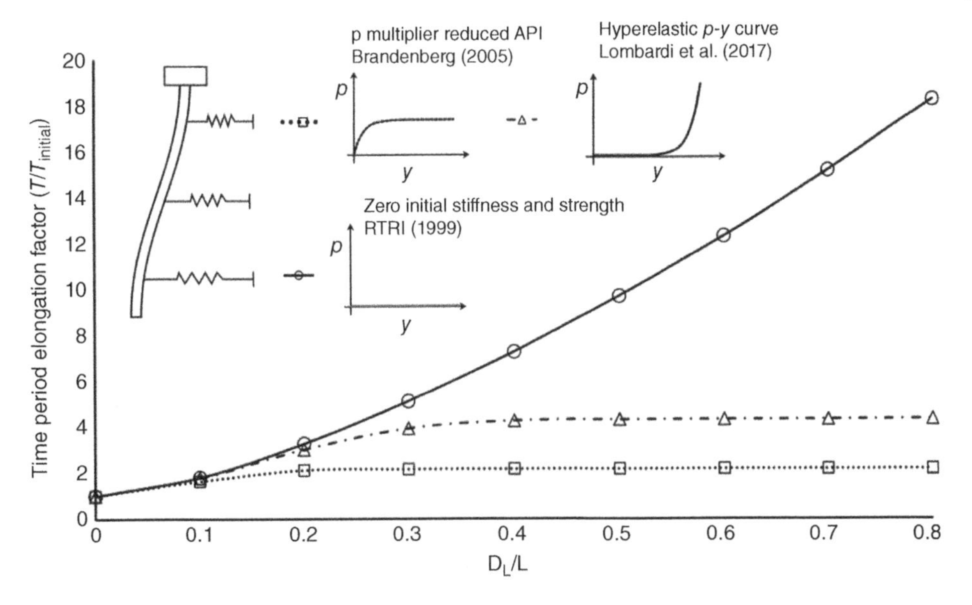

Figure 2.30 Elongation of a period of a pile due to different liquefiable depths for a given pile geometry and ground profile.

2.11.5 Analysis of the Foundation

Analysis of foundations is usually carried out using beams on nonlinear Winkler foundation approach as shown in Figure 2.31. Various details of modelling are discussed in Dash et al. (2010). Once the DL is estimated, it is important to use the Winkler or p-y springs of appropriate shape as shown in the figure. The shape of non-liquefied and

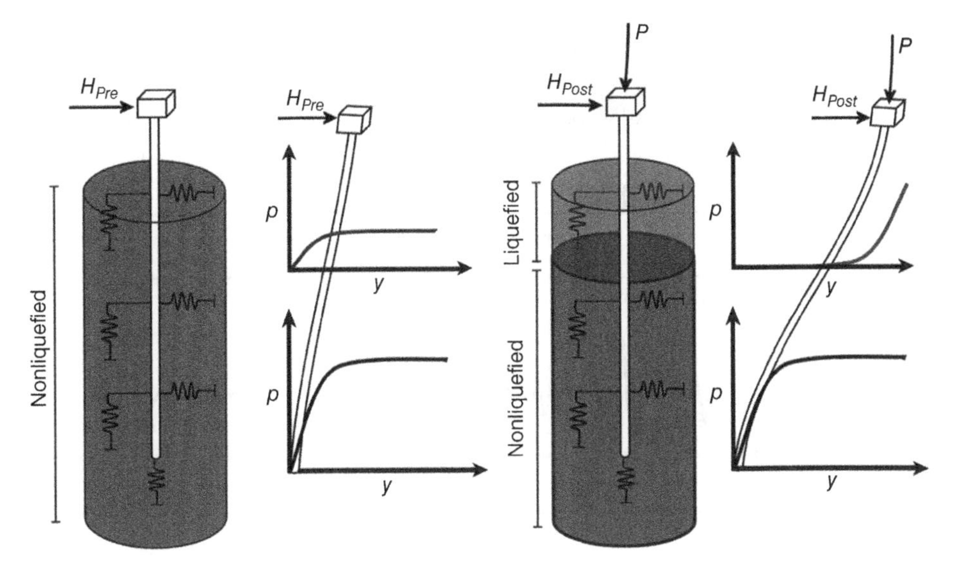

Figure 2.31 Beam on nonlinear winkler model.

liquefied p-y curves are very different. The seismic forces, H_{Pre} (pre-liquefaction) and H_{Post} (post-liquefaction), depend on the time period of the whole structure and can be estimated based on the response spectra of the site.

The aim of this section is to further the discussion carried out in Section 1.1 of Chapter 1 on the good performance of a near-shore wind farm. Calculations were carried out based on the available information of the ground profile and the turbine characteristics. In the absence of data, engineering judgements were used to carry out the analysis in order to draw broad conclusions. The focus of the investigation is the behaviour during the 2011 Tohoku earthquake seismic shaking and during the tsunami run-up, discussed in the following examples.

Example 2.7 ***Kamisu Wind Turbine Structure*** The wind farm consists of 2 MW OWT structure located in Kamisu wind farm and in the study two types of earthquake strong motion were used: 2011 Tohoku earthquake (Mw = 9.0) and 1995 Kobe (Mw = 6.9). The foundation is a monopile 3.5 m in diameter and 25 m long. The subsurface is sandy soil with varying relative density, ranging from 40% to 76%. Acceleration time series as shown in Figure 2.32 is used for the analysis. As can be observed, peak acceleration for the 2011 Tohoku earthquake is 0.22 g, while it is 0.29 g for the Kobe earthquake (1995). The Tohoku earthquake was of long duration (250 seconds) while the Kobe earthquake had a duration of about 20 seconds. Further details of the Tohoku earthquake can be found in Bhattacharya et al. (2011a), Goda et al. (2013). Tohoku earthquake is a subduction earthquake and Kobe is categorised as a shallow earthquake. PSD function for both earthquakes are also shown in Figure 2.33. Kobe earthquake is long-predominant-period earthquake while Tohoku earthquake is low-predominant-period earthquake. Macro-element method developed by Shadlou (2016), is used to carry out the analysis which also incorporates liquefaction.

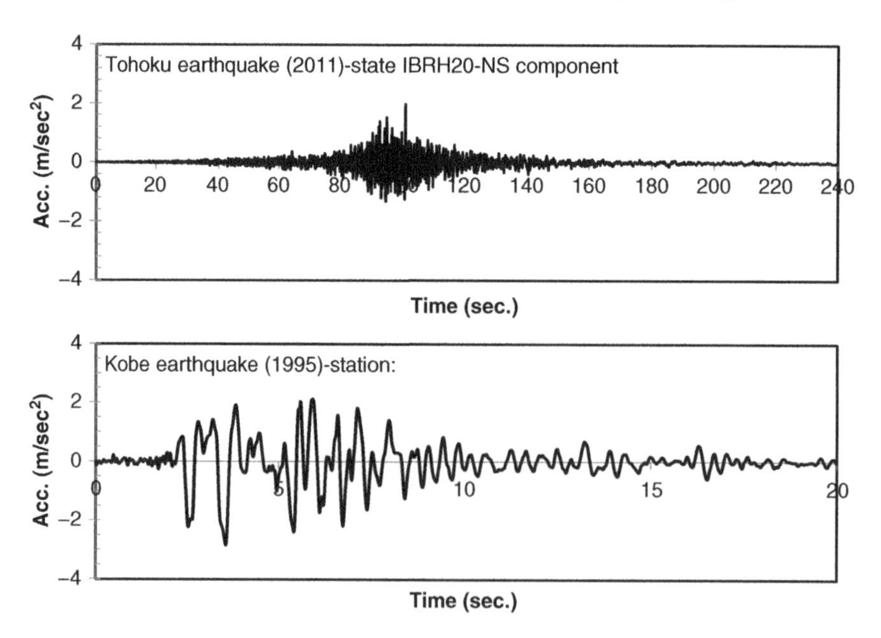

Figure 2.32 Acceleration input motions.

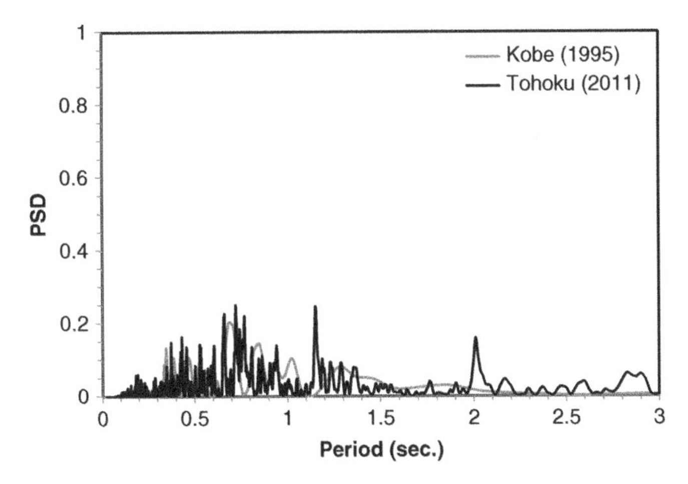

Figure 2.33 Power spectral density of the input motions in this research.

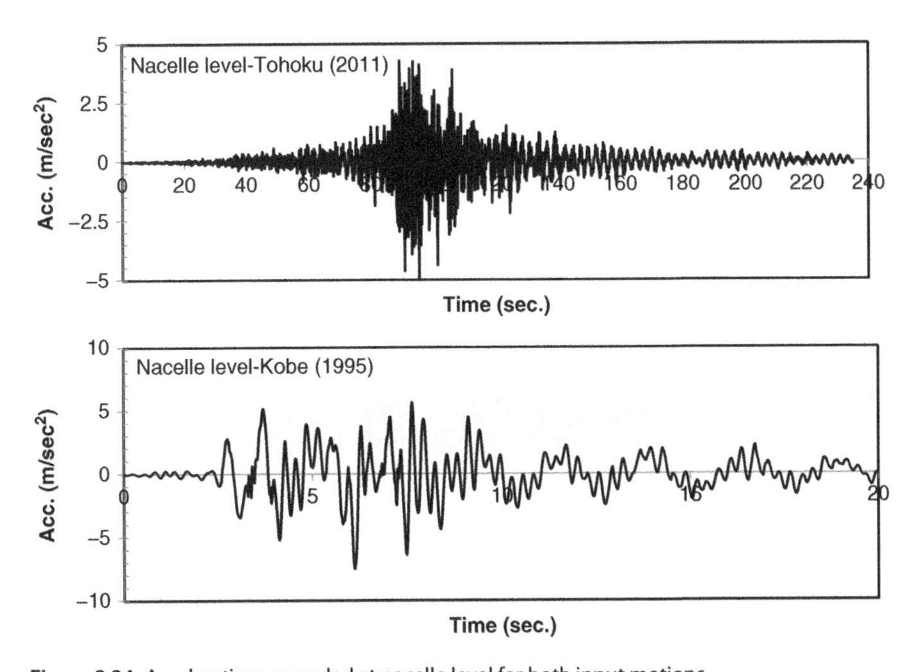

Figure 2.34 Accelerations recorded at nacelle level for both input motions.

Results of the nonlinear time series analysis using Macro-element model are shown in Figures 2.34–2.37. As can be observed, maximum acceleration observed at nacelle is lower in Tohoku earthquake than that of Kobe earthquake. Similar trends are also observed for displacement recorded at nacelle level for Tohoku and Kobe earthquakes. Nacelle displacement is estimated to be 30 cm during Kobe earthquake while it is 18 cm during Tohoku earthquake. Maximum nacelle rotation is about 0.2° and 0.5° for Tohoku and Kobe earthquake, respectively. This shows that the demand due to Kobe earthquake is higher than Tohoku. Recorded rotation at pile-head is 0.07° and 0.14° for Tohoku and

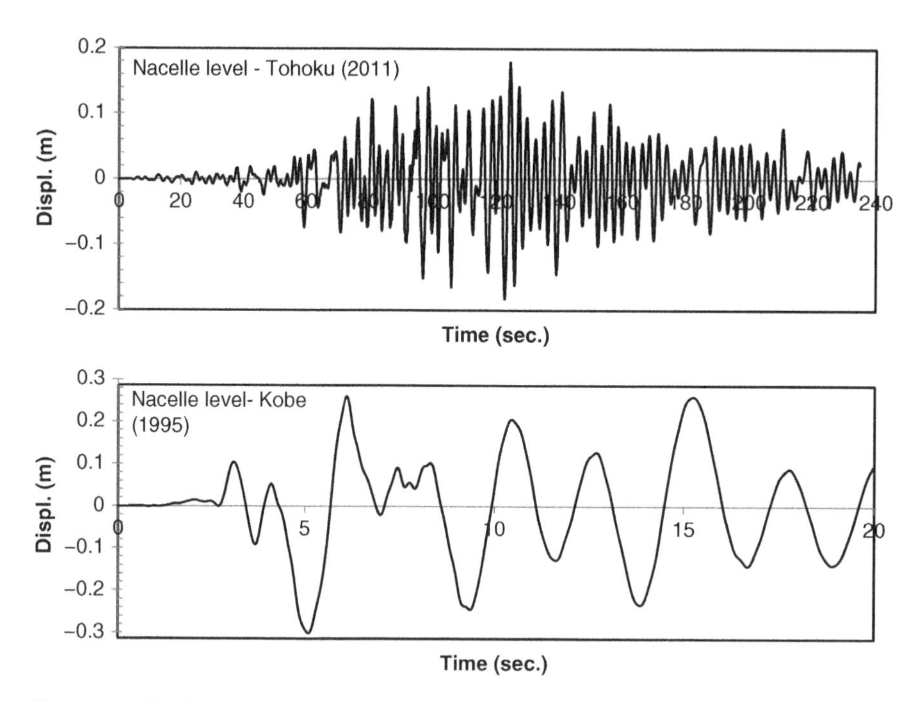

Figure 2.35 Displacement time series recorded at the nacelle level.

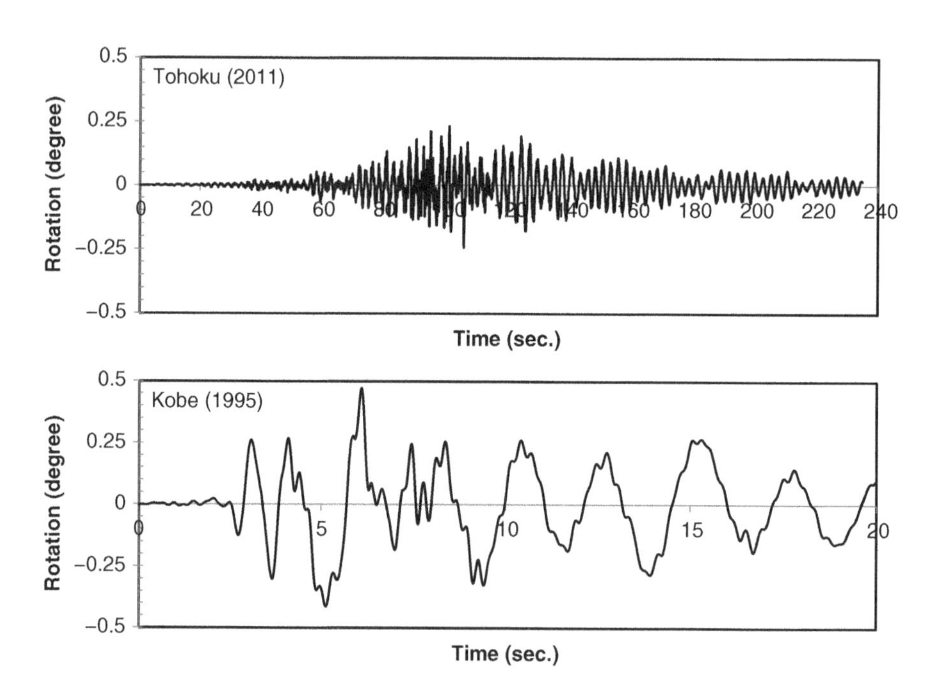

Figure 2.36 Rotation of the nacelle for both earthquakes.

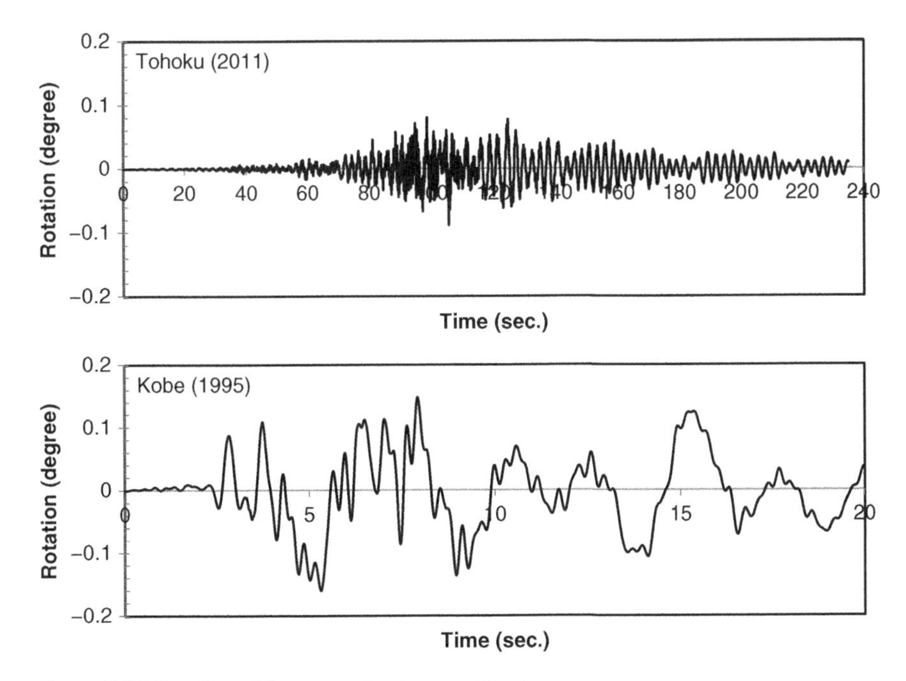

Figure 2.37 Rotation of the monopile at ground level.

Kobe earthquake, respectively. A permanent rotation is not observed at pile-head level during both earthquakes.

Two general points may be made:

1. The displacement and rotation at the nacelle level as well as foundation level depends on the strong motion characteristics and the ground profile.
2. If the ground liquefies, there can be a tendency for settlement as the shaft resistance needs to be transferred to the lower layers.

Example 2.8 ***Tsunami Run-Up Analysis*** To simulate behaviour of the Kamisu wind farm due to Tsunami run-up different technique may be used. In this book, ASCE 7-2016 is used to model hydrostatic and hydrodynamic effects. In simplified method, tsunami loading can be estimated as a triangular distribution of hydrostatic pressure and a uniform distribution pressure representing hydrodynamic effects as shown in Figure 2.38. The hydrodynamic term Figure 2.38 is quantified by a variable factor, b. Hydrodynamic term is $b\rho_w g D_p$, where ρ_w is water density, D_p is diameter of the superstructure.

As the rotation of the monopile foundation is the criteria for SLS (details given in Chapter 3), the lateral force and bending moment is applied on the pile-head and the rotation is calculated using the impedance matrix presented by Shadlou and Bhattacharya (2016) and provided in Chapter 4. Results are shown in Figure 2.39 for different b values. Two areas have been marked on the figure corresponding to different wave heights: Fukushima province (location of ill-fated nuclear power plant) and Ibaraki province (location of Kamisu wind farm). The pile-head rotation is acceptable zone Kamisu wind farm even if b is 10. Tsunami loading is an active area of research and readers are referred to specialised and latest literature on hydrodynamic loading.

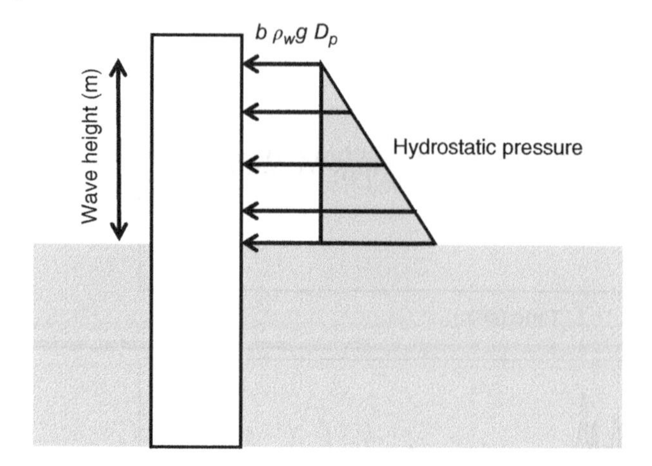

Figure 2.38 Statement of loading by tsunami run-up.

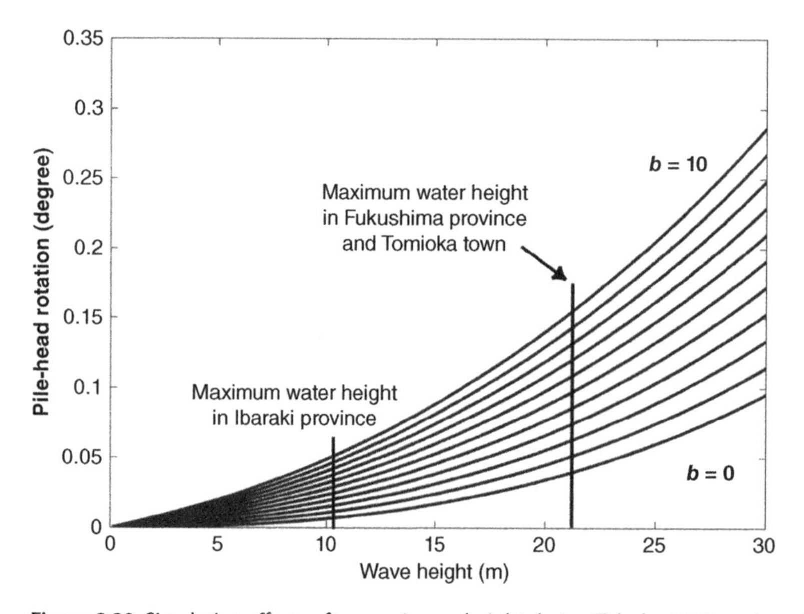

Figure 2.39 Simulating effects of tsunami wave height during Tohoku 2011 earthquake.

Shall We Combine Liquefaction and Tsunami Effects?

To understand effects of tsunami on OWTs, it may be useful to evaluate the cascading effect of Earthquake shaking + Soil liquefaction + Tsunami. The tsunami arrived circa 30 minutes following the earthquake, and therefore it is necessary to predict if there were complete dissipation of pore water pressure in soil stratum from the beginning of the seismic shaking till the tsunami run-up. Figure 2.40 shows the isochrones of excess pore water pressure generated by the shaking together with two extreme limits [hydrostatic condition and fully liquefied condition]. The analysis predicts that complete dissipation

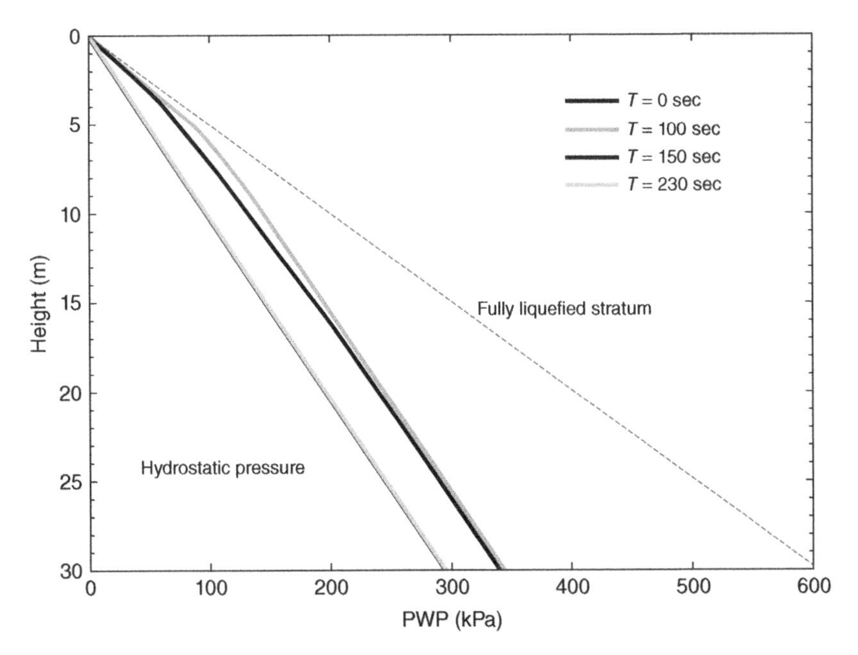

Figure 2.40 Isochrones of pore water pressure during earthquake showing the dissipation and the extreme values (hydrostatic and fully liquefied).

took place at about 230 seconds and the ground recovered most of its stiffness. In this particular case, tsunami effects may be evaluated alone.

2.12 Chapter Summary and Learning Points

(1) One of the first step in design of foundation is to fix the target frequency. To visualise the 'target frequency' it is necessary to construct the spectrum of the frequencies acting on the wind turbine structure, and this requires understating of wind and wave loading. The method proposed to visualise the target frequency is given in Figures 2.4 and 2.5, and it may be noted that the magnitudes of the spectral densities are normalised to 1. In reality, wind and wave will have few orders of magnitude higher energy than 1P and 3P. Order of magnitude calculations are shown in the chapter for some case studies.

(2) The motivation behind PSD formulation of the bending moment for the different loads is to provide a basis for a quick frequency domain fatigue damage estimation. This is particularly useful in the preliminary design phase of these structures, which is otherwise a very lengthy process and is usually done in time domain. This would also encourage integrated design of OWTs incorporating the dynamics and fatigue analysis in the early stages of structural design. Examples of frequency domain methods are the Dirlik, Tovo-Benasciutti, $\alpha_{0.75}$ methods.

(3) The formulations and methods presented had an objective to use the minimum information possible to obtain the loads such that simple spreadsheet-based

methods can be used to estimate the loads. Information such as the control parameters, the blade design, aerofoil characteristics, and generator characteristics are not included in this formulation. These can be carried out in aero-hydro-servo-elastic analysis such as HAWC, FAST, BLADED.

(4) For soft-stiff design, where the target frequency is placed between upper bound of 1P and lower bound of 3P, any change in natural frequency will enhance the dynamic amplifications, which will increase the vibration amplitudes and thus the stresses and fatigue damage on the structure.

3

Considerations for Foundation Design and the Necessary Calculations

Learning Objectives

Wind turbine structures are subjected to dynamic loads and modes of vibration of these structures play a key role in design considerations. Therefore, the chapter starts with a section on modes of vibration. The chapter then discusses the design consideration guided by limit states: ULS (ultimate limit state), SLS (serviceability limit state), and FLS (fatigue limit state). Issues related to installation are also discussed. This chapter will also discuss the calculations that needs to be carried out the designers: Ultimate capacity of the foundation, natural frequency of the whole system, deflection, and rotation of the foundation. This also includes allowable rotation and deflection, long-term tilting of the foundation, and allowable change in natural frequency of the whole system, installation of foundations including driveability and long-term issues such as soil erosion (scour), tilt, etc. For the various sections, fundamental, theory, and methodology will be described along with some solved examples. References will be provided to existing textbooks and codes of practice.

3.1 Introduction

Offshore wind turbines are complex machines, and the structure is dynamically sensitive. Furthermore, these are located in challenging offshore environment and, therefore, there are several types of design considerations. Before the design considerations are described, it is important to highlight some of the complexities and interdisciplinary nature of the issues. Construction of offshore structures in ultra-deep water is not new, and there are abundance of references and experience. However, what is new is the large-scale offshore wind turbine structures where a heavy rotating mass is placed at the top of the slender tower. Figure 3.1 shows the future of offshore wind turbines where heavier turbines are increasingly being placed in taller towers. One of the important design drivers is the dynamics of these systems. Therefore, the next section discusses the modes of vibration of these structures and how they affect the overall design considerations.

ASIDE

It is necessary to remind ourselves of the outcome of a design process, using as examples monopile and multiple-foundation supported structures. For monopile

Design of Foundations for Offshore Wind Turbines, First Edition. Subhamoy Bhattacharya.

Companion website: www.wiley.com/go/bhattacharya/offshorewindturbines

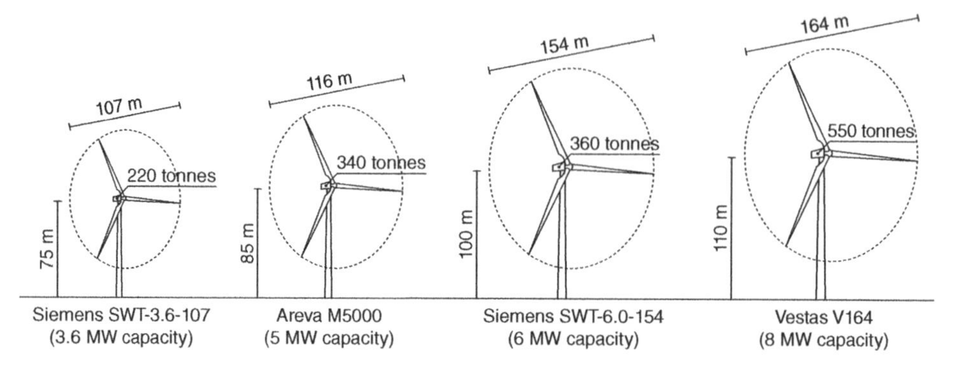

Figure 3.1 Future of offshore wind turbines.

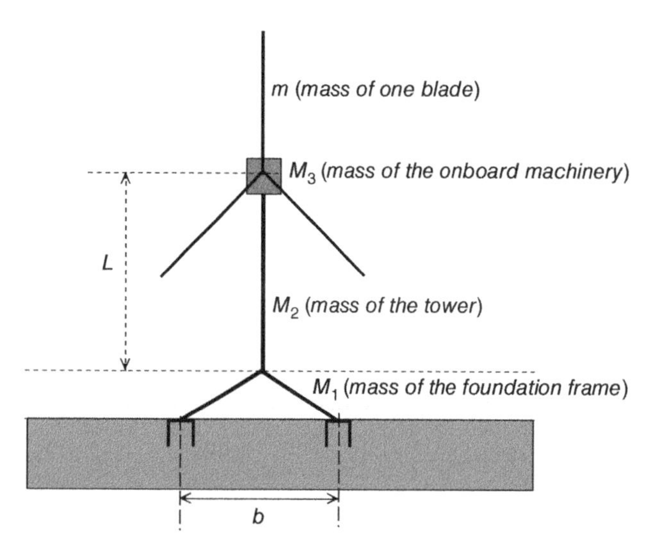

Figure 3.2 Definition of terms for design.

design, a designer needs to find three attributes to finalise a pile design – length, diameter, and thickness. On the other hand, for a multiple supported structure such as one shown schematically in Figure 3.2, the designer needs to attribute not only the dimensions of the foundation but also the numbers of such foundations and spacing between them.

3.2 Modes of Vibrations of Wind Turbine Structures

The modes of vibration depend on the combination of the foundation system (i.e. single foundation such as mono caisson or monopile or a group of piles or a seabed frame supported on multiple shallow foundations) and the superstructure stiffness and mass distribution. The fundamental modes of vibration for these structures can be mainly of two types:

1. *Sway-bending modes.* This consists of flexible modes of the tower together with the top rotor-nacelle assembly (RNA) mass that is essentially sway-bending mode of the tower. Effectively in these cases, the foundation is very stiff axially when compared with the tower and the tower vibrates. The foundation provides stiffness and damping to the whole system.
2. *Rocking modes.* This occurs when the foundation is axially deformable (less stiff) and is typical of wind turbine generators (WTGs) supported on multiple shallow foundations. Rocking modes can also be coupled with flexible modes of the tower.

The next section describes the modes of vibration through some examples. These aspects were investigated experimentally, numerically, and analytically by Bisoi and Halder (2014), Nikitas et al. (2016, 2017), and Lombardi et al. (2013). In the experimental methods, the modes of vibration were obtained from snap-back test, also known as free-vibration tests.

3.2.1 Sway-Bending Modes of Vibration

Essentially, this form is observed when the foundation is very rigid compared to the superstructure. Wind turbines supported on monopiles and jackets supported on piles will exhibit such kind of modes. Figure 3.3 shows a schematic diagram of modes of vibration for monopile supported wind turbines and Figure 3.4 shows schematic diagram of a jacket supported wind turbine system. It is important to note that the first two modes are quite widely spaced – the typical ratio is about four to six times. One of the important

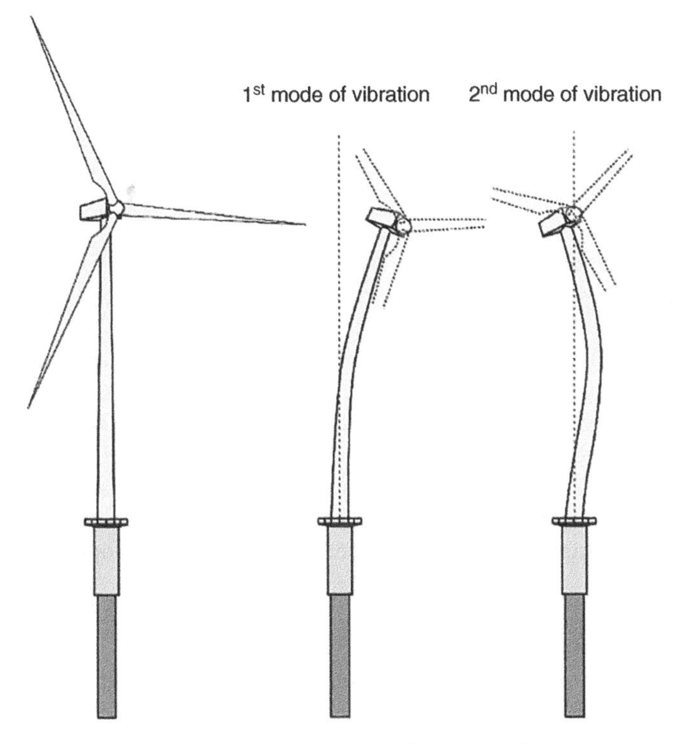

Figure 3.3 Modes of vibration for monopile supported wind turbines.

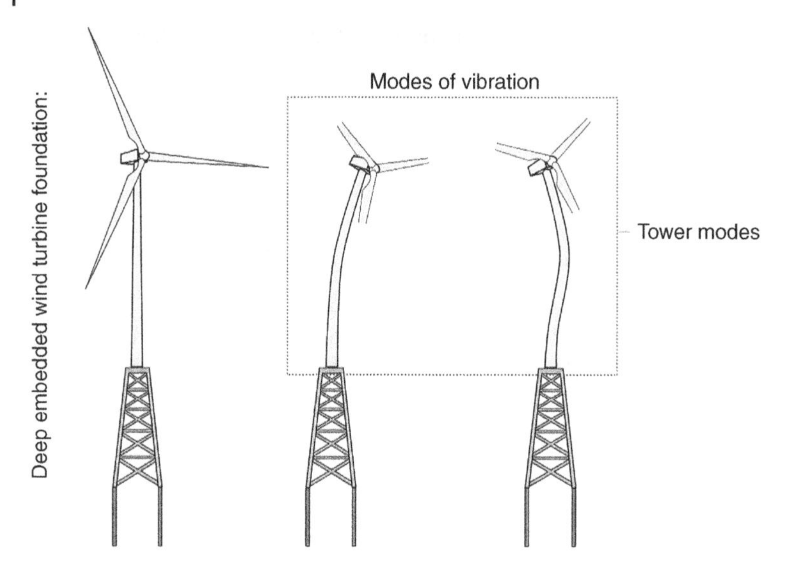

Figure 3.4 Schematic diagram of modes of vibration for jacket structures supported on piles.

points to note is that the foundation is axially very stiff. These analyses can be easily carried out using standard software.

It may be worth noting that the first three modes of vibration of a fixed-based cantilever beam are given by:

$$f_n = \frac{1}{2\pi}\alpha_n^2\sqrt{\frac{EI}{ML^4}} \tag{3.1}$$

where α_n is a mode number parameter having the value of 1.875, 4.694, 7.855 for the first, second, and third mode, respectively. EI is the bending stiffness of the beam having length L and M is the mass per unit length of the beam.

3.2.1.1 Example Numerical Application of Modes of Vibration of Jacket Systems

Figures 3.5 and 3.6 shows examples of numerical simulation of a jacket supported offshore wind turbine system where the modes of vibration may be appreciated. This is obtained through eigen solutions. For the twisted jacket (also known as inward battered guided structure [IBGS]), the frequency of first mode and second modes is 0.2678 Hz and the third mode is 1.049 Hz. It may be noted that the first two modes are identical and the task of the designer is to place this in the narrow band of allowable *soft-stiff* zone, discussed in Section 2.2 of Chapter 2. Similar observations were also noted for standard jackets on piles, see Figure 3.4. This figure also provides a plan view to visualise the mode shapes. Closed-form solution for obtaining natural frequencies of jacket is developed in Jalbi and Bhattacharya (2018).

3.2.1.2 Estimation of Natural Frequency of Monopile-Supported Strctures

Following the concept of *target natural frequency* (Section 2.2 in Chapter 2), in order to place the natural frequency of the whole system in the desired band, iterations are

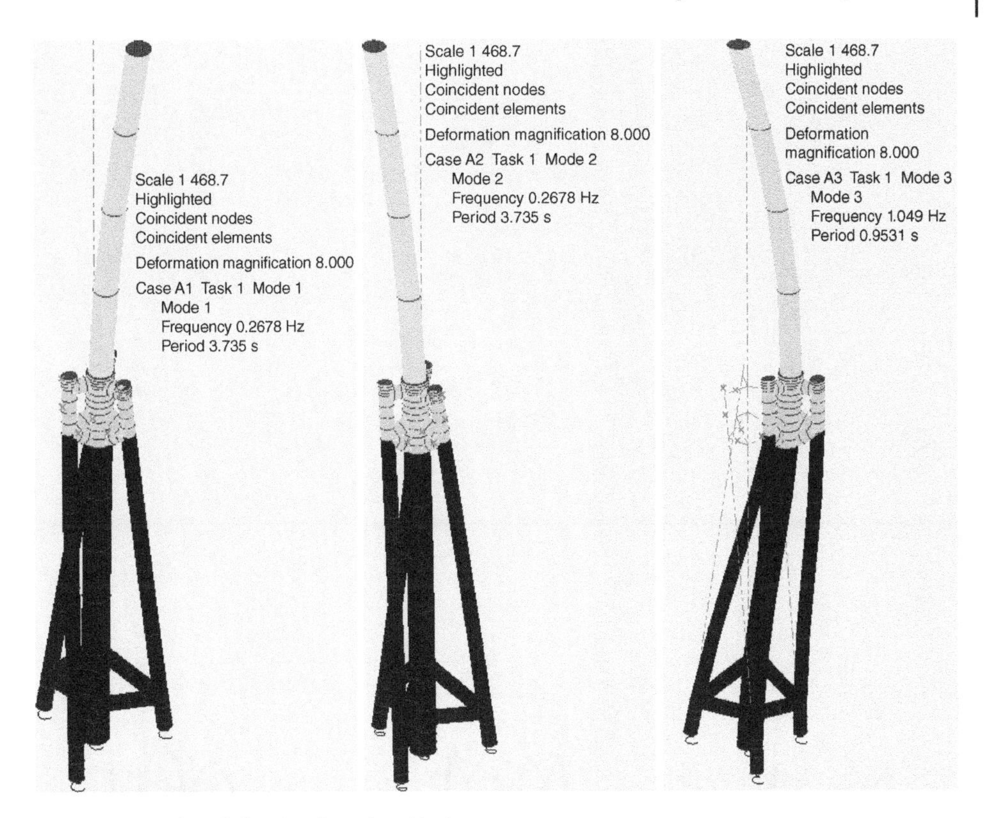

Figure 3.5 Modes of vibration for twisted jacket.

necessary whereby the different design parameters are altered. It is time consuming to carry out the optimization using software, and as a result, closed-form solutions are effective. Method to calculate natural frequencies are shown in Chapter 5 and worked-out examples are carried out in Chapter 6.

As an example, the natural frequency of a monopile supported wind turbine system can be estimated following the method developed by Arany et al. (2015a, b, 2016). This simplified methodology builds on the simple cantilever beam formula to estimate the natural frequency of the tower (fixed-base assumptions), and then applies modifying coefficients to take into account the flexibility of the foundation and the substructure. This is expressed as

$$f_0 = C_L C_R f_{FB} \tag{3.2}$$

where C_L and C_R are the lateral and rotational foundation flexibility coefficients, f_{FB} is the fixed-base (cantilever) natural frequency of the tower considering the varying stiffness along the length of the tower and transition piece (TP). Effectively, the formulation shows the flexibility provided by the foundation. The method is described in Chapter 5 and example applications are shown in Chapter 6. Arany et al. (2016) provides a detailed derivation of the method, along with validation and verification with 10 case studies of the wind turbine structure.

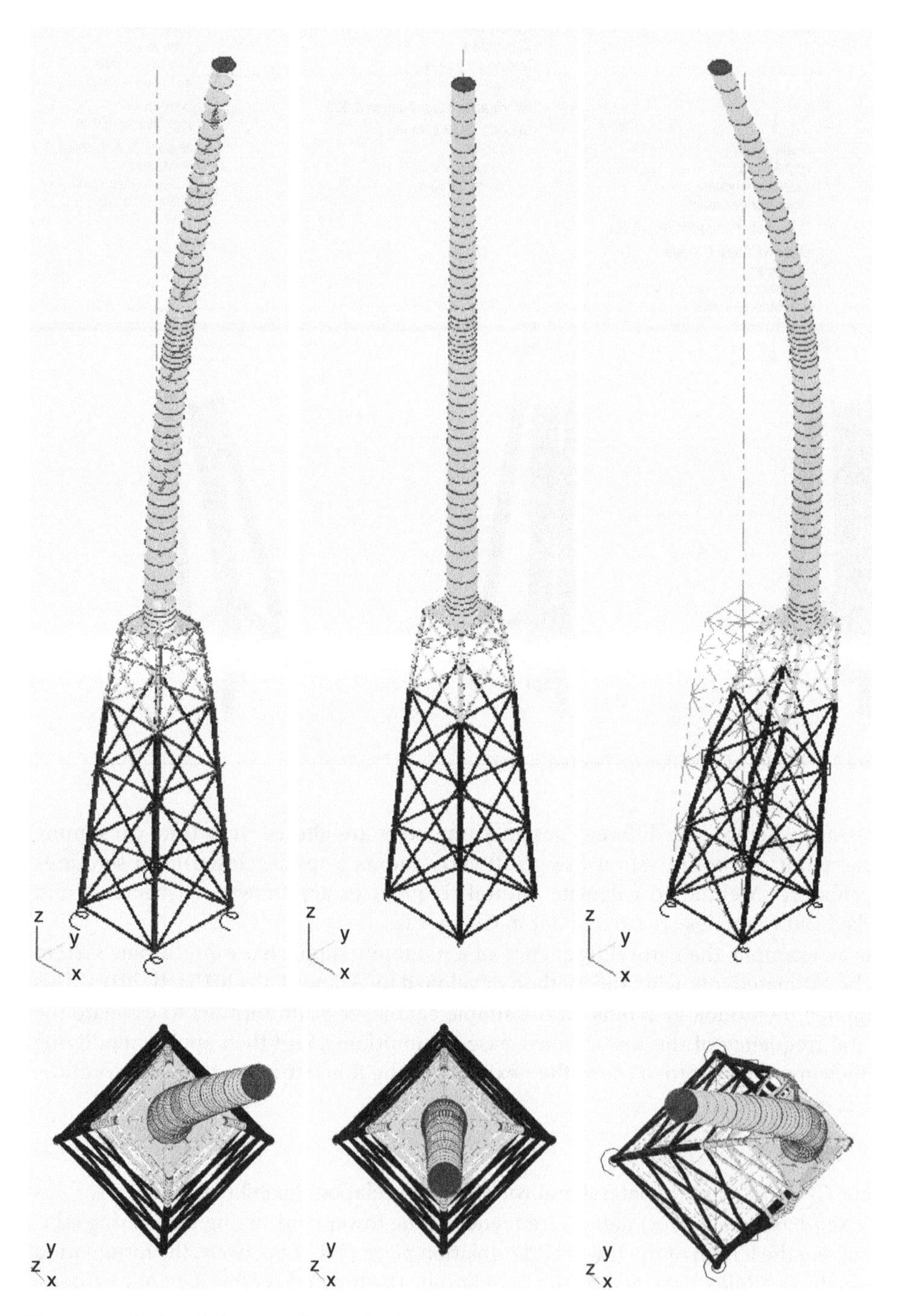

Figure 3.6 Modes of vibration of a standard jacket.

ASIDE

Modes of vibration for yet-to-be-built structures are best studied in scaled modes tests. Many of the proposed foundation types were therefore studied using scaled model tests in a stable laboratory floor. The experimental observations revealed interesting facts that are became important design considerations. The observations were later verified using analytical methods and numerical simulations.

3.2.2 Rocking Modes of Vibration

Rocking modes of foundation are typical of wind turbines supported on multiple shallow foundations; see, for example, Figure 3.7 where wind turbine structures are supported on multiple bucket-type foundations. Figure 3.8 shows three other types of scaled model tests where a seabed frame or jacket is supported on shallow foundations. A schematic representation of the rocking modes of vibration are shown in Figure 3.9 where the tower mode may get coupled with rigid modes of vibration. Essentially, there are two types of vibrations: (i) rocking mode of the jacket, which is very rigid; (ii) flexible modes of

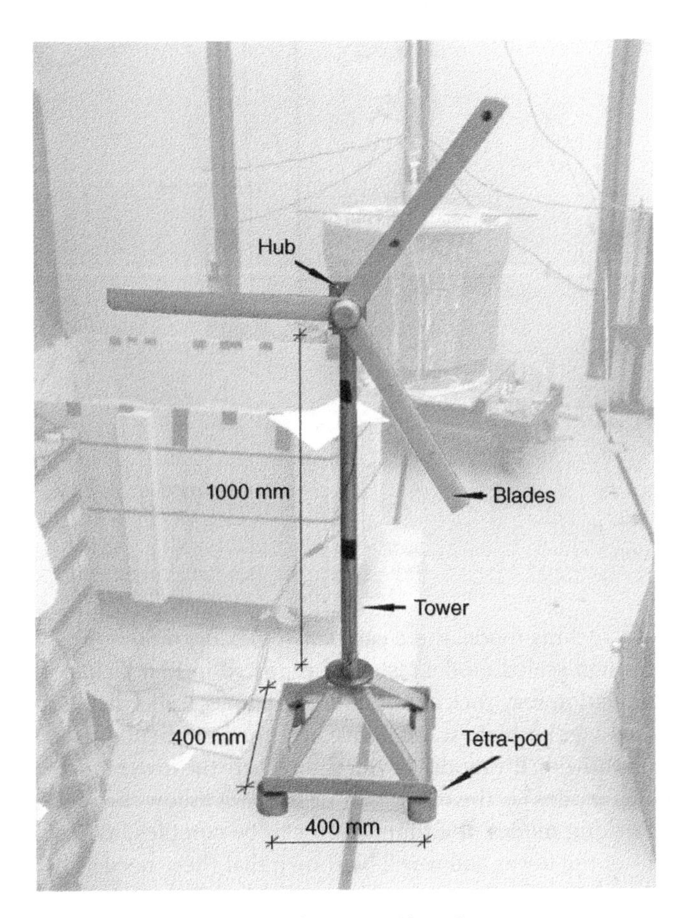

Figure 3.7 A scaled model of a tetrapod foundation.

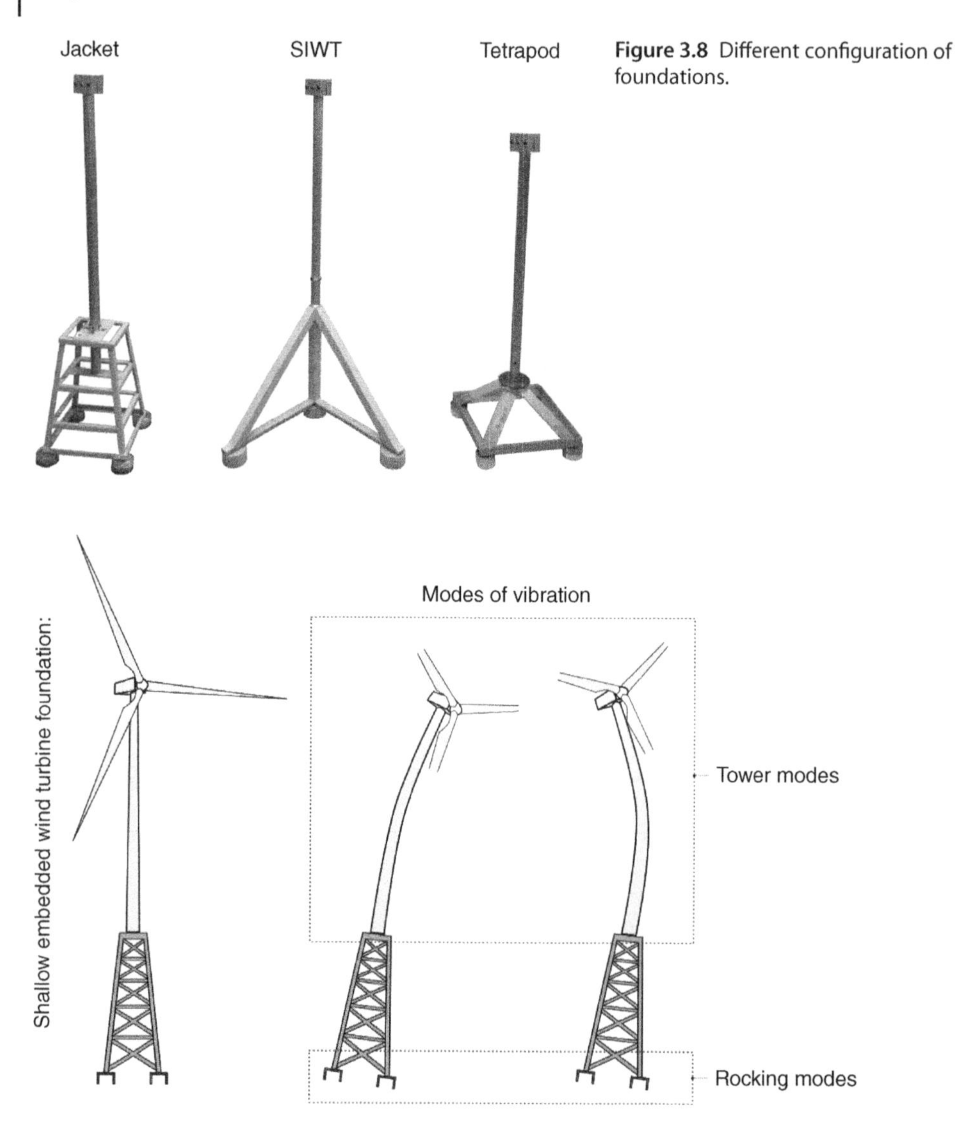

Figure 3.8 Different configuration of foundations.

Figure 3.9 Rocking modes of vibration.

the tower. Theoretically, for each rocking mode, there can be two flexible modes of the tower. These were observed through scaled model tests and reported in Bhattacharya et al. (2013a, b, 2017a). The foundation may rock about different planes and is dictated by the orientation of the principle axes i.e. highest difference of second moment of area. Figure 3.9 is a simplified diagram showing the modes of vibration where the tower modes can also interact with the rocking modes i.e. the tower may or may not follow the rocking mode of the foundation. Rocking modes of a foundation can be complex, as they interact with the flexible modes of the tower and it will be shown that these need to be avoided.

Three cases are discussed below:

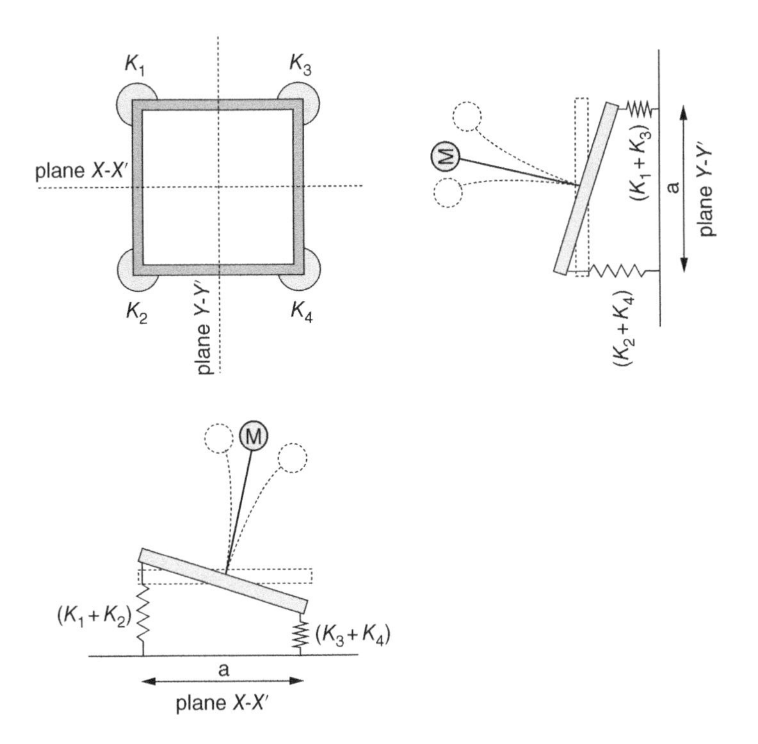

Figure 3.10 Rocking modes for a symmetric tetrapod about $X–X'$ and $Y–Y'$ plane.

1. Wind turbine supported on symmetric tetrapod foundations such as the one shown in Figure 3.7. A simplified model for analysis is illustrated in Figures 3.10 and 3.11. Research by Bhattacharya et al. (2013a,b) shows that even for the same foundations under each support, there will be two closely spaced vibration frequencies. This is due to different vertical stiffness of the foundation associated with variability of the ground. However, after many thousands of cycles of loading and vibration, these closely spaced vibration frequencies may converge to a single peak.
2. *Asymmetric tripod foundation.* Example is provided in Figure 3.12 inspired by the concept shown in Figure 1.31 (Chapter 1). Study reported in Bhattacharya et al. (2013a,b) showed that there will two modes of vibration with closely spaced frequencies but with millions of cycles of loading, and these two closely spaced peaks will not converge. This is because the foundation has two different stiffness in two orthogonal planes.
3. *Symmetric tripod foundation.* In a bid to understand the modes of vibration for a symmetric tripod, tests were carried out on a triangular foundation shown in Figures 3.13 and 3.14. Free vibration tests were carried out and a typical result is shown in Figure 3.15. The mode is like a 'beating phenomenon' well known in physics, which is possible for two very closed spaced vibration frequencies with low damping.

Taking into consideration Figures 2.4 and 2.5 (Chapter 2) where the design first natural frequency of the whole system is to be targeted between 1P and 3P, it is important

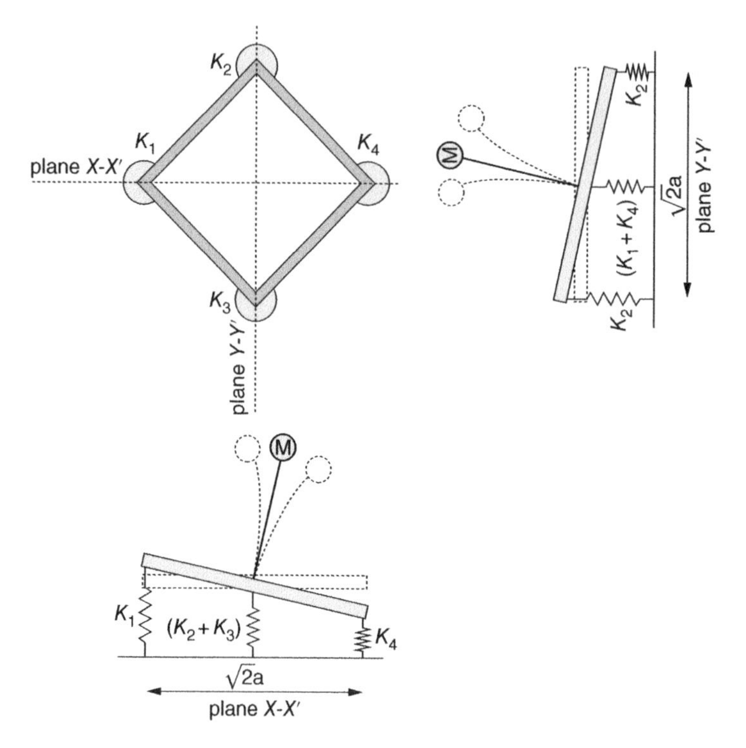

Figure 3.11 Rocking modes about diagonal plane.

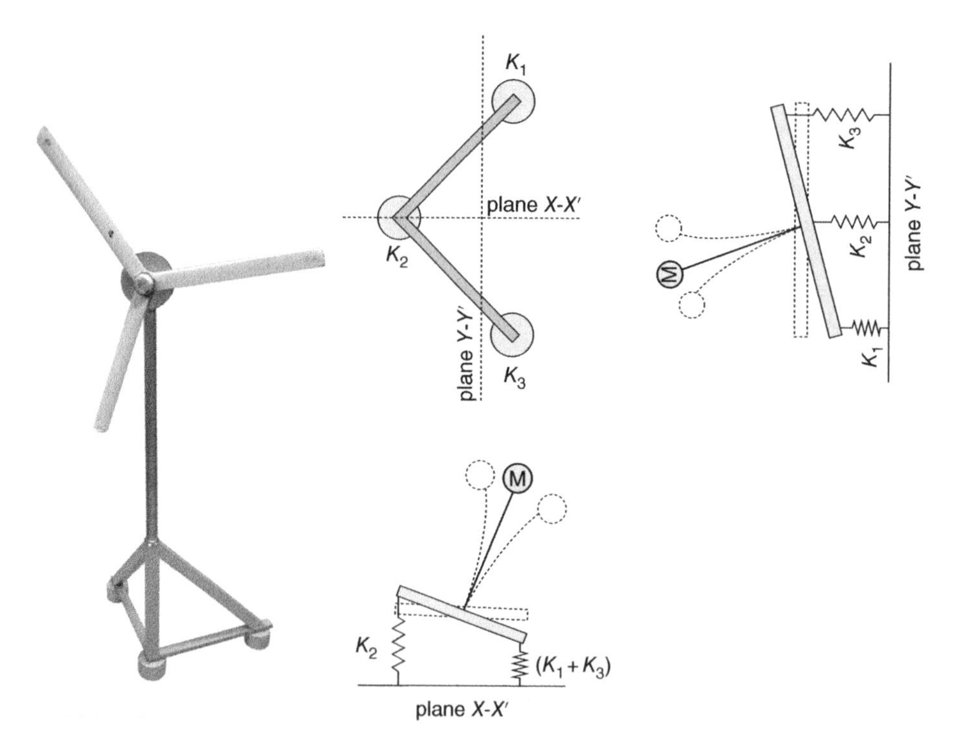

Figure 3.12 Modes of vibration for symmetric tripod.

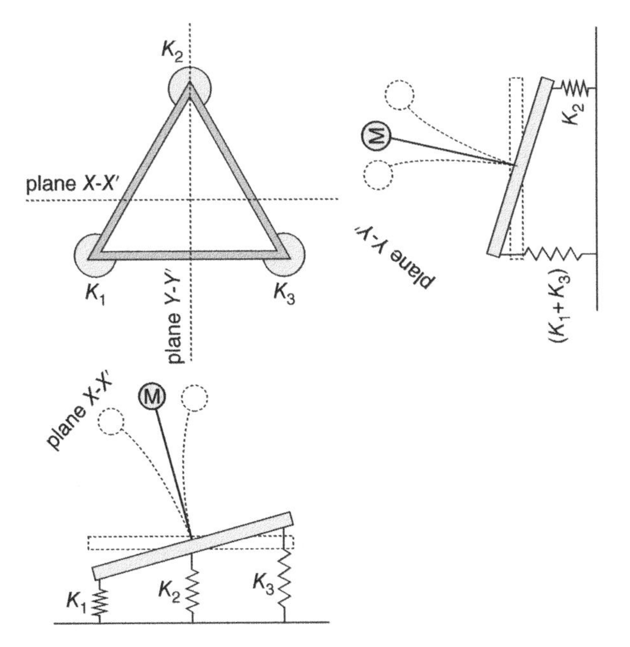

Figure 3.13 Symmetric foundation.

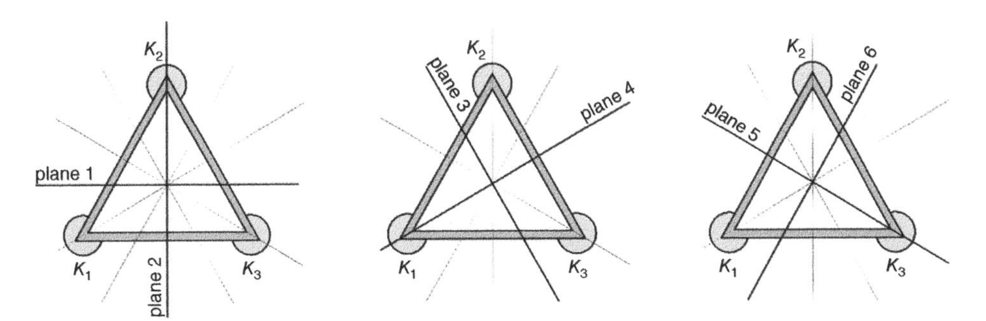

Figure 3.14 Planes of vibration.

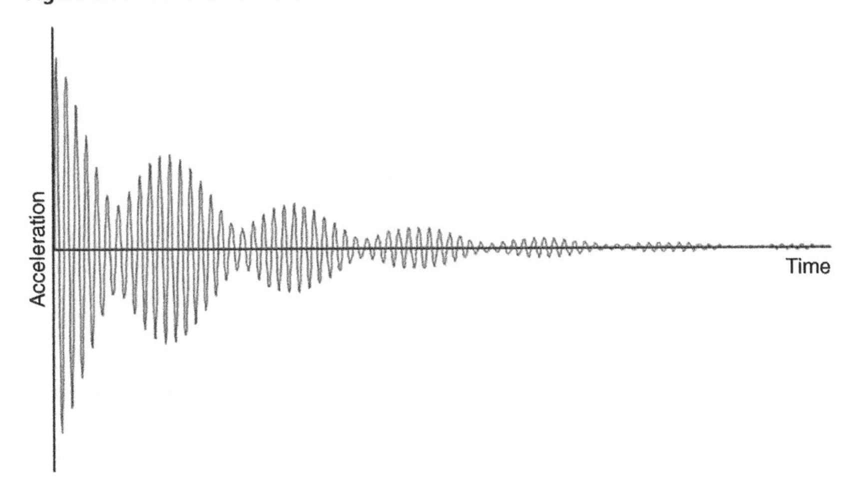

Figure 3.15 Free vibration acceleration response.

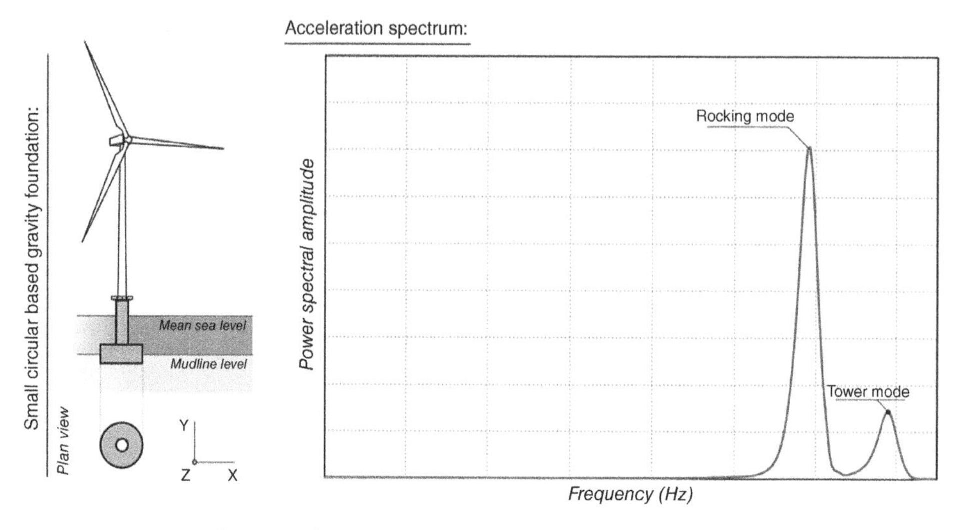

Figure 3.16 Modes of vibration for a small circular gravity-based foundation.

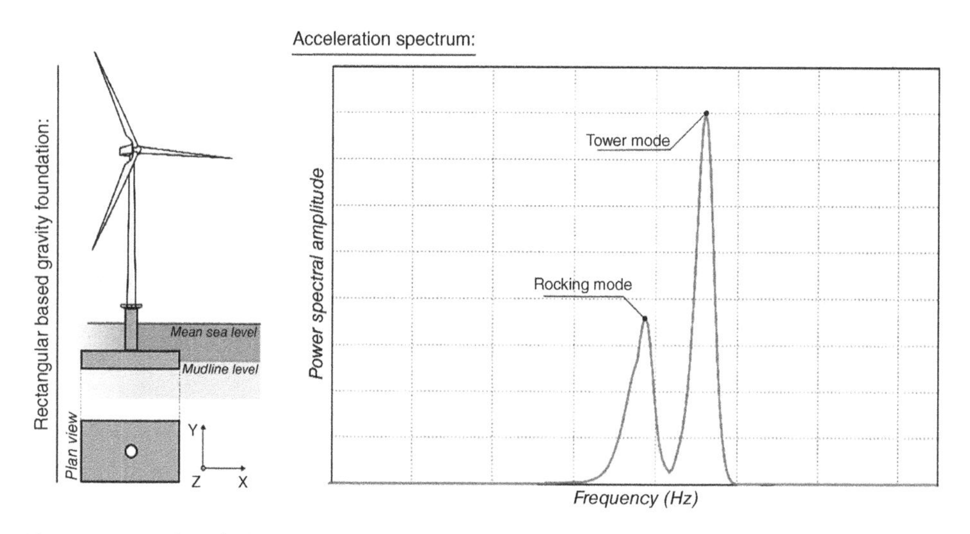

Figure 3.17 Modes of vibration in a rectangular-shaped GBS.

not to have two closely spaced modes of vibration. In practical terms, it is therefore recommended to avoid an asymmetric system. The previous case study also shows that a symmetric tetrapod is better than a symmetric tripod due to higher damping. It may be noted that the beating phenomenon is typical of low damping and two closely spaced modes. Gravity-based foundation will also exhibit rocking modes of vibration and it may also interact with tower flexible modes. Figures 3.16–3.18 show a schematic diagram of observed modes of vibration from a small-scale model test. Figure 3.19 shows the changing of natural frequency of foundations with cycles of loads observed in small scale testing.

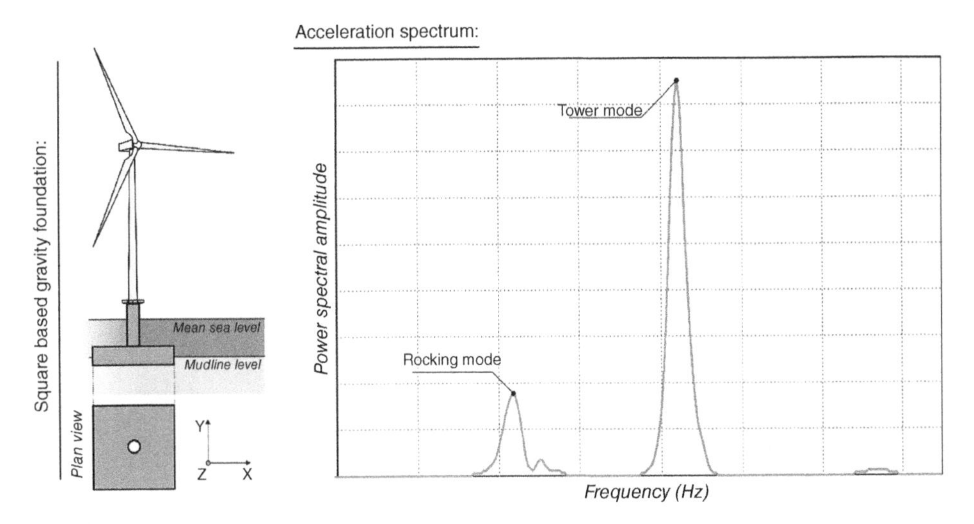

Figure 3.18 Modes of vibration in a square GBS.

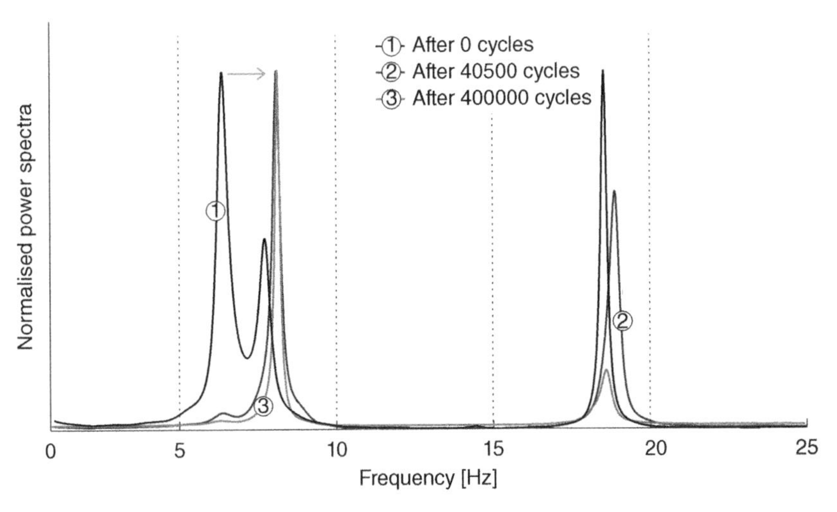

Figure 3.19 Changes in modes of vibration of a monopile supported wind turbine with cycles of loading.

3.2.3 Comparison of Modes of Vibration of Monopile/Mono-Caisson and Multiple Modes of Vibration

Deep foundations such as monopiles will exhibit sway-bending mode, i.e. the first two vibration modes are widely spaced – typical ratio is four to five. However, multiple pod foundations supported on shallow foundations (such as tetrapod or tripod on suction caisson) will exhibit rocking modes in two principal planes (which are, of course, orthogonal). Figure 3.20 shows the dynamic response of monopile supported wind turbine and tetrapod foundation plotted in the loading spectrum diagram obtained from scaled models tests.

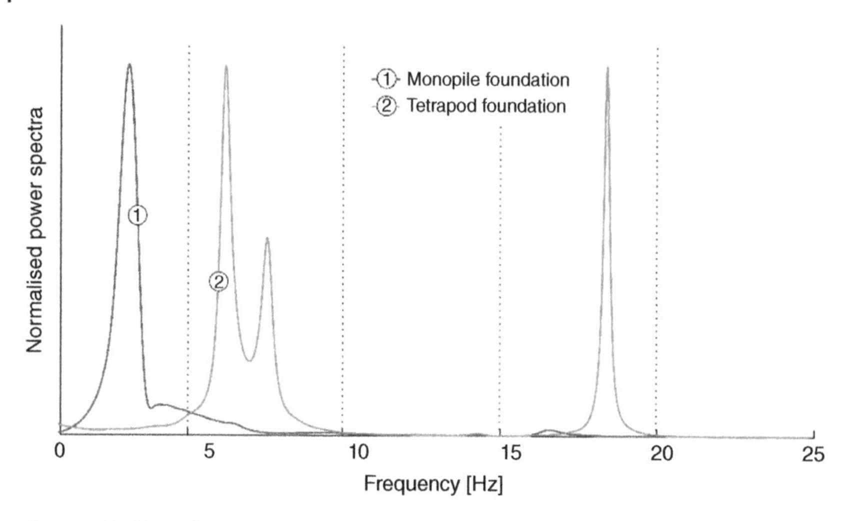

Figure 3.20 Free vibration response of tetrapod foundations on suction caissons (see Figure 3.8 for the foundation) and monopile.

3.2.4 Why Rocking Must Be Avoided

Foundations (symmetric or asymmetric) on multiple foundations will exhibit two closely spaced natural frequencies corresponding to the rocking modes of vibration in two principal axes. For a soft-stiff design, these need to be fitted in a narrow gap, as shown in Figure 3.21. Furthermore, as will be discussed in Chapter 5, owing to soil-structure

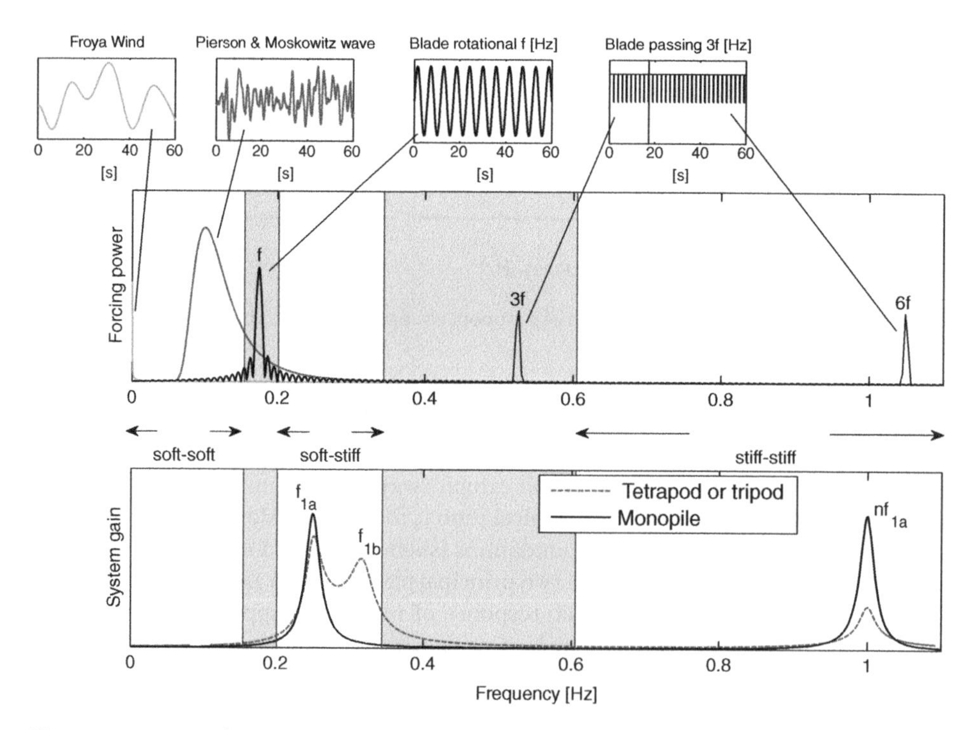

Figure 3.21 Fitting of two peaks in a narrow gap.

interaction (SSI), the two spectral peaks change with repeated cycles of loading. Also, for symmetric tetrapods (but not for asymmetric tripods), these two peaks will converge for sandy deposits. From the fatigue design point of view, the two spectral peaks for multipod foundations broaden the range of frequencies that can be excited by the broadband nature of the environmental loading (wind and wave) thereby impacting the extent of motions. Thus, the system lifespan (number of cycles to failure) may effectively increase for symmetric foundations as the two peaks will tend to converge. However, for asymmetric foundations the system life may continue to be affected adversely as the two peaks will not converge. In this sense, designers should prefer symmetric foundations to asymmetric foundations.

3.3 Effect of Resonance: A Study of an Equivalent Problem

Offshore wind turbines are a new type of structures, and it is important to learn from other disciplines. In this section, an example from helicopters is taken to show the importance of avoiding certain frequencies. Figure 3.22 shows still photographs from the well-known helicopter resonance problem known as ground resonance, the video can be accessed from YouTube. Effectively, due to the imbalance in the helicopter rotor the rotation-induced oscillations get in phase with the rocking frequency of the helicopter on its landing gears. This leads to collapse; the experiment is schematically shown in Figure 3.23. The helicopter starts rocking about the two landing pads (skids)

Figure 3.22 Ground resonance of a helicopter.

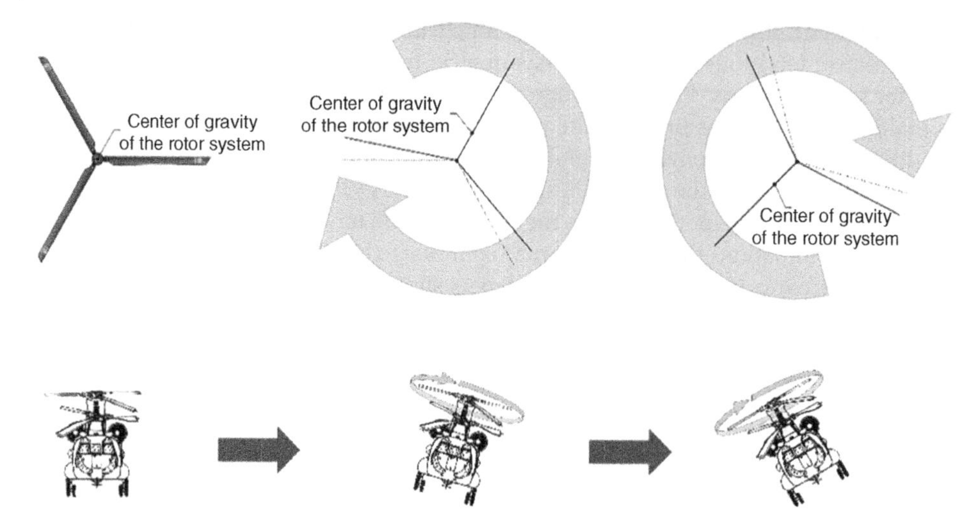

Figure 3.23 Rocking motion of a helicopter getting tuned with the RPM of helicopter rotor.

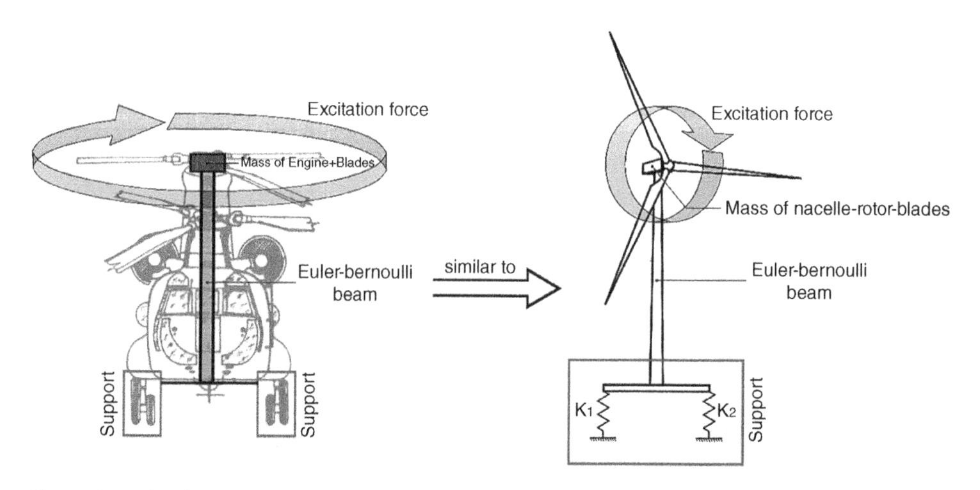

Figure 3.24 Similarities between a helicopter and offshore wind turbine structure.

until the stresses induced through resonance exceed the strength of the materials and connections, causing failure. There is a similarity between the helicopter ground resonance and jacket-supported offshore wind turbines, as shown schematically in Figure 3.24, in the sense that both support a heavy rotating top mass. As the motion under consideration is rocking, the vertical stiffness of the supports is a governing parameter. For a jacket structure, at the onset the vertical stiffness (kN m^{-1}) may not be identical. Therefore, they are shown as K_1 and K_2. It is clear that resonance must be avoided, and this emphasises the importance of understanding the subtle aspects of the dynamic behaviour of jacket-supported wind turbines. Resonance not only affects the FLS (fatigue limit state) and SLS (serviceability limit state) but will also impact O&M (operation and maintenance).

3.3.1 Observed Resonance in German North Sea Wind Turbines

Hu et al. (2014) reported resonance for wind turbines. This is discussed further in Chapter 5.

3.3.2 Damping of Structural Vibrations of Offshore Wind Turbines

As the natural frequency of offshore wind turbines is close to forcing frequencies, damping is critical to restrict damage accumulation and avoid premature maintenance. Therefore, discussion is warranted for issues related to damping, and for practical design purposes can be idealised in fore–aft and side-to-side vibrations of offshore wind turbines. The main difference between the sway-bending (or rocking) vibrations about two axes (X and Y) is that in the along-wind direction, higher damping is expected due to the high aerodynamic damping caused by the rotating blades interacting with the airflow. On the other hand, for side-to-side direction, the aerodynamic damping is orders of magnitudes lower.

A nonoperational (parked or idling) OWT has similar aerodynamic damping in the fore–aft as in the side-to-side direction. Since the wind load is acting in the along-wind direction, the highest load amplitudes are expected in fore–aft motion. This is because for most wind turbines in water depths less than 30 m, wind loading is the dominant load, while for very large diameter monopiles in medium to deep water, wave loading is expected to have equal or higher magnitude. It is worth noting that due to the yaw mechanism of the wind turbine, the along-wind and cross-wind directions are dynamically moving and are not fixed; therefore, the foundation is subjected to both cross-wind and along-wind loading in all directions during the lifetime of the OWT. One can conclude that analysing vibrations in both directions is important.

Studies considering the damping of the first bending mode either empirically or theoretically include Camp et al. (2004); Tarp-johansen et al. (2009); Versteijlen et al. (2011); Damgaard and Andersen (2012); Damgaard et al. (2013); Shirzadeh et al. (2013). Based on these studies and other estimates and in the absence of other data, the following assessment of damping ratio contributions is recommended:

- Structural damping: 0.15–1.5%. The value of structural damping depends on the connections in the structure (such as welded connections, grouted connections, etc) in addition to material damping (usually steel) through energy dissipation in the form of heat (hysteretic damping).
- Soil damping: 0.444–1%. The sources of damping resulting from soil-structure interaction (SSI) include hysteretic (material) damping of the soil, wave radiation damping (geometric dissipation) and, to a much lesser extent, pore fluid induced damping. Wave radiation damping and pore fluid induced damping are negligible for excitations below 1 Hz, and therefore hysteretic damping is dominant for the purposes of this study. The soil damping depends on the type of soil and the strain level.
- Hydrodynamic damping: 0.07–0.23%. Results from wave radiation and viscous damping due to hydrodynamic drag. In the low frequency vibration of wind turbines the relative velocity of the substructure is low and therefore viscous damping, which is proportional to the square of the velocity, is typically very low. The larger contribution results from wave radiation damping, which is proportional to the relative velocity.

- Aerodynamic damping: in the fore–aft direction for an operational turbine 1–6%, for a parking turbine or in the crosswind direction 0.06–0.23%. Aerodynamic damping is the result of the relative velocity between the wind turbine structure and the surrounding air. Aerodynamic damping depends on the particular wind turbine, and is inherent in the popular blade element momentum (BEM) theory for aeroelastic analysis of wind turbine rotors. The magnitude for a particular wind turbine also depends on the rotational speed of the turbine.

The total damping of the first mode of vibration is typically between 1–4% in side-to-side vibration, or parked or stopped or idling turbine. On the other hand, the total damping is between 2–8% for an operational wind turbine in the fore–aft direction.

3.4 Allowable Rotation and Deflection of a Wind Turbine Structure

Allowable rotation and deflection fall under SLSs i.e. operational tolerances imposed on the turbine system. These limits specify the total deflection, instantaneous deflection, and differential settlement/accumulation of rotation allowable throughout the lifetime of the turbine system. Excessive deflections or settlements can have an adverse effect on nonstructural components such as the generator and gearbox. This section of the chapter highlights these considerations. Possible reasons for the strict requirements from the point of view of the safe operation of the turbine, can be identified and are listed in Table 3.1:

1. One of the important issues with excessive rotation is the risk of tower strike by the blades and the difficulties with the yaw system of control to provide a constant power. Due to the pre-bend of the blades caused by mudline tilt, the blade-tower clearance may reduce, resulting in an increased risk of tower strike; see Figure 3.25. The capacity of the yaw motor may not allow the turbine to yaw into the wind when the rotor-nacelle assembly has to be turned 'uphill'. Furthermore, the yaw brakes may not be able to keep the RNA in the upward position.
2. There are also related issues such as reduced-power production, increased bending moment on the support structure, reduced lifetime of the bearings, and problems with cooling fluid levels or movement. These issues are becoming less importance with design improvements.

3.4.1 Current Limits on the Rotation at Mudline Level

In terms of SLS criteria, some specific requirements are to be met and the most important being the maximum deflection and rotation (tilt) at the foundation level/pile head (mudline) and at nacelle level. When assessing the total rotation, the initial tilt as well as the accumulated permanent rotation resulting from cyclic and dynamic loading throughout the design lifetime of the offshore wind turbine (OWT) should be considered.

Table 3.1 Possible reasons for strict verticality requirements.

Possible problem area	Description of the problem
Blade – tower collision	The tilt of the turbine may cause reduction of blade-tower clearance due to initial deflection of the blade. These effects for an upwind machine may increase the risk of tower strike at the extreme deflection of a blade in the downward pointing position as shown in Figure 3.25. Blade-tower collision is a serious risk that needs to be taken into account. In order to avoid tower strike, wind turbine manufacturers apply tilt angles to the main shaft of the rotor in the tune of 5–6°.
Reduced energy production	The tilt at mudline causes the wind to hit the rotor 'at an angle' and reduces the total energy production by the wind turbine.
Yaw motors	The limited yaw motor capacity may stop the turbine from yawing into the wind when the rotor has to be turned upwards against gravity.
Yaw brake	The yaw brake may not be able to keep the rotor in the upwards pointing position.
Yaw, pitch, and main bearings	The tilt at the nacelle level changes the direction and magnitude of the loading of bearings which may reduce their fatigue life or affect movement criteria.
Fluid levels and cooling fluid movements	Tilt at the nacelle level might interfere with the cooling system of the turbine.
Increased bending moments	The tilt of the turbine causes higher bending moments in the tower, grouted connection of the monopile and the transition piece, and the monopile itself.

The Det Norske Verita (DNV) code (DNV 2010a,b) gives typical limits of 0.25° and 0.5° for allowable rotation and states that 'the deformation tolerances are typically derived from visual requirements and requirements for the operation of the wind turbine'.

Serviceability criteria appear to be 'turbine manufacturer requirements' for the operation of the wind turbine, which originates from onshore wind technology and is very similar to a typical tall structure (such as tall buildings). The very low tilt angle requirement for monopile supported structures appears to be especially overcautious in light of floating OWT technology, where tilt angles up to 7° are acceptable.

ASIDE

(1) Tilting will be an important design consideration in seismic areas with liquefaction susceptibility. Typically, 8–16 m soil may liquefy in loose to medium dense deposit under moderate to strong earthquakes. This will reduce the rotational stiffness of the foundation (K_R), causing higher tilting as well as settlement.

(2) Scour can substantially reduce the embedded length of the foundation. It may also reduce the rotational stiffness (K_R) of the foundation.

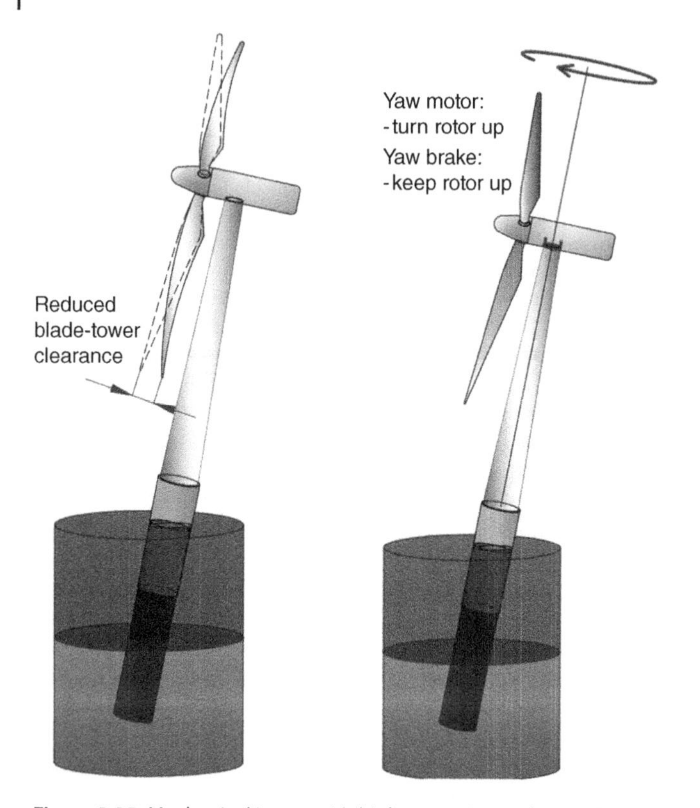

Figure 3.25 Mechanical issues with higher rotation at the mudline (Part 1).

3.5 Internationals Standards and Codes of Practices

The considerations necessary for foundation design depend on its type (grounded or floating system), site location (wind and wave climate and subsurface conditions), availability and expertise of marine contractors, economics, type of contract and investors (governmental or private), and type of economy (developed or developing). However, the aim of this chapter is to discuss the scientific considerations with emphasis on dynamics requirements. Where applicable, general considerations for offshore installation and constructions will be briefly discussed and further reading will be suggested (Figure 3.26).

The design criteria and considerations are typically established based on the following:

1. *Design codes.* The most important ones are: International Electrotechnical Commission (IEC) regulations, IEC61400-3 and IEC61400-22; DNV – guidelines on 'Design of Offshore Wind Turbine Structures' (DNV 2014), Germanischer Lloyd (GL) Windenergie's 'Guideline for the Certification of Offshore Wind Turbines', ISO standards, American Petroleum Institute (API) codes, Eurocodes, British Standards (BSI), and BSH (German) codes.
   a. The British Standards Institute (BSI) is a business service provider originating out of London, England. In addition to providing standards for the majority of

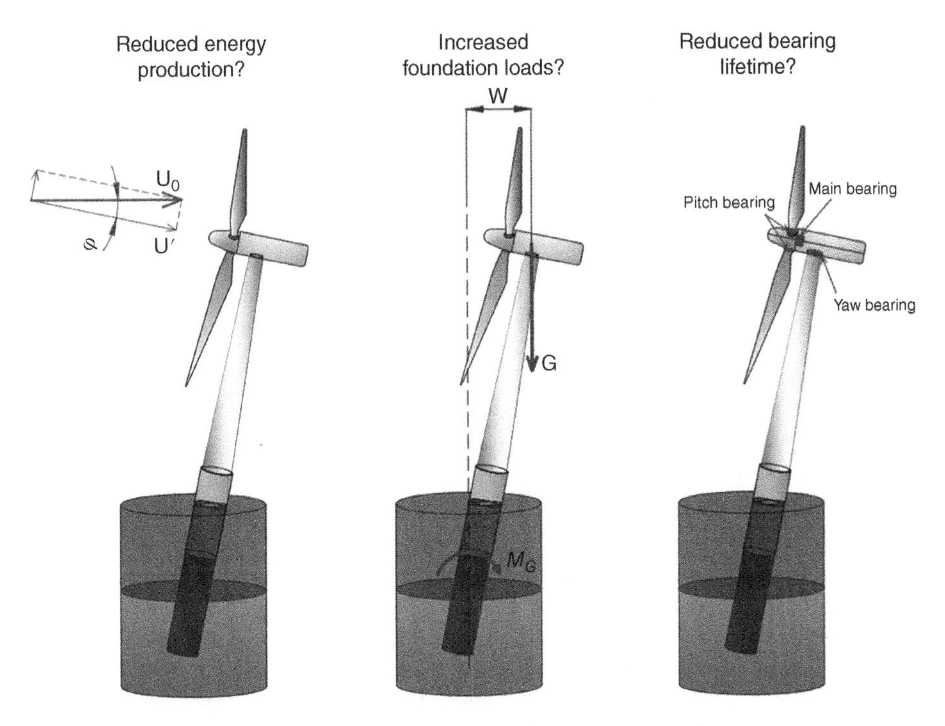

Figure 3.26 Mechanical issues with higher rotation at the mudline (Part 2).

UK industry, it has also produced a series of publications for use in the design and construction of offshore wind turbines. These regulations are BS EN 61400-3 and BS EN 61400-22, and they draw heavily from and elaborate on the IEC regulations already mentioned (BSI – British Standards Institution 2009; BSI – British Standards Institution 2011).

b. DNVs is a Norwegian classification society dating back to 1864 that has produced numerous standards and guidelines for the offshore industry. With the emergence of offshore wind power in the North Sea, DNV started producing standards for the renewable sector, such as DNV-OS-J101 and DNV-OS-J201 (Det Norske Veritas 2009, 2013).
c. GL is a German classification society based out of Hamburg. Similar to DNV, GL also produces a series of regulations providing guidance for the design and construction of offshore wind turbines (Germanischer Lloyd 2012). Recently, DNV and GL merged to form a classification society; however, to date the standards produced by each organisation remain separate.
d. The API is a trade association that has provided regulatory services to the oil and gas industry. Despite no specific regulations regarding the design of offshore wind turbines, their guidance for conventional offshore programs is worth noting (American Petroleum Institute 2007).

2. *Certification body*. Typically, a certification body allows for departure from the design guidelines if the design is supported by sound engineering and sufficient evidence/test results.

3. *Client*. Occasionally, the client may pose additional requirements based on appointed consultants.
4. *Turbine manufacturer*. The manufacturer of the wind turbines typically imposes strict SLS requirements or natural frequency requirements. In addition, the expected hub height is also a requirement for the turbine type and the site. The tower dimensions are also often inputs to foundation design and are normally provided by the turbine supplier.
5. For specialised design such as seismic, ISO standard and Eurocode 8 may be used. The readers are also referred to text book titled *Seismic Design of Foundation: Concepts and Applications* by Bhattacharya et al (2018) for guidance on seismic design.

ASIDE

These are new type of structures, and major innovations are underway for efficient design and construction and lower the O&M costs. However, there are limited track records of long-term performance, and as a result, the current design considerations are often very conservative. Due to this lack of track record or limited monitored data, it is necessary to learn from other disciplines where a similar problem is being encountered. Future trends suggest that turbines are getting larger/heavier and must be placed at a higher height, and so they will need taller towers. Figure 3.1 shows the future of offshore wind turbines in terms of sizes. By far, these are the largest rotating machine in the world. It is important to discuss the special considerations very much relevant to this problem, i.e. under what conditions the turbine will be considered to have failed in terms of *performance-based design*. In this context, it is important to remember the definition of an ideal foundation, discussed in Chapter 1.

3.6 Definition of Limit States

A limit state is a condition beyond which the structure-foundation assembly will no longer satisfy the specified performance requirements. For offshore wind turbines applications, they are:

(i) the Ultimate Limit State (ULS); (ii) the Serviceability Limit State (SLS); (iii) the Fatigue Limit State (FLS); and (iv) the Accidental Limit State (ALS).

3.6.1 Ultimate Limit State (ULS)

The ULS is related to the maximum load-carrying resistance, and can be reached for several reasons: (i) excessive yielding and/or buckling (i.e. loss of structural resistance); (ii) a failure of a component (e.g. brittle fracture of connections); and (iii) a loss of static equilibrium of the structure (whole or part) with a consequent mechanism (e.g. rigid body behaviour, overturning, and capsizing).

As the main aim of a foundation is to transfer all the loads, during its design life, from the wind turbine structure to the ground safely and within the allowable deformations. The design calculations should ensure that the maximum loads on the foundations are much lower than the capacity of the chosen foundation. This calculation is most dependent on the ultimate strength of the soil, i.e. this is a *strength type calculation*. The first

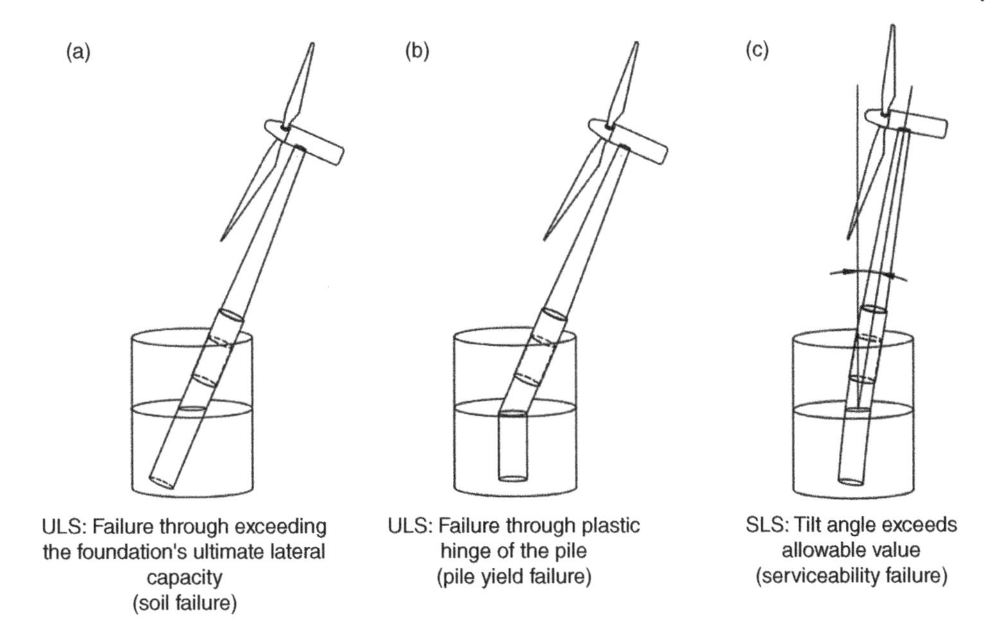

Figure 3.27 Examples of ULS and SLS failure.

step in design is to estimate the maximum loads on the foundations (predominantly overturning moment, lateral load and the vertical load) due to all possible design load cases – this is covered in Chapter 2. Subsequently, the loads need to be compared with the capacity of the chosen foundation. This calculation is necessary to avoid failure of foundation; see, for example, Figure 3.27a,b, which shows two cases of ULS failure for monopiles. In the case of Figure 3.27a, the foundation fails due to soil failure around the foundation and eventually through uprooting of the foundation. On the other hand, Figure 3.27b shows the case where the pile fails by forming a plastic hinge where the overturning moment in the monopile exceeds the plastic moment carrying capacity of the pile. For a jacket on small-diameter piles, this is also very similar. Figure 3.28 shows some of the ULS cases for anchor piles supporting a floating wind turbine, as discussed in Chapter 2. On the other hand, Figure 3.29 shows the ULS cases for anchor piles where two types of failure mechanisms (overturning i.e. rigid body rotation and sliding/translation) are depicted.

3.6.2 Serviceability Limit State (SLS)

The SLS corresponds to tolerance criteria associated with the regular and normal use of the wind turbine, including: (i) excessive deflection leading to the second order effects modifying the distribution of loads between supported and supporting structures; (ii) excessive vibration jeopardising the functioning of the nonstructural components; (iii) displacements that exceed the limitation of the equipment; (iv) differential settlements of foundation and soil causing intolerable tilt of the wind turbine; and (v) temperature-induced excessive deformations.

SLS is also directly linked to target natural frequency (eigen frequency) as discussed in Chapter 2 due to possible amplification due to dynamics. For these calculations,

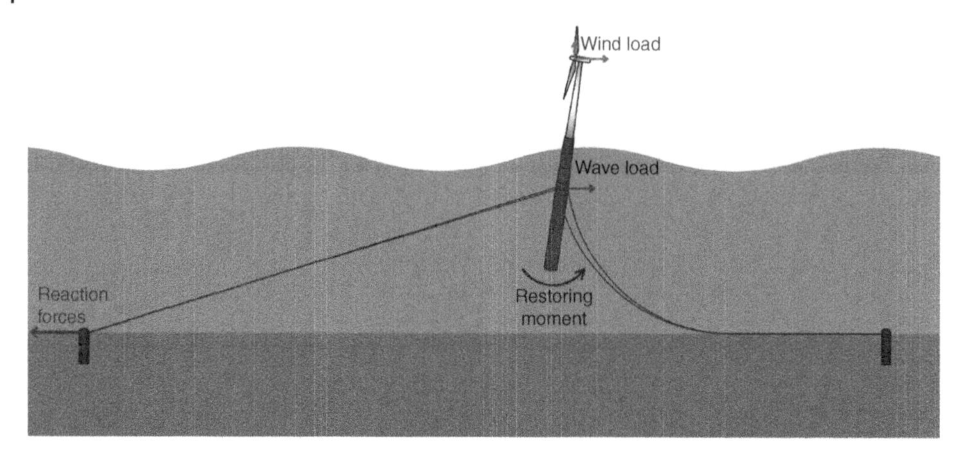

Figure 3.28 ULS loads for spar type floating wind turbine structure.

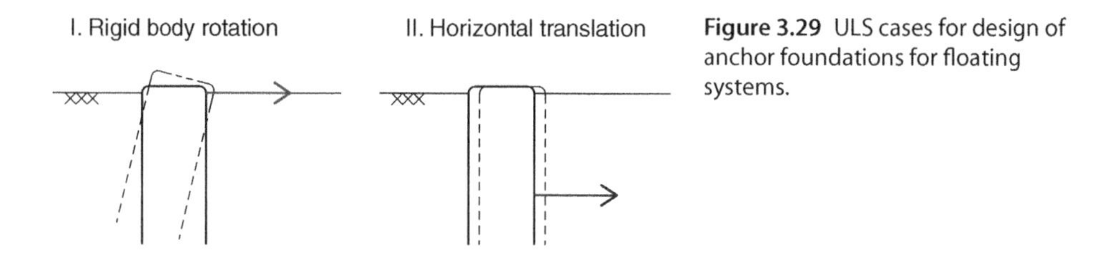

Figure 3.29 ULS cases for design of anchor foundations for floating systems.

prediction of the natural frequency of the whole system (eigen frequency) is necessary. As natural frequency is concerned with very small amplitude vibrations, linear eigen value analysis will suffice and prediction of initial foundation stiffness is necessary. Therefore, determination of stiffness of the foundation is an important design step.

3.6.3 Fatigue Limit State (FLS)

The FLS is related to cumulative damage due to repeated loads. For the case of monopile, this would require predicting the fatigue life of the monopile, as well as effects of long-term cyclic loading on the foundation. Again, this step requires stiffness of the foundation.

3.6.4 Accidental Limit States (ALS)

ALS considers potential accidental or unexpected loads (e.g. vessel impact) that can lead to loss of global or local structural integrity.

3.7 Other Design Considerations Affecting the Limit States

This section discusses design considerations affecting limit states, including scour, corrosion, and marine growth.

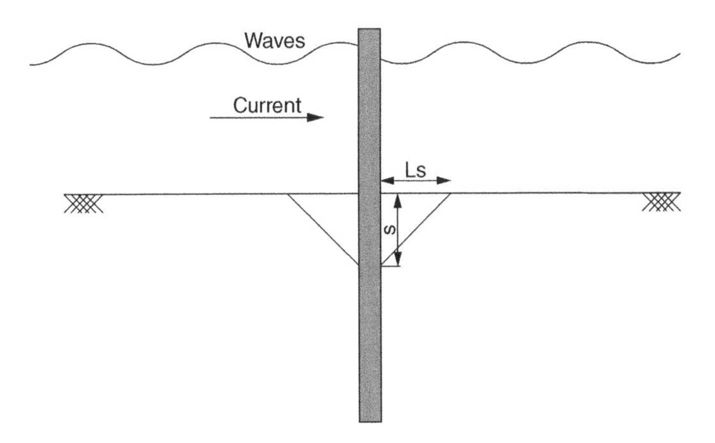

Figure 3.30 Definition of scour terminology.

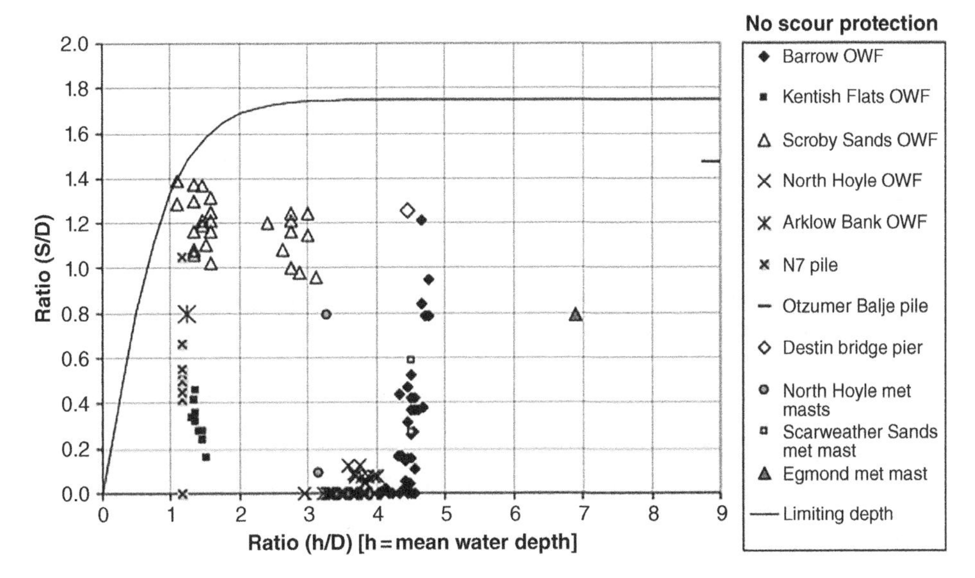

Figure 3.31 Observed scour depth for field case records, Whitehouse et al. (2011). (Source: Reproduced with permission of Elsevier).

3.7.1 Scour

Scour, the manifestation of which is a local lowering of seabed around foundations, is essentially sediment transport or erosion of soil and is caused by waves and current. Scour can affect the structural stability of foundations by making the foundations unsupported up to the scour depth and is an important design consideration. Scour is usually quantified in terms of its equilibrium depth S and the scour extension Ls as shown in Figure 3.30.

The scour depth depends on the current speed and there is a threshold current speed at which undisturbed sand far from the pile just starts to move. If the current speed is lower than the threshold, scour does not develop and the seabed remains stable. However, if the current speed increases scour starts to develop. Table 3.2 summarises

Table 3.2 Formulas for scour depth.

Publication	Flow condition	Relationships
Sumer et al. (1992) and adopted in DNV-OS-J101 (2014)	Steady Current KC → ∞	$\frac{S}{D_P} = 1.3$
Sumer et al. (1992) adopted in DNV-OS-J101 (2014)	Waves KC ≥ 6	$\frac{S}{D_P} = 1.3\{1 - \exp[-0.03(KC - 6)]\}$

empirical formulas on scour depth based on literature and design standards. Caution must be exercised while using these expressions for large diameter monopiles as they may not have been calibrated.

The equations given in Table 3.2 are governed by Keulegan-Carpenter (KC) number given by the following equation:

$$KC = \frac{U_{\max} T}{D_P}$$

where $U_{\max}$ is the maximum value of the orbital velocity at the bed, T is wave period, and D_P is the pile diameter. DNV code recommends using linear theory to find the orbital velocity at the bed.

KC number is a nondimensional number representing the relative importance of drag forces over inertia forces for objects in an oscillatory fluid flow. Scour protection measured are often taken to avoid unplanned maintenance.

Example 3.1 ***Observed Scour Depth in Wind Turbine Foundations*** Whitehouse et al. (2011) performed a comprehensive study on 115 reading from 10 wind farms where multiple readings were taken from each wind farm at different times and locations. The results are presented in a nondimensional form in terms of S/D and h/D where D is the diameter of the pile in figure 3.31. Figure 3.32 shows scour survey from three sites at different times following the completion of installation of bucket foundation. In the plots S is the maximum scour and D is the diameter of the bucket.

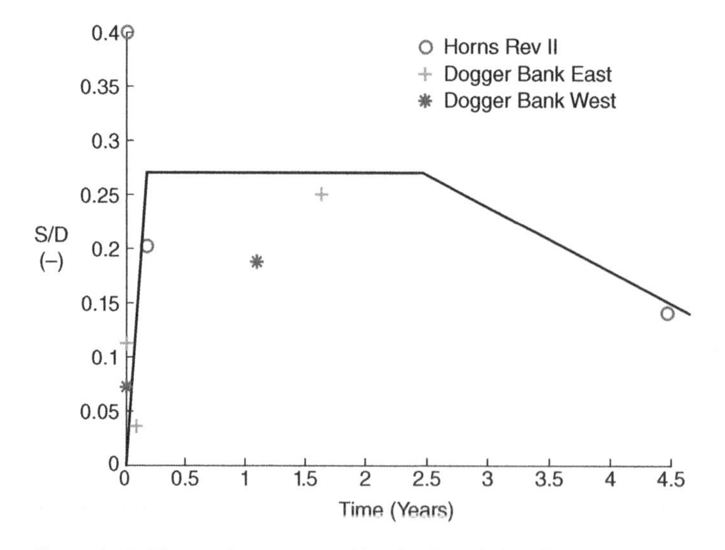

Figure 3.32 Observed scour around bucket foundation, Stroescu et al. (2016).

Table 3.3 Marine growth.

Depth below mean water level (m)	Marine growth thickness (mm)	
	Central and north sea	Norwegian sea
−2 to 40	100	60
>40	50	30

3.7.2 Corrosion

Monopiles located in offshore marine environments are susceptible to corrosion, which has implication on long-term performance and fatigue life. Current design standards such as the DNV and literature suggest different methods to account for in design:

1. Corrosion allowance, i.e. extra wall thickness, added during design to compensate for any reduction in wall thickness during design life. The corrosion allowance is based on the corrosion rate, which is a minimum of 0.1 mm per year following DNV.
2. Use a coating system such as zinc paint.
3. Cathodic protection is a technique to prevent the corrosion of steel surface by making the surface to act as a cathode of an electrochemical cell.

3.7.3 Marine Growth

The flora and fauna (such as bacteria, plant, and animal life) surrounding an offshore submerged structure may causes marine growth, see Figure 3.33. This will have implications on the structural design as it adds weight, influences the geometry and surface texture, and affects the corrosion rate. Marine growth can be due to tubeworms, barnacles, corals, and sea anemones, which colonise on a structure shortly after installation. The thickness and rate of growth depend on numerous factors such as salinity, oxygen content, pH, current, and temperature, age, orientation, and depth of the structural component below sea level. Structurally, the extra weight needs to be addressed relative to the total mass the foundations and when performing dynamic analysis of the whole system. The increased thickness will also increase the hydrodynamic loads. In the absence of field data, the DNV recommendations provided in Table 3.3 may be used as a guidance. The values are based on Norwegian and UK waters and are provided to give the reader an indication about the size and impact of marine growth on the structural performance of offshore wind turbine foundations.

3.8 Grouted Connection Considerations for Monopile Type Foundations

Additional considerations must be considered:

- Connection design between monopile, TP, and the tower
- Long-term tilt considerations in light of 30 years of uncertain wind and wave conditions

For monopile type of foundations, transition pieces are meant to transmit high bending moments from the superstructure to the pile. Grout were adopted for this

Figure 3.33 Marine growth around piles [Sources: Foundation Zone].

connection for most of the initial European projects for prior oil and gas experience, speedy construction, and apparent cost savings. Subsequently, it has been observed that these grouts have settled, cracked, and failed to deliver the intended performance for monopiles. Back analysis revealed that most of these designs excluded reinforcing shear keys and resulted in expensive retrofitting on many European projects and in some cases legal battles.

In this context, it must be mentioned that for decades oil & gas (O&G) platform jackets used API designed grouted connections. For O&G jackets, the grout connection transfers mainly axial load. In contrast, OWT structures are subjected to cyclic overturning moments apart from the axial loads and the stresses in the grout are therefore a combination of axial, cyclic bending, and shear. Behaviour of grouted connections is still an area of active research and industry best practice, and codes are therefore constantly being updated.

Other connection options that are being considered are:

- Bolted flange connections (e.g. in Scroby Sands Wind Farm)
- Use conical TP sections as a solution

3.9 Design Consideration for Jacket-Supported Foundations

Jackets or seabed frames supported on multiple shallow foundations are currently being installed to support offshore wind turbines in deep waters ranging between 23 and 60 m – see, for example, Borkum Riffgrund 1 (Germany, water depth 23–29 m), Alpha Ventus Offshore (Germany, water depth 28–30 m), Aberdeen Offshore wind farm (Scotland, water depth 20–30 m) (4C Offshore Limited). Typical design is three- or four-legged jackets supported on either deep foundations (piles) or shallow foundations (suction caissons). The height of the jacket currently in use is between 30 and 35 m where height is governed by water depth and wave height. However, it is expected that future offshore developments will see jacket heights up to 65 m to support larger

Figure 3.34 Schematic of a three-legged jacket supported on a suction caisson. [Source: DONG Energy (now known as Orsted) and Carbon Trust Offshore Wind Accelerator Project.]

turbines (12–20 MW) in deeper waters. Figure 3.34 shows a schematic of a three-legged jacket inspired by some recent offshore developments.

A jacket needs to be engineered towards a no-rocking solution by optimising two parameters: (i) ratio of vertical stiffness of the foundation to lateral superstructure stiffness; (ii) aspect ratio of the jacket-tower geometry. A low value of vertical foundation stiffness values together with a low aspect ratio will promote a rocking mode of vibration. On the other hand, a high vertical stiffness of the foundation with higher aspect ratio (broader base of the tower) will encourage a sway-bending mode. It can be shown that the transition from rocking to sway-bending is nonlinear and depends not only on the aspect ratio but also on the ratio of vertical stiffness of the foundation and lateral stiffness of jacket-tower configuration.

Other aspects are design of the joints for the jacket, together with the fatigue design. Welds must be periodically checked.

3.10 Design Considerations for Floating Turbines

Floating offshore wind farms (FOWFs) are emerging as a solution with many demonstration projects over the last decade. Design considerations include the selection of the anchoring system, which has a significant impact as it influences the mooring radius, seabed penetration, required anchor size, length of the forerunner, as well as the anchoring (transportation and installation, [T&I]) costs. There is a set of design choices for floating wind turbine platforms, with the most important ones being the stabilisation type (spar, semi-sub, tension-leg platform, semi-spar), the mooring system type (catenary, taut, or tension-leg), and the anchor type (gravity anchors, drag anchors, pile anchors, suction caissons). Considerations that are also important are the forerunner and mooring chain types (chain, rope, cable, etc.).

The most common approach in the mooring system design is to decouple the sea-keeping analysis (i.e. hydrodynamic motions analysis of the platform, including the mooring system) and the anchor system design. In other words, the expected anchor loads may be obtained and an anchor chosen without accounting for any effect of the anchor on the seakeeping of the platform.

Catenary mooring lines have so far been the most common choice for tested floating wind turbine concepts such as Hywind. In the case of catenary mooring lines, a so-called theoretical anchor point (TAP) is determined on the seabed through hydrodynamic motions analysis of the platform. The TAP is a point where no motion of the mooring line is expected over the life of the platform. The mooring line tension can be determined for this point, and thus the anchor loads can be calculated and the anchor chosen accordingly. This requires the assessment of the following for various types of anchors:

- Load transfer process
- Failure mechanisms
- Anchor size and weight
- Required seabed preparation
- Required seabed area
- Impact of soil conditions on site and the available knowledge about it in the early phases of the project
- T&I
- Forerunner type and length
- Long-term behaviour
- Overall anchoring costs

3.11 Seismic Design

Aspects of seismic design are provided in Chapter 2. However, there are two important considerations:

1. The response of the ground to the expected strong motion. Ground may respond in one of the two ways: either by amplifying the motion (site amplification) or some layers may liquefy. As a rule of thumb, loose- to medium-dense sandy soil liquefies and clay layers typically amplify the motion.
2. The change in properties of the supporting soil (strength and stiffness) will alter the response of the wind turbine structure. These need to be evaluated and allowance must be made so as to not have permanent deformation of the structure.

3.12 Installation, Decommission, and Robustness

These considerations will ascertain that the foundation can be installed and taken off the ground and that there are adequate redundancies in the system. Figure 3.35 shows an installed wind farm.

3.12.1 Installation of Foundations

Installation of large monopiles is currently carried out in the following ways: (i) hydraulic or vibration hammers from above water; (ii) underwater hydraulic hammers; (iii) in different ground conditions, drill-drive-drill operation; (iv) drill and grout. In comparison

Figure 3.35 London array wind farm after installation.

to drilling operations, driving a pile is faster and is relatively less weather sensitive. However, driving may damage the pile head and there may be issues with verticality. As expected, drill-drive is slower than driving and is used if the driving operation didn't reach the required penetration depth. There are some calculations that engineers must carry out, and these are discussed next (Figure 3.36).

3.12.1.1 Pile Drivability Analysis

One-dimensional wave equation analysis is used to simulate the pile response to the driving equipment. The analysis is based on the assumption that from the moment of impact the ram starts to transmit its energy to the pile cap and pile head, and an energy wave starts to travel through the pile at high velocity. The wave is assumed to be one-dimensional, acting longitudinally down the pile axis. The amplitude of the wave depends on the energy transmitted. Due to the energy losses in the whole system of hammer, pile, and soil, the amplitude of the wave decreases during travelling down. If sufficient energy is left once the wave has reached the pile tip, the pile starts to penetrate into the soil. As soon as all the energy is consumed, the pile stops penetrating and a permanent set is reached. This set for one blow can be computed by a program, and this is an iterative process. The program needs an initial guess of soil resistance to driving (SRD). Usually, the static-bearing capacity data are used to determine the initial guess of SRD. In the next steps of the analysis, SRD is varied, keeping all the remaining pile and hammer parameters constant, and the corresponding permanent sets are computed. When the so-obtained permanent sets are plotted versus the corresponding SRD, a blowcount resistance curve is obtained. A blowcount resistance curve shows the relation between blowcounts and SRD for certain pile, hammer, and soil conditions. There are many such program (GRLWEAP, CAPWAP) that can carry out the analysis. The

Figure 3.36 Driving of piles in London Array wind farm Courtesy: London Array.

output of the analysis is a prediction of driving stresses, hammer performance, and total driving time. These studies allow the designer to optimise the hammer requirements for a certain pile and soil condition.

3.12.1.2 Predicting the Increase in Soil Resistance at the Time of Driving (SRD) Due to Delays (Contingency Planning)

During pile installation, often driving has to be stopped and restarted, for various reasons such as changes of cushion or hammer or the addition of a follower. Such delays typically last from a few hours up to a few days. At the design stage, it is necessary to estimate both the SRD during continuous driving and also the amount of set-up (increase in pile driving resistance) that may occur during delays, to ensure that the hammers taken offshore are sufficient to meet all eventualities. This is a critical design consideration for offshore pile installation projects. Empirical relationships have been proposed for quantifying set-up – see Table 3.4 for one such case. Set-up is primarily dependent on three mechanisms or physical processes – consolidation, stress relaxation, and soil ageing. Theoretically, all these mechanisms/processes start acting as soon as the pile penetrates the ground. It is, however, still uncertain which of the above dominate in a particular soil condition, how long each process continues, and the contribution of each component to the observed overall set-up. There is widespread evidence showing that the capacity of most driven piles increases with time after driving; see Bhattacharya et al. (2009b).

3.12.1.3 Buckling Considerations in Pile Design

Piles can never be assumed to be straight and free from residual stresses due to construction processes. Apart from out-of-line straightness (see Figure 3.37), a pile may be thin walled and therefore vulnerable to local buckling (see Figure 3.38). This has been

Table 3.4 Empirical formulas for predicting pile capacity with time.

Reference	Equation	Type of soils	Comments (all times are in days)
Skov and Denver (1988)	$Q_t = Q_0\left\{A.\log_e\left(\frac{t}{t_0}\right)+1\right\}$ where Q_t and Q_0 are the pile capacities at time t and t_0 respectively.	Sand and clay	For sand: $A = 0.2$, $t_0 = 0.5$ For clay: $A = 0.6$, $t_0 = 1.0$
Huang (1988)	$Q_t = Q_{EOD} + 0.236\{1 + \log_e(t)(Q_{\max} - Q_{EOD})\}$	Soft ground soil of Shanghai	Q_t = Pile capacity at time t $Q_{\max}$ = Maximum pile capacity Q_{EOD} = Pile Capacity at the end of driving (EOD).
Guang-Yu (1988)	$Q_t = Q_{EOD}(0.372S_t + 1)$	Piles driven in soft soils.	Q_{14} = pile capacity at 14 days and S_t is the sensitivity of soil.
Svinkin (1996)	$Q_t = A.Q_{EOD}.t^{0.1}$	Sand	For upper bound; $A = 1.4$ For lower bound; $A = 1.025$

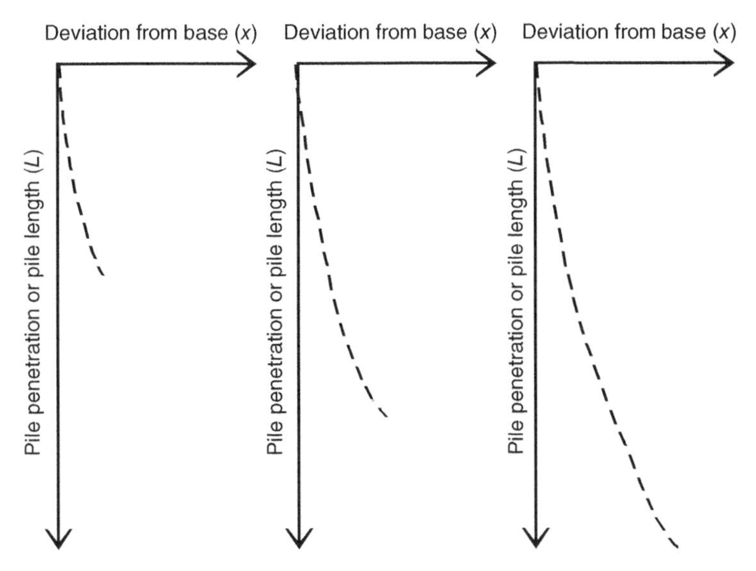

Figure 3.37 Imperfections in driven pile.

observed in a number of cases where offshore piles have collapsed during driving due to progressive closure of the internal dimensions – the initiating mechanism being local buckling. Thus, the design method has also to consider the interaction between local and global buckling. Figure 3.39 shows a typical offshore installation and Figure 3.40 shows the pile stick-up. Once the pile is in the sleeve, it is important to check the buckling potential of the pile under the action of the lateral forces due to the wave loading and the hammer weight. Often attachments are used for transportation and

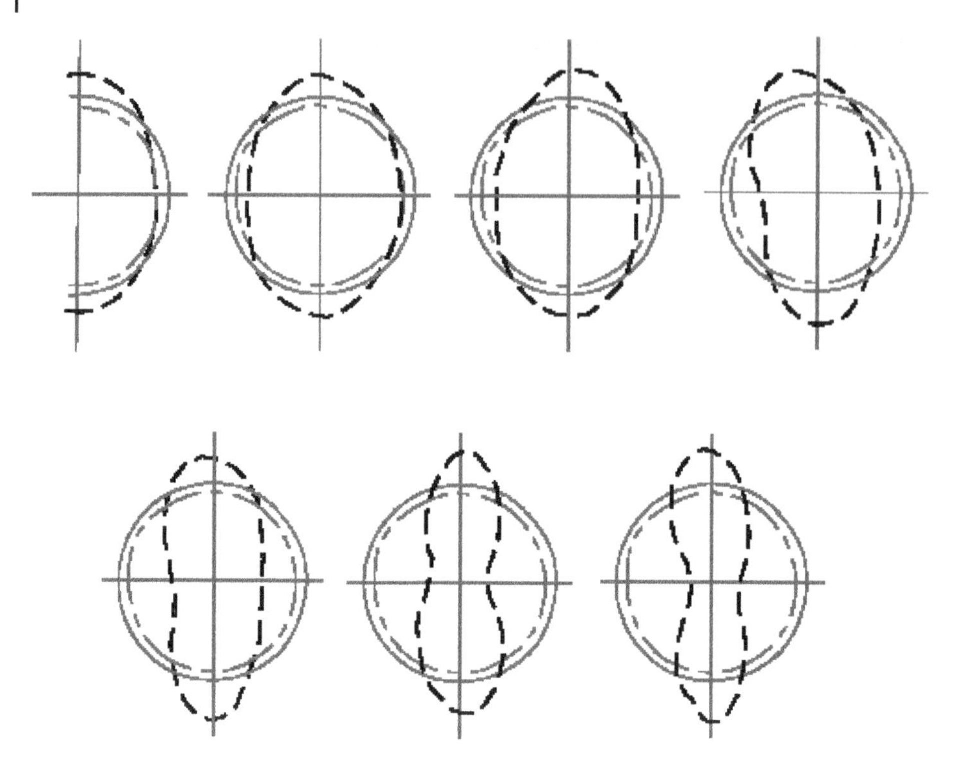

Figure 3.38 Pile tip buckling. The figure shows changes in cross-sections leading to progressive pile collapse during driving following an initial deformation of the pile tip.

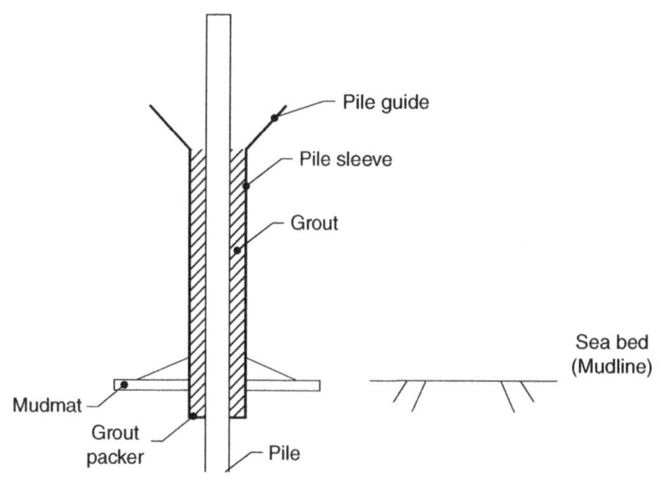

Figure 3.39 A typical offshore pile installation.

lifting, see Figure 3.41, and it is necessary to evaluate the impact of such attachments on pile driving. Details of buckling considerations can be found in Bhattacharya et al. (2005) and Aldridge et al. (2005).

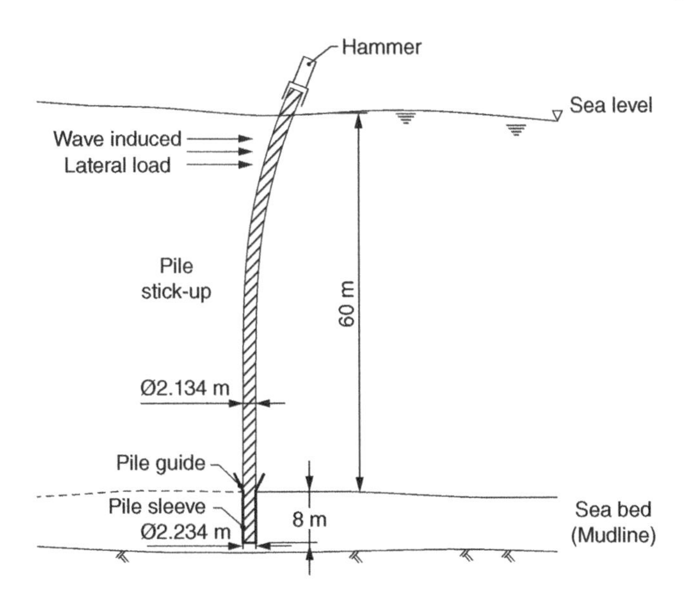

Figure 3.40 Pile stick-up.

Figure 3.41 Attachments in pile for transportation and lifting.

Example 3.2 ***Analysis of Driving Records from North Sea Sites (Beatrice & Amethyst Site)*** Bhattacharya et al. (2009b) analysed 53 pile-driving records for steel pipe piles driven at two sites in the North Sea, where the soil conditions consist of overconsolidated sand and clay strata and contain significant amounts of hard to very hard clay with undrained shear strengths typically 500–800 kPa. The main conclusions are as follows:

1. The driving records indicate clearly that the soil resistance increases (set up) during delays in the driving of steel pipe piles in very dense sands and very hard overconsolidated clays.

2. The observed increase in total soil resistance for delays with piles tipping between 35 and 50 m below mudline is of the order of 20–60% after 24 hours delay.
3. The observed increase in total soil resistance for delays with piles tipping at about 17–27 m metres below mudline is of the order of 60% at 24 hours.
4. There is more scatter in the calculated set-up factors for short time delays and lesser penetration, with set-up factors between 40% and 240% being observed for delays of less than 10 hours.

The above information may be useful in selecting the suite of hammers to be used when planning pile installation for North Sea projects.

3.12.2 Installation of Suction Caissons

The installation of suction caissons occurs in two distinct stages.

3.12.2.1 First Stage

In this stage, the buoyant self-weight of the caisson and attached structure pushes the caisson into the marine bed. This buoyant weight is resisted by a combination of skin friction and annulus-bearing force. The caisson will continue to sink into the sea bed until a sufficient portion of the caisson skirt is embedded in the sediment, creating an equivalent resistance to the self-weight of the structure. During this installation, valves are left open in the caisson lid to allow water to flow out of the space encompasses by the caisson. If sufficient water pressure develops in the caisson, outward piping failure can occur in the sediment surrounding the caisson. One of the design considerations is to allow sufficient drainage capacity within the caisson lid to prevent this effective from having more drainage valves.

3.12.2.2 Second Stage

In this stage of installation, pumping the water (contained in the cavity that exists between the caisson lid and the soil plug) is carried out. The act of pumping water out of this void is twofold: the first is to increase the downward force, effectively pulling the caisson into the sea bed. The second is to create an upward hydraulic gradient in the soil mass that loosens the soil around the caisson annulus and on the inside of the caisson skirt. This suction needs to be limited to avoid boiling of the sediment on the inside of the caisson. This will still further install the caisson over that achieved through self-weight installation. In some instances, caissons may be installed without applying suction (stage one only); this is achieved by using a sufficiently large surcharge. A simplified figure showing the installation process is shown in Figure 3.42.

The major reference for the installation load for suction caisson have come from a series of papers produced by Houlsby and Byrne (2005a,b). These estimations are based on theory and corroborated against experimentally obtained data. Both papers provide a conclusive guide as to caisson installation in sand and clay.

3.12.3 Assembly of Blades

The blades can be assembled in two ways: (i) each blade is individually attached see Figure 3.43 (ii) star type: All the blades are assembled and lifted and installed. Figure shows from London Array wind farm see Figure 3.44.

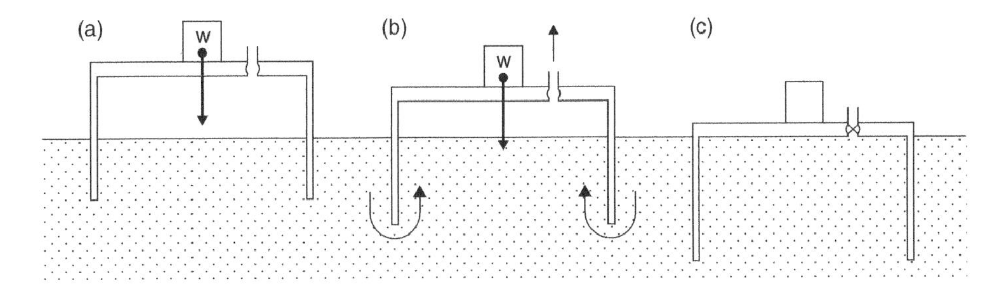

Figure 3.42 Stages of suction caisson installation, (a) self-weight, (b) suction assisted, and (c) installed.

Figure 3.43 London Array installation.

For jackets or seabed frame supported on multiple piles, there are two methods:

1. Subsea templates are first laid in the seabed, and piling is done at the correct location using the template (also known as pre-piling). The jackets are then transported and the jacket leg-pile connection grouted.
2. Jackets are transported and then laid in the seabed using mudmats. The piles are then driven.

Figure 3.45 shows installation photograph for Germany's Alpha Ventus offshore Wind farm in deeper waters.

3.12.4 Decommissioning

Decommissioning of energy infrastructure is increasingly becoming an important design consideration from the view of sustainability. This section of the chapter

Figure 3.44 UK – Ormond Site [Photo Courtsey: A2SEA].

Figure 3.45 Installation using a template. [Source: Alpha Ventus Offshore Wind Farm Project.]

provides a case study of the decommissioning of Lely Wind Farm (Netherlands) where the monopile is also extracted from the ground. All the self-explanatory photos are provided. In this context, it is necessary to provide an overview of the current legislation on offshore installations.

For the United Kingdom, decommissioning of offshore structures is regulated by UK law and by the OSPAR Decision 98/3 on the Disposal of Disused Offshore Installations (OSPAR Commission 1998). However, the removal of offshore piles is not required under legislation. Typically, the piles are cut below the seabed, at a suitable distance appropriate, often around three metres below the natural seabed.

Example 3.3 ***Decommissioning of Lely Wind Farm*** Lely wind farm (four 2 MW turbines in the IJsselmeer located approximately 800 m) built in 1992 generated electricity for 22 years and was fully decommissioned in 2017. Crane barges and tugs were used to

Figure 3.46 Offshore vessel and crane positioned near the WTG installation. Source and Courtesy: Dieseko Group - ICE, PVE & Woltman and Vattenfall.

Figure 3.47 Lift of nacelle and rotor. Source and Courtesy: Dieseko Group - ICE, PVE & Woltman and Vattenfall.

dismantle the turbines and tower sections; see Figures 3.46–3.51. The monopiles were extracted using a vibratory hammer; see Figure 3.52–3.55.

3.13 Chapter Summary and Learning Points

Offshore wind turbines are new types of offshore structure characterised by low stiffness (as a result, flexible and having low natural frequency) and therefore sensitive to the

Figure 3.48 Lift of tower, first section. Source and Courtesy: Dieseko Group - ICE, PVE & Woltman and Vattenfall.

Figure 3.49 Lift of tower, second section. Source and Courtesy: Dieseko Group - ICE, PVE & Woltman and Vattenfall.

dynamic loading imposed on them. The design guidelines available for offshore oil and gas installation foundations cannot be direct extrapolated/interpolated to offshore wind turbine foundation design.

3.13.1 Monopiles

Monopiles have been predominantly used to support WTGs in water depths up to 30 m. However, there are discussions with regard to the use of monopiles in deeper water

Figure 3.50 Superstructure on the vessel. Source and Courtesy: Dieseko Group - ICE, PVE & Woltman and Vattenfall.

Figure 3.51 Vibratory hammer used to remove the monopile from the ground. Source and Courtesy: Dieseko Group - ICE, PVE & Woltman and Vattenfall.

depths termed as 'XL' monopile. Preliminary calculations suggest that 10 m diameter monopiles weighing 1200 t may be suitable for 45 m water depth, of course dependent on ground conditions. However, the use is uncertain due to the following:

- No codified cyclic design to predict long term tilt;
- Lack of redundancy in foundation system and therefore chance of single-point failure;

Figure 3.52 Power pack used to along with the vibratory hammer.

Figure 3.53 Removal of the monopile.

- Installation costs and lack of adequate specialised vessels;
- Connection between foundation, transition piece, and the tower.

Some of these aspects are described below in further details.

1. *Lack of redundancy*. Monopiles are 'overturning moment' resisting structures, and there are two main components: (i) overturning moment arising from the thrust acting at the hub level; (ii) overturning moment due to the wave loading. Also, these two moments can act in two different planes and will vary constantly, depending on

Figure 3.54 Lift of the monopile on the vessel.

Figure 3.55 Monopile on the vessel.

the time of the day and time of the year. Monopiles are rigid piles and the foundation collapse can occur if the soil around the pile fails, i.e. there would be rigid body movement. If the foundation starts to tilt, it is very expensive to rectify.

2. *Cyclic (rather dynamic) design of monopile.* The response of monopile under cyclic/dynamic load is not well understood and there is a lack of guidance in codes of practice. If cyclic design is incorrect, monopiles can tilt in the long term. If the tilt is more than the allowable limit, the turbine may need a shutdown. Monopile design is usually (also wrongly) carried out using API design procedure calibrated for flexible pile design, where the pile is expected to fail by plastic hinge.

3. *Issues related to installation of monopiles.* Large monopile installation requires suitable vessel availability, as well as specialised heavy lifting equipment. Other issues are noise refusals, buckling of the pile tip, drilling out, and grouted connections. If the site contains weak rock (siltstone/sandstone/mudstone) and where the local geology shows bedrock or hard glacial soils at shallow depths, drive-drill-drive techniques may be required, with subsequent increases in cost and schedule. It must be mentioned here that driving reduces the fatigue life.

3.13.2 Jacket on Flexible Piles

There has been a considerable interest in jacket-type structures for deeper-water applications but is perceived being expensive due to the amount of steel required. However, jackets supported on piles can be considered as a safe solution due to excellent track record of good performance in offshore O&G industry. Offshore O&G industry has been using long flexible piles (diameters up to 2.4 m), which are easy to drive, and necessary vessels are readily available (relatively as opposed to vessels to install monopiles). This aspect will drive down the TIC (time in construction) costs regarding piling, and also, large vessels are not required for pile installation. However, there are costs associated with jacket installation. One of the requirement is the optimisation of the jacket so as to consume minimum steel. There are two types of jacket – normal jacket or twisted jacket. The advantage of the twisted jacket over normal jacket is fewer number of joints and therefore less fatigue issue.

3.13.3 Jackets on Suction Caissons

Jackets on suction caissons need to be designed so that rocking modes of vibration are avoided.

4

Geotechnical Site Investigation and Soil Behaviour under Cyclic Loading

Learning Objectives

This chapter discusses the site investigation and the soil testing required for obtaining the parameters for carrying out the foundation design. The chapter will also discuss the advanced soil testing apparatus that may be used to obtain the design parameters. There are plenty of text books on site investigation and soil testing and therefore those topics are not discussed in detail in this chapter. However, references are provided for those topics for guided learning.

4.1 Introduction

As discussed in Chapter 2, foundation design requires not only ground profile but also information on wind, waves, and site-specific geohazards, including natural hazards (e.g. seismic). However, this chapter is devoted to ground investigation, i.e. data and ground/soil parameters necessary for foundation-related design. Site Investigations related to ground are necessary for the following:

1. Foundation design for the turbine support structure and offshore substation structure. See Chapter 1 for a typical wind farm layout.
2. Cable route. Chapter 1 shows the layout of a wind farm, and there are two types of cables (inter-array and export cables) to transmit the electricity generated to the onshore grid. Investigations are necessary along the cable route for design purposes. The major considerations are: scour prediction, cable protection, and turbidity changes in the water. Turbidity changes in water will have implications in the fish population, thereby impacting the environment.
3. Planning and analysis of installation of foundations, together with the seabed anchoring of installation vessels. For example, jack-up vessels are often used to install the foundation, and a detailed ground profile is necessary for jack-up vessel leg penetration analysis. Other examples are suction pressure required to install foundations or vibratory/impact energy for driving piles.
4. Geo-hazards at the wind farm site, which include slope stability (submarine slides) or presence of gassy soils. Seismic analysis would require prediction of liquefaction susceptibility of the ground and site-response analysis.

Design of Foundations for Offshore Wind Turbines, First Edition. Subhamoy Bhattacharya.

Companion website: www.wiley.com/go/bhattacharya/offshorewindturbines

Detailed coverage on all these aspects are beyond the aims and scope of the book. There are many books and specialised reports available on offshore site characterisation and ground investigations, and they will be referenced here for further reading. The aim of this chapter is to discuss soil parameters that are necessary for preliminary and detailed design calculations, including soil-structure interaction (SSI) analysis. The testing methods through which these parameters are obtained are discussed and where possible further reading are suggested.

Arguably, ground-related issues are one the greatest project risks due to the uncertainties in the predicted costs due to installation and the problems associated with it. For example, delays in installation will have a direct impact on capital expenditure (CAPEX) cost, project schedule, and therefore profitability. Often, construction methodologies can lead to health and safety issues and also to environmental damage, which could, in turn, lead to penalty as well as negative publicity of the developer. All the above can be avoided as far as practicable using knowledge of the site. The aim of the discussion is to highlight the importance of thorough site investigation.

4.2 Hazards that Needs Identification Through Site Investigation

Ground-related hazards depend on the location of the wind farm. SUT (2014) classifies hazards into three types:

1. *Man-made hazards*. Examples include existing infrastructure such as oil and gas pipelines either on the seabed or buried, communication cables, shipwrecks, jack-up footprints, unexploded ordnance (UXO), etc.
2. *Natural seabed hazards*. These include, as the name suggests, rock outcrops, seabed topography, large diameter boulders, and unstable and steep slopes, etc.
3. *Subsurface geological hazards*. Examples include rapidly changing stratigraphy, buried infilled channels, igneous intrusions near seabed, tectonic faults, gas hydrates zone, and presence of gassy soils, etc.

The readers are referred to SUT (2014) for a comprehensive list of hazards. Once the hazards are identified, it is necessary to assess the risk and impact of those to the projects.

4.2.1 Integrated Ground Models

Essentially, the integrated ground model is a way to manage and sort information in a meaningful way so that ground risks can be captured and communicated to various stakeholders of the projects. Examples of stakeholders include investors, due diligence personnel, insurers or underwriters, certifiers, asset owners, etc. SUT (2014) defines a ground model as a database of information that includes structural geology, geomorphology, sedimentology, stratigraphy, geohazards, and geotechnical properties of the site.

The development of ground model and managing geotechnical risk is an iterative process that involves collecting relevant data and then interpreting and presenting in a meaningful way. Ground models are constantly updated as more information is available. Obtaining offshore ground data (geological, geophysical, and geotechnical) is expensive, as it involves mobilising a vessel. Therefore, ground models are not usually finalized until after the business decision of wind farm development is made. Development of the ground model is a continuous process and may begin through the use of satellite images and desk study (i.e. geological information of the site).

ASIDE

Ground modeling is often the deciding factor for development of offshore wind farm. For example, the Atlantic Array (United Kingdom) wind farm project didn't go ahead because of ground-related uncertainties. Although wind farm developers often avoid carrying out expensive site investigation before a project is secured, risk must be incorporated into the bidding process for offshore development, with some effort made to predict the CAPEX and OPEX cost. The Netherlands government carries out extensive site survey at the cost of the taxpayer before the start of the bidding process. In this way, some certainty is guaranteed to the bidders.

4.2.2 Site Information Necessary for Foundation Design

A typical offshore wind farm consists of large number turbines widely spaced and therefore change in water depth and geology may be expected. Therefore, the investigation will cover a large amount of area together with the plausible route for cables. To carry out feasibility studies as well as create a detailed design, various information and data are necessary. Table 4.1 lists the essential ones.

ASIDE: Chinese Offshore Wind Farm Development

Figure 1.2c in Chapter 1 shows the offshore wind farm development in China. China has abundant offshore wind energy resources. Junfeng (2012) reports that the coastal area of China has a wind power capacity of 200 GW at an elevation of 50 m above sea level and within a water depth of 5 to 25 m. However, the challenges are typhoons, earthquakes and soft seabed conditions. The eastern coast of China is prone to experience earthquakes and source can be located in the continenal shelf extending from the coastal land to offshore undersea, creating additional hazards for offshore wind farms. In such cases, probabilistic seismic hazard analysis (PSHA) as explained in Chapter 2 is a rational approach to quantify the uncertainty.

Wind information (i.e. wind speed at the hub level) will provide an estimation of the energy yield from the turbine. Wave information will be required for estimating the loads on the structure. On the other hand, ground information i.e. the variability of the ground over the wind farm is necessary for designing the foundations and for the construction phase.

ASIDE: Unexploded Ordnance in Horns Rev Wind Farm

Unexploded explosive devices have not been fully detonated. Examples include air-dropped bombs, grenades, anti-aircraft missiles, and projectiles. During desktop studies, it was noted from the 1944 German logs that there were mines, and two of the mines were in the vicinity of Horns Rev 3, were from WWI, and were quite small.

Table 4.1 Site information and why it is needed.

Type of data	Information required	Remarks on the use of data for design purpose
Metocean data: This involves meteorological and oceanographic condition at the wind farm location	The main information typically gathered are: (a) Wind speed and direction (Wind Rose) at selected heights above MSL. The data are structured in 1 h mean, 10 min mean, and 3 s gust. (b) Significant and maximum wave heights and the corresponding periods. Wave directions are also recorded. (c) Profile of current speed and direction. (d) Water depth and changes in levels caused by tides and surge.	Wind speed, wave height and period, current speed, and water depths are needed for foundation load calculations. Current data are required for scour prediction and design of the cable route. Wave and wind (magnitude and frequency) data are necessary for fatigue life calculations. Wind and wave data are required to understand the nature of the load in terms of one-way or two-way.
Seismic and tsunami hazard	Location and nature of faults around the site which includes historical occurrences. For further details see Chapter 2.	This are needed for PHSA (probabilistic seismic hazard assessment) and DSHA (deterministic seismic hazard assessment). The output will be seismic hazard parameters for seismic load estimation.
Slope stability and seabed mobility	Slopes in the seabed around the area. Any history of subsidence, i.e. presence of any oil and gas reservoir under the seabed or gassy soils.	Estimation of the ground subsidence (if any) around the foundation. Seabed mobility will determine the extent of unsupported length in the lifetime of the project and the scour protection measures needed to avoid such problem.
Site characterisation, which is a combination of geological, geophysical and geotechnical	The strength and stiffness profile of the ground along the depth. For simplified calculations of monopile stiffness as shown in Chapters 5 and 6, two parameters are required: (a) Stiffness of the ground at one diameter depth; (b) Variation of the ground stiffness with depth.	Strength of soil is required for ULS calculations, i.e. capacity of the foundations. Stiffness of the ground is required for prediction of foundation stiffness, which is necessary for natural frequency calculations and deformations.
Other considerations	Marine growth studies, corrosion, and site-specific considerations. For example, in European waters there are still plenty of UXOs. Environmental studies are likely needed.	Long-term prediction of capacity and load demand.

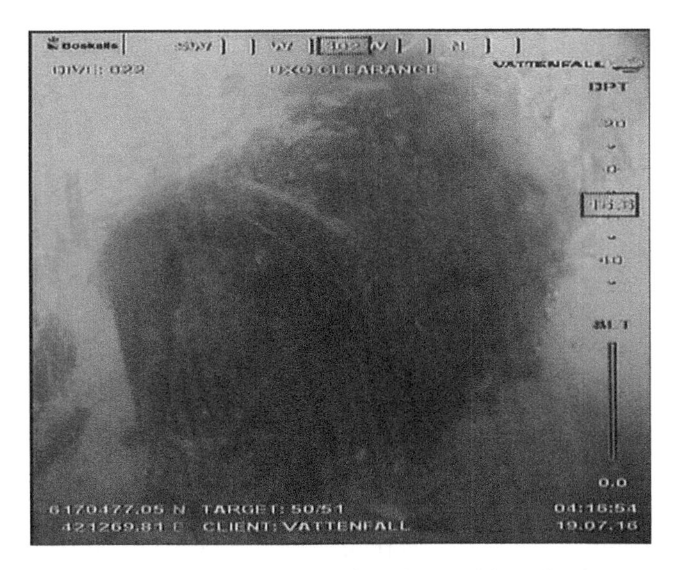

Figure 4.1 UXO in Horns Rev wind farm (Source: [Photo Courtesy: Vattenfall]).

A third mine was a so-called LMB mine from WWII and had an explosive charge of 700 kg. These were handled by specially forces of the Naval Staff division of Danish Defence Command (DDC), whereby the mine was detonated in 2016 in a controlled way using a wire and a detonator. Figure 4.1 shows the UXO and further details can be found in Vattenfall (2016).

4.2.3 Definition of Optimised Site Characterisation

The ground engineering requirement covers a large area and there are significant costs associated with mobilisation of an offshore site investigation vessel. Therefore, at the planning stage, it is necessary to develop understanding of the project's needs and requirements so that necessary data and soil samples for the project can be obtained from one mobilisation of a vessel. For the purpose of foundation design, the main requirement is the development of *design ground profile* and assigning *engineering parameters* to each of the soil strata. This is one of the most important and critical steps that demand expertise. Section 4.3 provides typical examples of design ground profile.

4.3 Examples of Offshore Ground Profiles

4.3.1 Offshore Ground Profile from North Sea

This section provides examples of typical ground profiles from a few offshore locations. Figure 4.2 and Table 4.2 shows a design ground profile from North Sea site following Bhattacharya et al. (2009). The location of the site is provided in Figure 4.3.

Other ground profiles from European wind farms are discussed in Chapter 5.

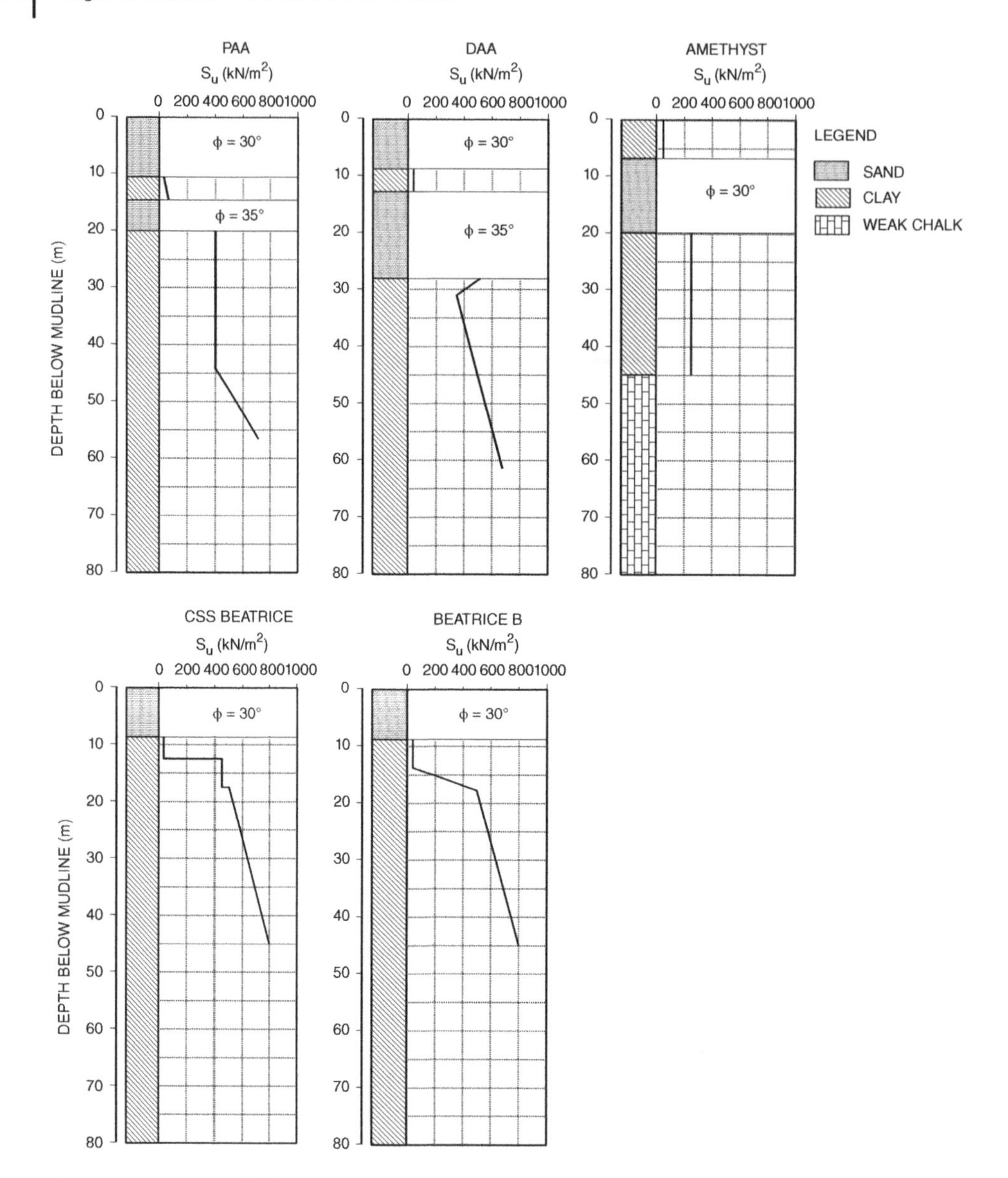

Figure 4.2 Example of offshore ground (Design profile) from North Sea. For location of these sites, see Figure 4.3.

4.3.2 Ground Profiles from Chinese Development

There are four main sea areas in Chinese water where offshore wind development is in progress: (i) Bohai Sea (enclosed sea); (ii) Yellow Sea; (iii) East China Sea; (iv) South China Sea. Generally, the ground conditions in China seas are soft and layered, in contrast to North Europe.

For example, the Bohai Sea is a low-lying land settled by North China Plain, where the soil consists of sediments brought from mountains and Yellow River and is soft, silty clay or clayey silt. Borehole is usually a widely used method for underground geology

Table 4.2 Soil profile at the sites.

Location and time of the test	Relevant soil Layer (typical)
Beatrice field, PAA jacket site	0.0–9.0 m sand (ϕ = ~30°) 9.0–12.5 m soft to firm clay; (40–50 kPa) 12.5–17 m/29 m sand (ϕ = ~35°) >17 m/29 m very hard clay (400–700 kPa)
Beatrice field, DAA jacket,	0.0–8.5 m sand (ϕ = ~30°) 8.5–12.5 m soft to firm clay (40–50 kPa) 12.5–29.0 m sand (ϕ = ~35°) 12.5–15.0 m stiff to hard clay (40–500 kPa) >29 m very hard clay (400–700 kPa)
Amethyst field	0.0–7.0 m stiff clay (40–150 kPa) 7.0–20.0 m medium dense sand (ϕ = ~30°) 20.0–45.0 m stiff clay (~250 kPa) >45.0 m very weak chalk
Beatrice field (North Sea), block 11/30,	0.0–8.6 m medium to dense sand (ϕ = ~30°) 8.6–12.6 m soft to firm clays (~ 40 kPa) 12.6–17.6 m HARD CLAY (400–500 kPa) >17.6–45 m hard to very hard clay (500–800 kPa)
Beatrice field (North Sea), block 11/30, B jacket (adjacent to CSS)	0.0–9.0 m loose very silty fine sand to dense fine sand (ϕ = ~30°) 9.0–14.0 m soft sandy very silty clay (~ 40 kPa) 14.0–18.0 m stiff to hard clay (40–500 kPa) >18.0 m hard to very hard clay(500–800 kPa)

survey. There are many records of boreholes in Bohai Sea, and a typical profile (from top to bottom) is as follows:

1. *Soft clay silt (Layer 1)*. The depth of this layer varies but is typically 0–7.4 m in the western Bohai.
2. *Mucky clay to silty clay (Layer 2)*. In this layer, the clay is changing from soft to stiff, with materials of mucky clay, silt, and loose sand. The depth of the clay layer varies (thickness 2.5–9.0 m in western Bohai), centre Bohai (thickness 4.0–10.0 m), Jinzhou (thickness 2.7–14.8 m) and Suizhong (thickness 0.8–5.8 m).
3. *Combination of silty clay, clay, and loose sand (Layer 3). Again, the thickness of this layer varies* – 2.5–7.0 m in western Bohai.
4. *Fine sand and silt*. In this layer, the main material is fine sand, combined with silty sand and clay. The stiffness is that of medium dense sand. The thickness in western Bohai is sea is 4.0–15.0 m, in centre Bohai is 4.0–13.0 m, in Jinzhou is 3.0–15.0 m, and in Suizhong is 0–14.2 m.
5. *Clay*. In this layer, the clay is quite stiff and hard, combined with some silty clay. The thickness in western Bohai is 2.0–18.0 m, in centre Bohai is 3.0–15.0 m, in Jinzhou is 3.0–17.0 m, and in Suizhong is 0–15.0 m.
6. Fine sand and silt

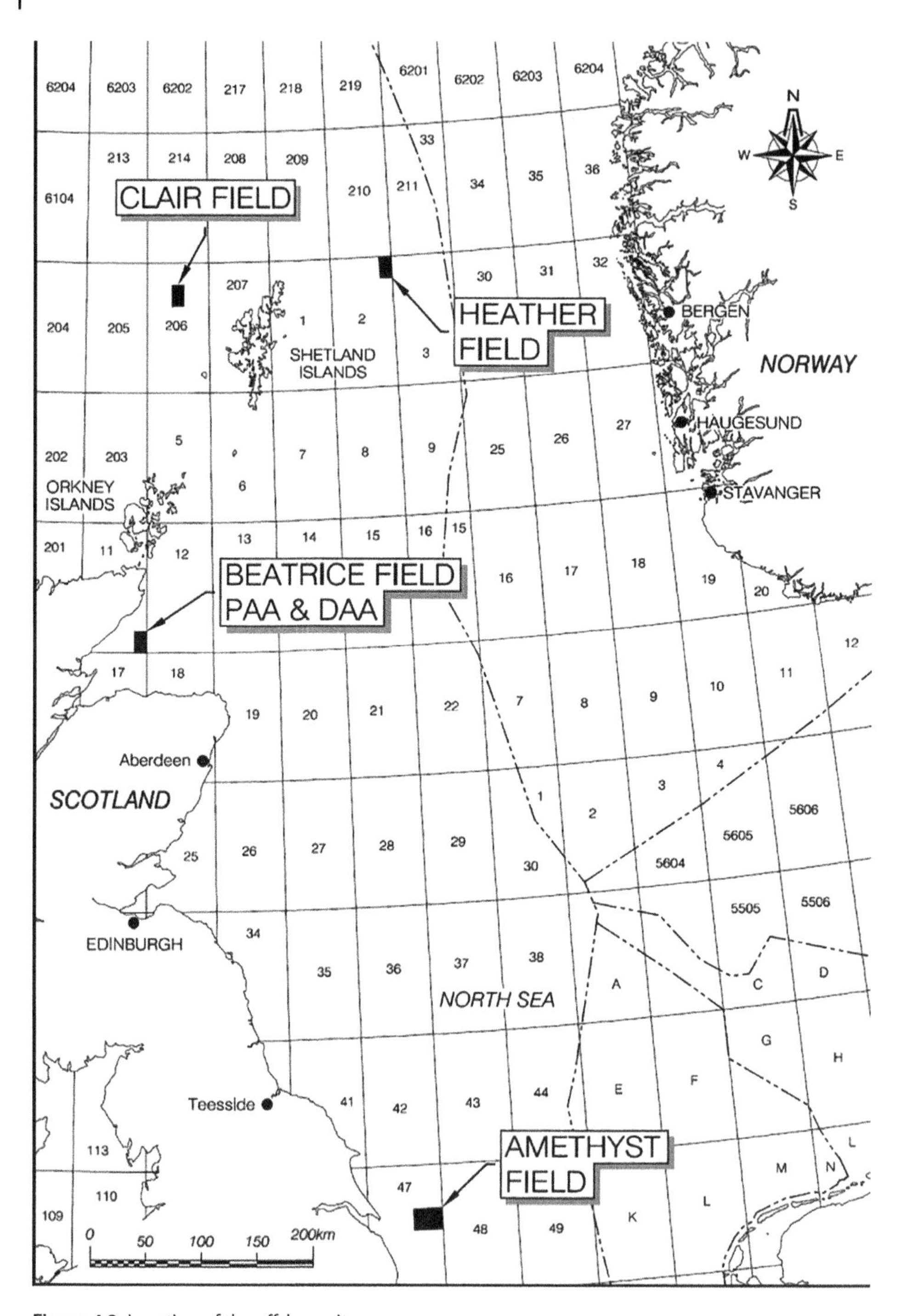

Figure 4.3 Location of the offshore sites.

In this layer, the main material is fine sand, combined with silty clay where the layer is thin. The stiffness is quite dense. The thickness in western Bohai is 5.0–11.0 m, in centre Bohai is 6.0–15.0 m, in Jinzhou is 7.0–15.0 m, and in Suizhong is 12.0–18.0 m (see Figure 4.4).

By contrast, Figure 4.5 shows ground profile for the Yellow Sea, and it is evident that the ground profile is variable, which calls for detailed site investigation to minimise risks

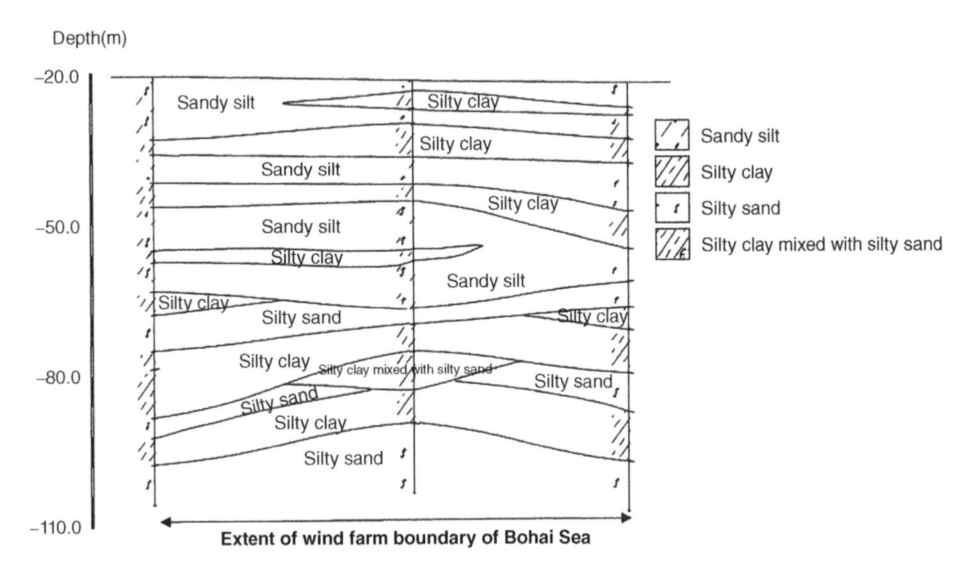

Figure 4.4 Ground profile in Bohai Sea wind farm.

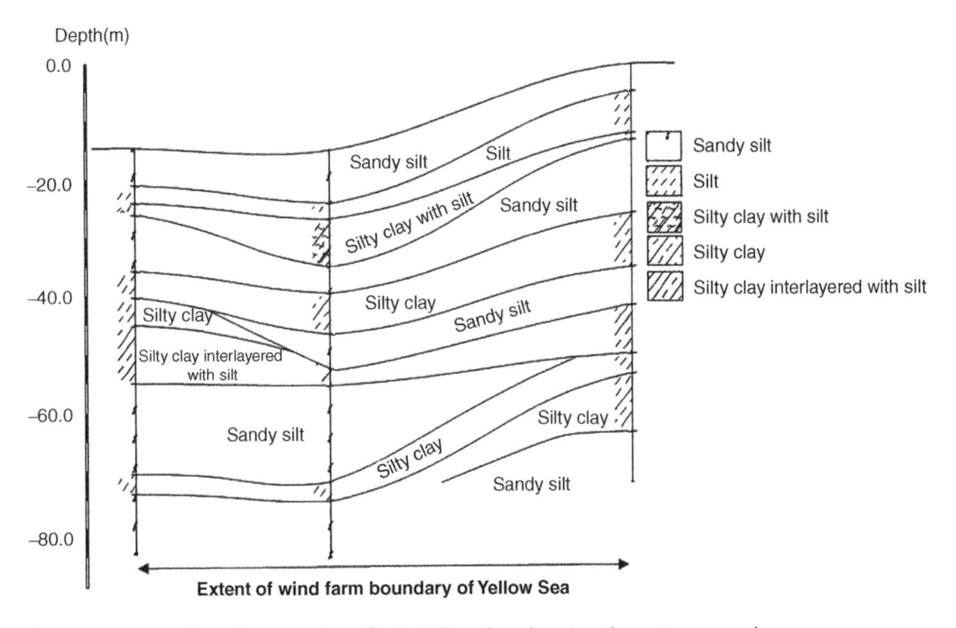

Figure 4.5 A section of a ground profile in Yellow Sea showing the extreme geology.

during construction. Figures 4.6 and 4.7 show the ground profiles for other two Chinese seas.

Table 4.3 shows the ground profile in South China Sea. A general observation can be made. The ground profile in Chinese waters is variable, and it shows the importance

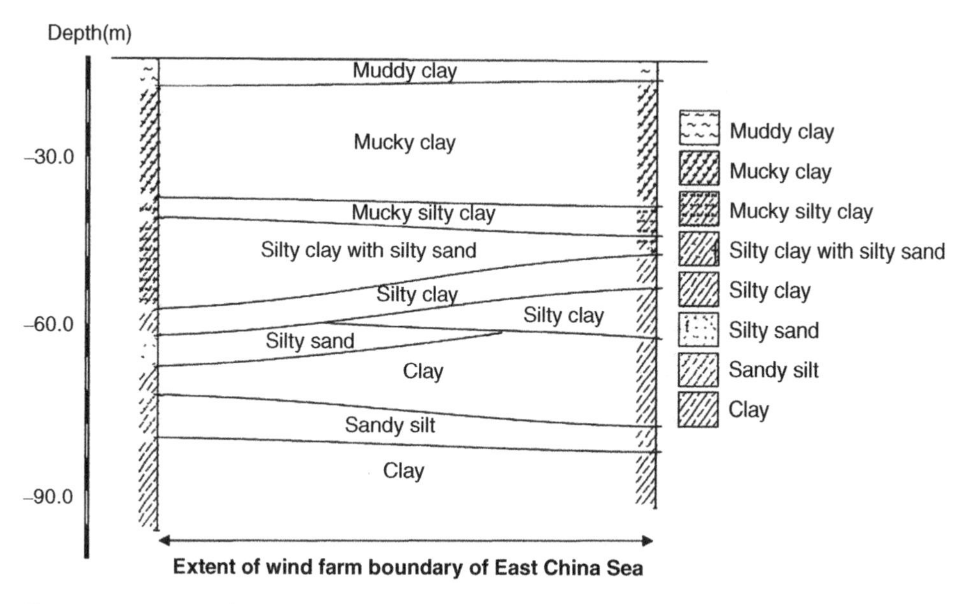

Figure 4.6 A section of the ground profile in East China Sea.

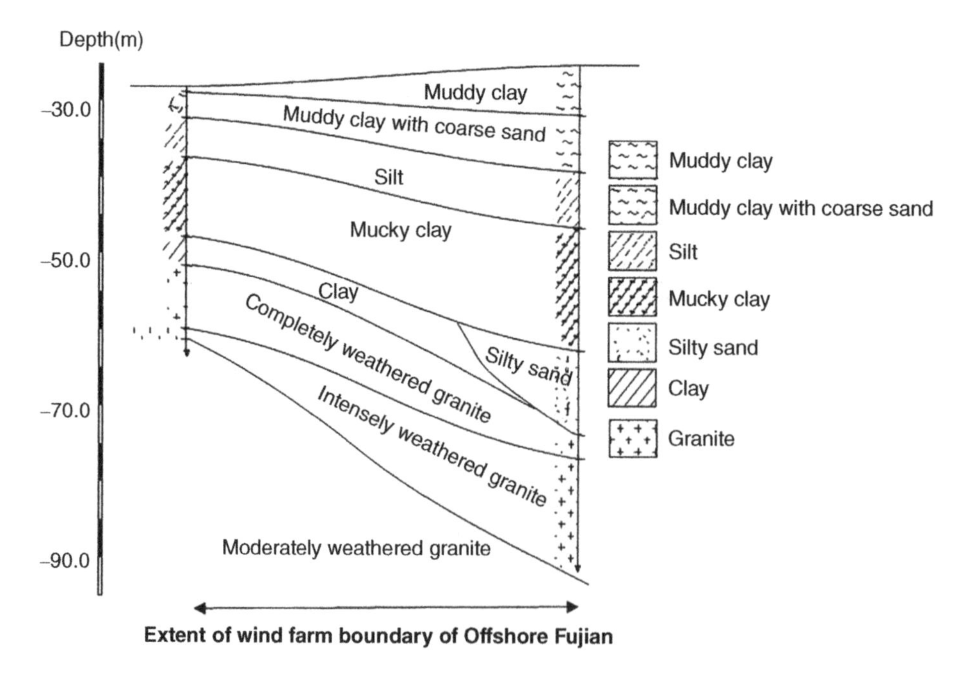

Figure 4.7 Ground profile in offshore Fujian Sea (Taiwan Strait).

Table 4.3 Ground profile in South China Sea.

Position no.	Depth (m)	Soil
1	0–4.2	Silty clay
	4.2–19.4	Laminated soil
	19.4–40.0	Silty clay
2	0–3.0	Sandy soil
	3.0–7.4	Silty clay
	7.4–27.0	Laminated soil
	27.0–35.3	Silty clay
3	0–1.9	Silty clay
	1.9–2.6	Silty sand
	2.6–5.5	Silty clay
	5.5–9.0	Sandy soil
	9.0–15.0	Laminated soil
	15.0–39.8	Silty clay
4	0–1.5	Sandy soil
	1.5–4.9	Silty clay
	4.9–7.0	Silty sand
	7.0–13.7	Laminated soil
	13.7–25.4	Silty clay
	25.4–40.6	Sandy soil

of a comprehensive site investigation (Table 4.3). Further details can be found in Bhattacharya et al. (2017).

4.4 Overview of Ground Investigation

The site investigation for offshore wind farms can be divided into three categories: geological, geophysical, and geotechnical. These are used to develop a comprehensive integrated ground model.

4.4.1 Geological Study

Chapter 1 showed that an offshore wind farm must cover a large area in order to maximise the energy yield. As shown, in some cases, the area is as large as 20 × 6.5 km. A study based on geological history of formation can be the basis of planning site investigation. Geological study is typically a desk study and is used to create the first ground model.

4.4.2 Geophysical Survey

This involves mobilising specialist vessels to map the seabed surface as well as subsea floor and the data generated can be used to update the ground model. A geophysical

Table 4.4 Types of geophysical survey.

Geophysical equipment	Design data	Remarks
Multi-Beam Echo Sounder (MBES)	Water depth and seabed topography of the whole wind farm and the cable route is obtained.	The bathymetry survey system and is generally hull-mounted with sensors to compensate for vessel movement. Survey line spacing depends on the site
Side scan sonar (SSS)	Seabed features are recorded	Provides an acoustic image of the seabed
Sub-bottom profiler	Ground profile and geology	Depending of the sub-bottom resolution (e.g. pinger, sparker, boomer), they are required for shallow seabed for inter-array or export cables or correlation of deeper layers in the geological model
Magnetometer/ Gradiometer	Unexploded ordnance (UXO)	These are used to map the change in magnetic field around a ferrous object

survey can be conducted alongside geotechnical site investigation. Table 4.4 shows the list of geophysical equipment for geophysical survey and design data/information obtained. Geophysical surveys are often carried out before the final business decision is taken.

DNV-OS-J101 suggests that line spacing of survey at the seabed location should be sufficient to detect all soil strata of significance for the design and installation of the wind turbine structure. It is also recommended that the survey needs to detect any buried channels with soft infill materials.

An example of an alignment chart used in a cable route investigation is given in Figure 4.8. The chart shows bathymetry, seabed features, sub-bottom profile, magnetometer, route planning, and service chart.

4.4.3 Geotechnical Survey

The geotechnical site investigation required is governed by the foundation concept likely to be used and is, in turn, dependent on water depth, geology of the area, and the environmental conditions. In this process, a vessel is chosen depending on the requirements of the project for *in-situ* testing and sample recovery from boreholes. Typically, *in-situ* investigation involves pushing a cone (also known as cone penetration test, CPT) at different location to measure the soil resistance to penetration, see Figure 4.11 for an example bore log. Selection of vessel depends of various factors, and a summary of the important ones are given in Table 4.5.

Further details on the types of vessels can be found in Dean (2010), SUT (2014), and Rattley et al. (2017). Geotechnical investigation will provide the engineering parameters for the following analysis and design of: (i) foundations and its installation; (ii) inter-array and export cables along a defined route; (iii) long-term prediction in terms of serviceability limit state (SLS) and fatigue limit state (FLS). Some examples are:

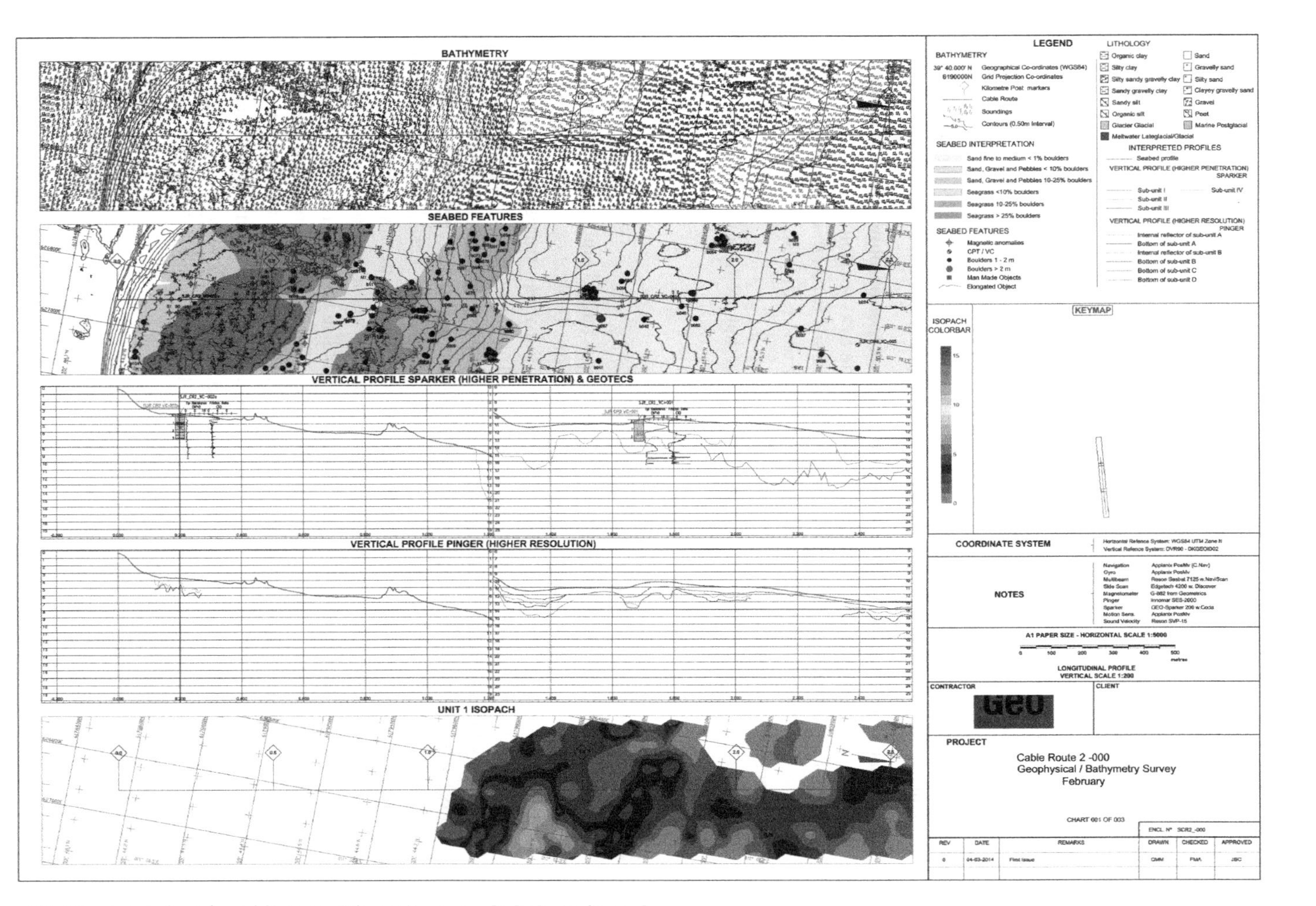

Figure 4.8 Alignment chart for cable route. [Photo Courtsey: GEO, Copenhagen].

Table 4.5 Vessel selection.

Types of site	Vessels
Shallow water depth (typically less than 15–20 m) and good seabed conditions	Jack-ups are suitable and standard onshore equipment can be used.
Water depth more than 20 m	Barges or ship with dynamic positioning system.
Site with high current	Anchored vessels with heave compensators

axial capacity of piles in tension and compression, p-y curves for laterally loaded pile design, pile driveability characteristics, and bearing capacity of shallow foundations.

Typically, a geotechnical site investigation campaign involves the following:

1. CPT at many locations to a depth depending on the types of foundations. A tentative guidance on the depth of penetration is given in the next section.
2. High-quality samples are obtained at selected locations through boreholes.
3. *In-situ* measurement of small strain stiffness may involve the use of a pressuremeter, dilatometer, or seismic cone.
4. Laboratory testing (Table 4.6) provides a list of tests that are necessary. However, this is not an exhaustive list.

ASIDE:

Offshore site investigation is challenging, requiring specialised equipment and vessels. This is very expensive and requires constant innovation. Typical cost of a comprehensive offshore investigation for a 1 GW wind farm in European waters is \$3 million to \$5 million. It is therefore necessary to be well planned and directed, taking into account the most likely foundations and cable routes. At the planning stage, geology-based desktop study can be helpful. The geophysical methods have the advantage of wide coverage and helps to build up a big picture and reinforce the geology-based study. Stratification showing strong contrasts of soil stiffness can be detected. *In-situ* testing with CPT can provide quantitative and continuous measurement of soil strength. Sample recovery and testing has no substitute, as it is necessary to visualise the materials and carry out essential characterisation. Advanced methods such as X-ray diffraction, carbon-14 dating, and SEM (scanning electronic microscope) are often used to find fabric in the soil, mineralogy, ageing of the sample, etc. Tests on disturbed samples can be easily carried out based on re-constitution. However, undisturbed sample recovery is very difficult in offshore, especially if there is a very soft deposit. Often, dissolved gas comes out of soil as the pressure is relieved, destroying the fabric.

4.5 Cone Penetration Test (CPT)

CPT initially developed in the 1950s in Holland, also known as the Dutch Cone Test, is one of the most used site investigation tools. The test consists of pushing an instrumented cone (together with a friction sleeve), with the tip facing down, into the ground

Table 4.6 Engineering parameters that are required for various calculations.

Description and Index properties of the soil	Atterberg limit, maximum and minimum void ratio, water content of the sample	This is a fundamental property and many engineering correlations are based on these properties. These tests are often carried out in-situ in the exploration vessel.
Strength properties of the soil	Friction angle (peak as well as critical state angle), undrained strength of soils. The following apparatus may be used: direct shear box, triaxial apparatus (e.g. CU/UU tests, consolidated / unconsolidated undrained)	These parameters are necessary for ULS calculations. The stress–strain graph is required for SLS and FLS calculations
Stiffness properties of the soil	In-situ stiffness may be measured. For laboratory testing, the following apparatus may be used: resonant column, triaxial or cyclic triaxial with bender element, cyclic simple shear apparatus	Stiffness at small strain (G_{max}) is one of the important parameters for natural frequency estimate
Permeability and consolidation parameters	Triaxial tests can be used or permeameters or CPT	Permeability information is required for assessing if the SSI interaction is drained or undrained.
Liquefaction potential	This is required for earthquake studies. Cyclic triaxial tests can be carried out to find the cycles to liquefaction and post-liquefaction behaviour.	The analysis can predict the depth of liquefaction. A multi-stage triaxial test (first cyclic load and then monotonic load) can also be carried out for post-liquefaction behaviour.
Thermal conductivity	Thermal needle method may be used. Details of method can be found in Mitchell and Kao (1978), Farouki (1981, 1982)	This is required for cable insulation and protection design.
Chemical composition	To find the content of gassy soils and other chemicals that can cause corrosion to different buried components	Cable protection design
Damping properties	The following apparatus may be used: Resonant column and cyclic loading apparatus (cyclic triaxial and cyclic simple shear).	Damping measurement and are necessary for dynamic analysis of the whole system.

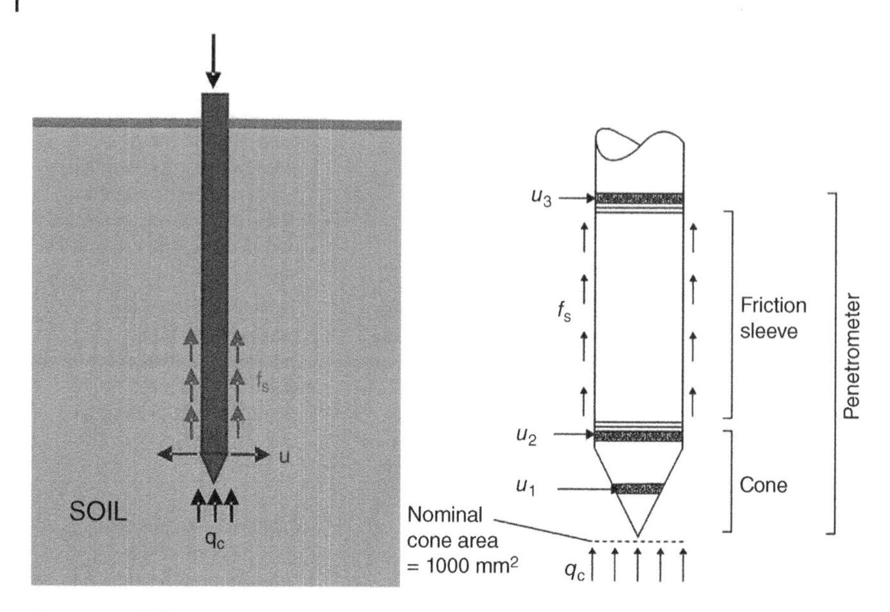

Figure 4.9 CPT test.

at a controlled rate (typically 15–25 mm s^{-1}). The standard cone has an apex angle of 60° and a cross-sectional area of 10 cm^2 corresponding to a diameter of 3.56 cm; see Figure 4.9. Electrical transducers measure the soil resistance on the cone tip as it penetrates into the soil. The frictional resistance along the sleeve is also measured. Another variation of CPT, known as CPTU, measures simultaneously tip resistance (q_c), the sleeve friction (f_s), and the pore pressure generated during penetration (u). In saturated soils, especially in fine-grained soils, the development of excess pore water pressure reduces the value of tip resistance q_c. Different proportions of sleeve friction f_s and tip resistance q_c for measured for different soils. Coarse-grained soils will have high q_c and low f_s. On the other hand, fine-grained soils will have high f_s and low q_c. Standard correlations and calibration charts exist for classifying soils; see Figure 4.10. Specialist knowledge is required for interpreting CPT data, and the readers are referred to Lunne et al. (1997) and Robertson et al. (1986).

Table 4.7 provides a preliminary guide for the relation between cone resistance (q_c) and the angle of internal friction φ(°) in fine sand. It should be noted that this guide is applicable to unaged, uncemented sands up to about 10–15 m depth.

The results from the CPT test can also be used to estimate soil strength and stiffness.

To estimate the soil strength and assess the undrained shear strength of a soil (S_u), the net resistance from the CPT q_{net} (corrected cone resistance minus the vertical stress) is divided by a cone factor (N_{kt}).

$$S_u = \frac{q_t - \sigma_v}{N_{kt}} \tag{4.1}$$

where typical N_{kt} factors for the North Sea are 15–20.

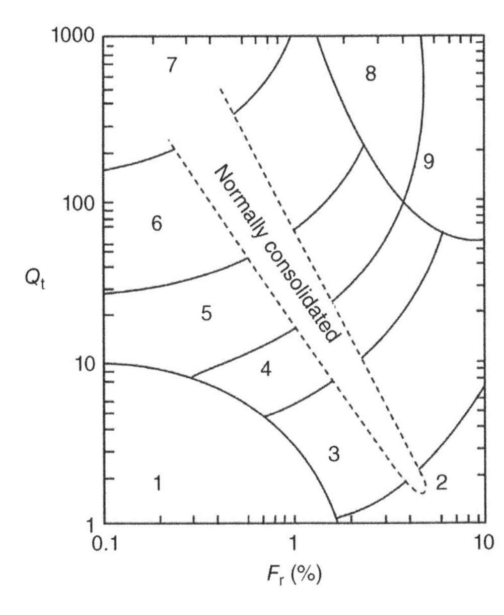

Zone Soil behaviour type

1. Sensitive, fine grained
2. Organic soils–peats
3. Clays–clay to silty clay
4. Silt mixtures; clayey silt to silty clay
5. Sand mixtures; silty sand to sand silty
6. Sands; clean sands to silty sands
7. Gravelly sand to sand
8. Very stiff sand to clayey sand
9. Very stiff fine grained

$$Q_t = \frac{q_t - \sigma_{vo}}{\sigma_{vo}}$$

$$F_r = \frac{f_s}{q_t - \sigma_{vo}} \; 100\%$$

Figure 4.10 Soil classifications based on CPT.

Table 4.7 Correlations of cone resistance and angle of internal friction.

Cone resistance, q_c (MPa)	Compaction of fine sand	Relative density, D_r (%)	Angle of internal friction, φ(°)
<2	Very loose	<15	20–25
2–5	Loose	15–35	25–30
5–15	Medium dense	35–65	30–35
15–30	Dense	65–85	35–40
>30	Very dense	85–100	>40

In terms of stiffness, empirical equations suggested by different researchers may be used to estimate shear velocity and shear modulus. A few examples are given below:

The cone resistance can be used to determine the shear modulus of different sands using either of the following relationships Rix and Stokoe, (1991), Mayne and Rix, (1993):

$$G_o = 1634(q_c)^{0.25}(\sigma_v')^{0.375} \tag{4.2}$$

$$G_o = 406(q_c)^{0.695}(e)^{-1.13} \tag{4.3}$$

where e is the soil void ratio:

The cone resistance can also be used to determine the shear velocity (V_s) as given.

$$V_s = 277(q_c)^{0.13}(\sigma_v')^{0.27} \tag{4.4}$$

ASIDE

Readers are reminded on the use of correlations. The correlations should not be used beyond its range of validity.

4.6 Minimum Site Investigation for Foundation Design

Many codes of practice and best practice guides, such as the DNV, American Petroleum Institute (API) and SUT (2014), provide guidance for the minimum site investigation requirements and this section summarises them. For wind turbine structures in a wind farm, a tentative minimum soil investigation program may contain four aspects:

1. One CPT test per turbine location for a grounded system based on monopile. For suction buckets, and other foundations such as floating/anchored foundations, one CPT for each suction bucket or anchored location will be needed.
2. Soil boring to sufficient depth in each corner of the area covered by the wind farm for recovery of soil samples for laboratory testing. This will provide a basis for variability of the ground.
3. An additional soil boring in the middle of the area will provide additional information about possible non-homogeneities over the area.
4. In cases where soil conditions are highly varying within small spatial areas and foundations, more than one soil boring or more than one CPT per wind turbine position may be needed.

In-situ testing such as CPT and soil borings with sampling should be carried out to sufficient depths. The following are recommended:

- For slender and flexible piles in jacket type foundations, a continuous CPT from seabed to the anticipated depth of the pile plus three times pile diameters is recommended.
- For monopile with larger diameters, a continuous CPT from seabed to the anticipated depth of the pile plus 0.5 times the pile diameter is recommended.
- For gravity-based system, a sample borehole or deep continuous CPT borehole at the centre of the proposed structure. The number and depth of borings will depend on the soil variability in the vicinity of the site. This is necessary for settlement predictions.
- For jack-up installation vessels, the CPTs performed for the wind turbine generator (WTG) foundation are often used. Otherwise, depth equal to at least one and a half spudcan diameter are recommended.

4.7 Laboratory Testing

Once the offshore samples are recovered, laboratory tests must be carried out to obtain as much information as possible. It will be shown in Chapter 5 that soil–structure interaction plays an important role in the long-term performance of these structures and, as a result, advanced soil tests are necessary. The types of tests depend on the types of foundations and the nature of loading. It must be mentioned that the cost of laboratory tests is negligible when compared with the cost required to obtain them.

4.7.1 Standard/Routine Laboratory Testing

The most commonly used element testing methods are direct shear and ring shear tests, simple shear tests, triaxial tests, and consolidation tests. The readers are referred to textbooks on element testing for details and methodology. Some examples are Dean (2010), Mitchell and Soga (2005), Head (2006), and Randolph and Gourvenec (2011). Table 4.8 provides a summary on the soil parameters that can be obtained from such tests.

4.7.2 Advanced Soil Testing for Offshore Wind Turbine Applications

As discussed in Chapter 2, loads on the foundation are cyclic as well as dynamic. It is therefore necessary to understand the behaviour of soils under such loading conditions. From practical view of view, three types of apparatus readily available and widely used

Table 4.8 Routine element tests.

Name of the test	Description of the test	Requirement in design
Direct shear test	This apparatus is used to determine shear strength parameters. It consists of horizontally split, open metal box where one part of the box can move relative to the other. Soil is poured in the box at a particular density and one part of the box is moved relative to the other. The soil is designed to fail in a thin zone along the interface of the two part box. This apparatus is commonly known as shear box.	Drained tests are generally conducted in a Shear Box. Angle of friction (φ), Angle of dilation Undrained strength (cohesion). Shear Box cannot prevent drainage i.e. pore pressure cannot be measured. Undrained shear strength can be estimated approximately by shearing the soil very fast, i.e. running the test very quickly) and the sample fails at the specific plane and not necessarily at the weakest plane.
Triaxial tests	Triaxial Test: A cylindrical soil sample is tested where confining pressure can be applied together with different drainage condition i.e. drained/undrained. Deformation of the soil along with volume change/ pore water pressure change can be measured. The main advantage is variety of stress and strain paths that a soil experiences in the field can be simulated.	We can obtain: Stress-strain characteristics of a soil, shear strength, and volume change together with pore water pressure Permeability of soil samples can also be measured. The following parameters may be obtained: Angle of friction (φ), cohesion/undrained strength (Su) Shear modulus (G) from the slope of the stress strain curve.
Oedometer test	This involves applying compression to a disk of soil vertically to obtain the time-dependent characteristics.	Compressibility of the soil
Ring shear test	In this test an annular soil sample is twisted to induce a circular shear plane.	Residual friction angles for clay

in engineering practice: cyclic triaxial with bender element, cyclic simple shear (CSS), and resonant column apparatus.

4.7.2.1 Cyclic Triaxial Test

The triaxial test is the most widely used and versatile shear strength test. As the name suggests, the soil specimen is subjected to compressive stresses in three directions by fluid pressure. To fail the soil sample, additional vertical stresses are applied on the top of the sample, axially (either monotonically or cyclically) by a loading ram. Different types of tests can be carried out: drained, undrained, consolidated undrained, and multistage. The basic features of the triaxial test apparatus are presented in Figure 4.12. Figure 4.13 shows a cyclic triaxial set at SAGE (Surrey Advanced Geotechnical Engineering) laboratory. The readers are referred to textbooks for further details of triaxial test procedures. This apparatus is widely used in earthquake engineering practice to study the problem of cyclic liquefaction of saturated granular materials. The response of soils can be represented by plotting stress path graph ($p' - q$) where p' and q are the mean effective stress and deviator stress, respectively, and the definition are given below:

$$p' = \frac{\sigma_1' + \sigma_2' + \sigma_3'}{3} \tag{4.5}$$

$$q = \sigma_1' - \sigma_3' \tag{4.6}$$

where, σ_1' and σ_3' are the maximum and minimum principal effective stress, respectively. In triaxial tests, it is assumed that the intermediate and the minimum principal stress are the same (i.e. $\sigma_2' = \sigma_3'$).

Both compression and extension cycles in the effective triaxial plane (q–p') are usually performed. An example of a cyclic triaxial apparatus, under undrained conditions to simulate earthquake loading, is presented in Figure 4.14. The number of cycles necessary in undrained conditions for the material to reach a state of liquefaction is evaluated and the effective stress path is recorded. During the tests, both pore pressure within the sample and axial strain can be recorded (Figures 4.13 and 4.14).

Depending on the perceived imposed loading on the soil, loading stages can be generated or recreated. For example, Figure 4.15 illustrates the schematic diagram of the testing scheme adapted for the multistage soil element test to simulate an extreme case of soil liquefaction. In these tests, undrained stress-controlled sinusoidal cyclic loading with a certain frequency and amplitude was initially applied in order to liquefy the soil sample. Once the specimen liquefied, cyclic loading was stopped and strain-controlled monotonic load was then applied under undrained condition. This will provide the stress–strain curve of the liquefied sand. Figure 4.16 shows a result from the multistage test on Assam sand (India) to study liquefaction. Readers are referred to Rouholamin et al. (2017) and Dammala et al. (2017) for examples of multistage tests.

Bender element can be used to obtain the shear wave velocity of the sample through wave propagation technique. The cyclic triaxial text can be applied in two ways to issues related to offshore wind turbines:

1. Cyclic triaxial apparatus, together with the bender element, can be used to study the effect of an extreme storm on long-term performance. Cyclic loading may be applied to a soil sample corresponding to a storm to recreate the stress history. At the end

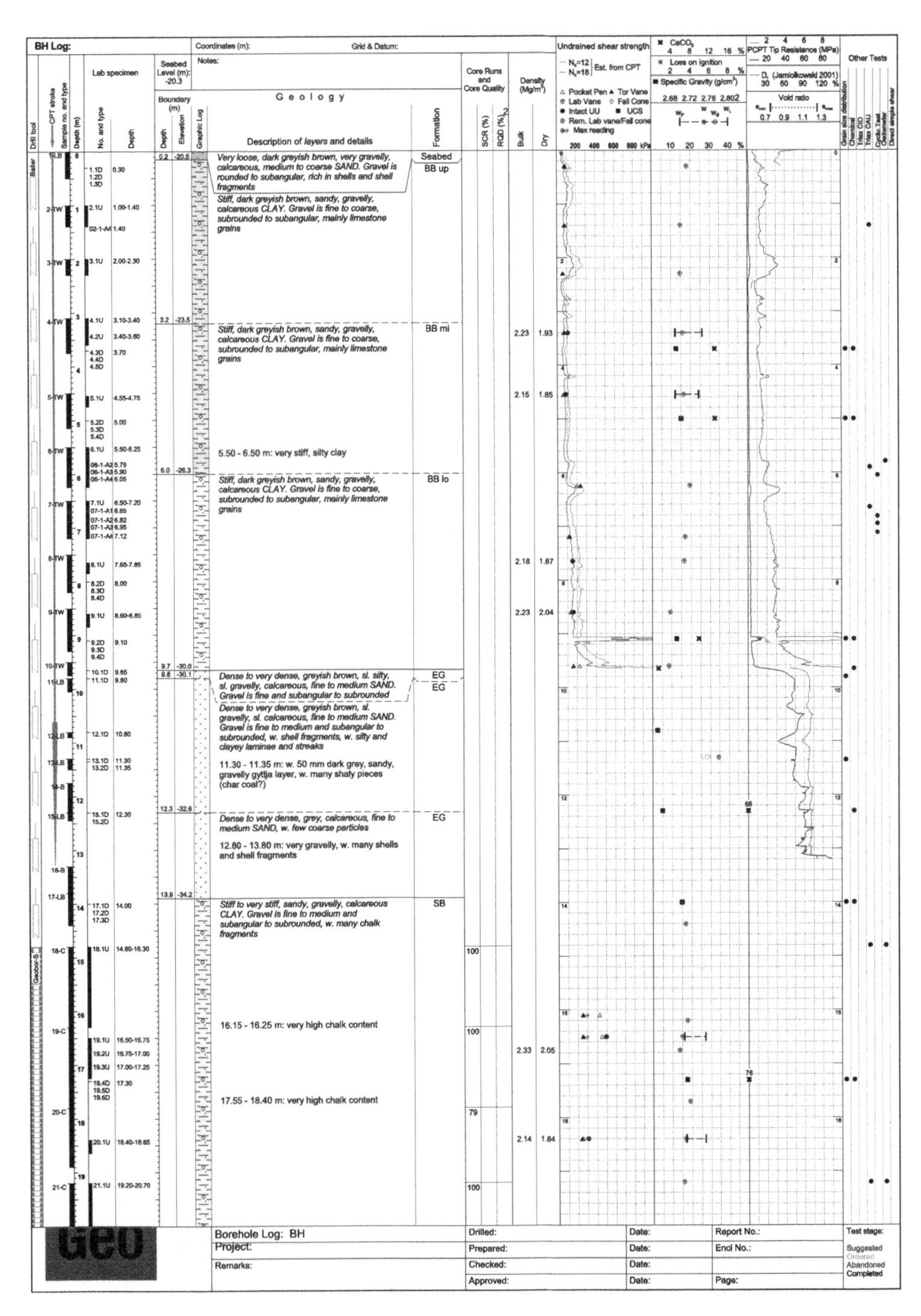

Figure 4.11 Typical example of a bore log and engineering parameters.

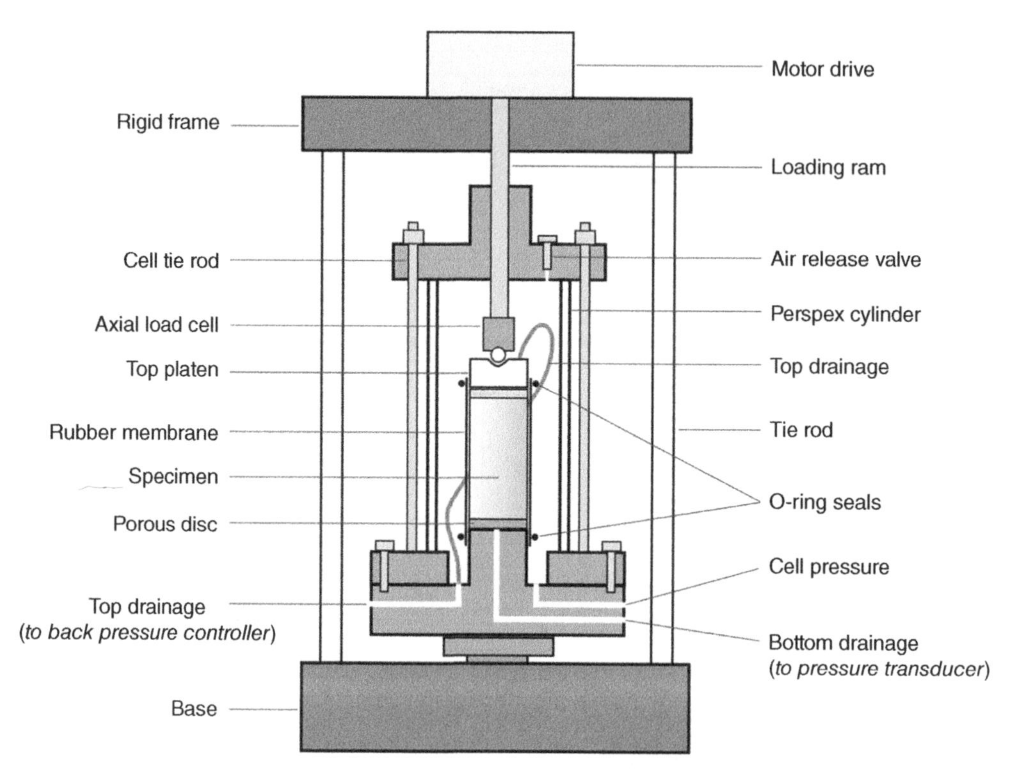

Figure 4.12 Schematic outline of a triaxial apparatus.

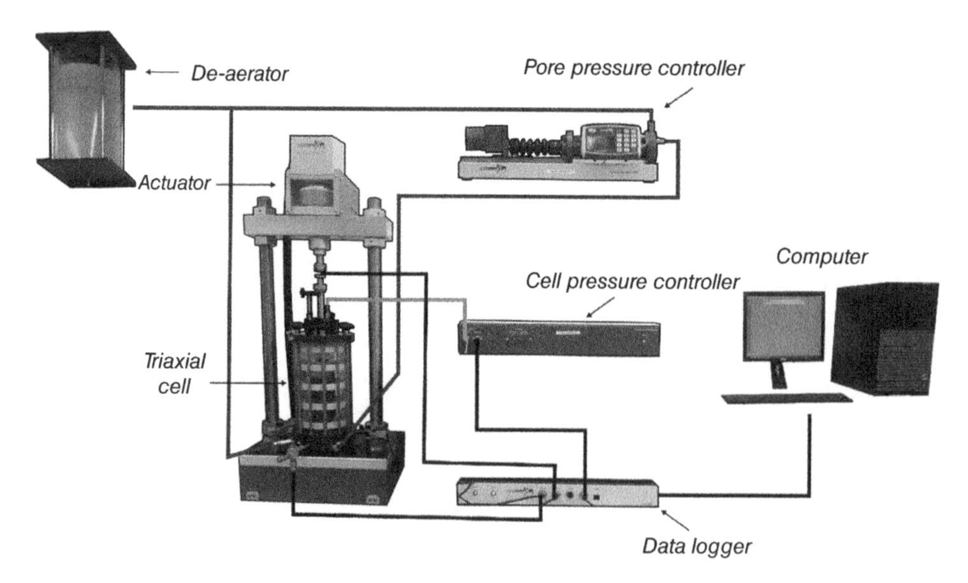

Figure 4.13 Schematic of the testing system from SAGE laboratory.

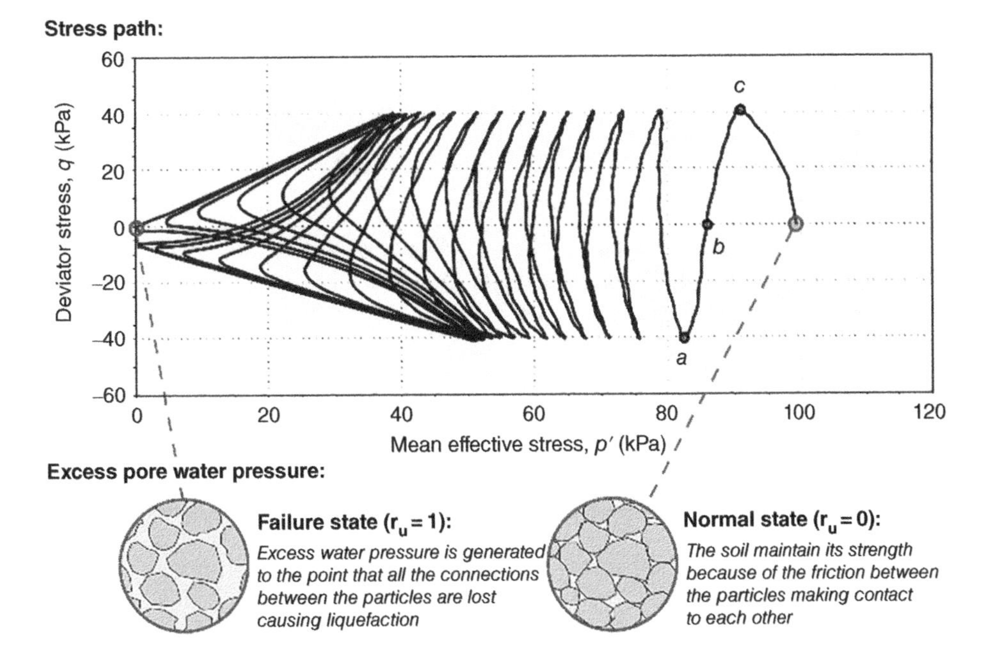

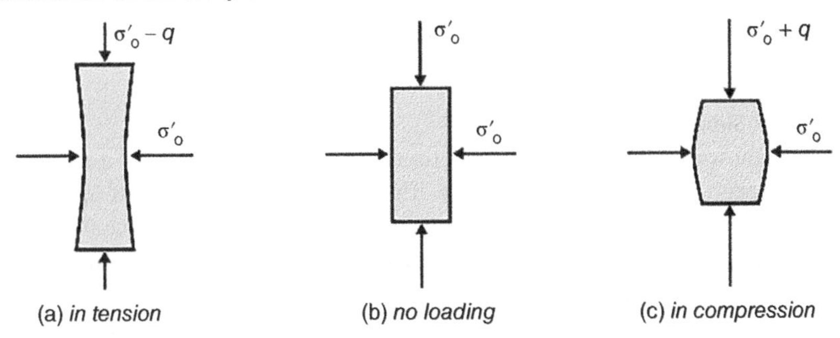

Figure 4.14 Typical result of a cyclic triaxial test, under undrained conditions.

of the loading, monotonic load may be applied to find the residual soil stiffness and strength.

2. Based on the literature, the behaviour of standard soils such as pure clay or pure sand can be predicted with some degree of confidence. It is, however, difficult to predict the behaviour of intermediate soils, i.e. sandy-silt or silty-sand, or clayey-silt, as the main question is whether the sample will behave like *clay-like* material or *sand-like* material. In other words, the behaviour of soils with varying amount of fines content is difficult to predict, and in such cases cyclic triaxial tests are useful. This will have a big impact on the prediction of long-term behaviour, as it is the accumulation of pore water pressure in the soil that will dictate the behaviour due to the effects of millions of cycles of loading.

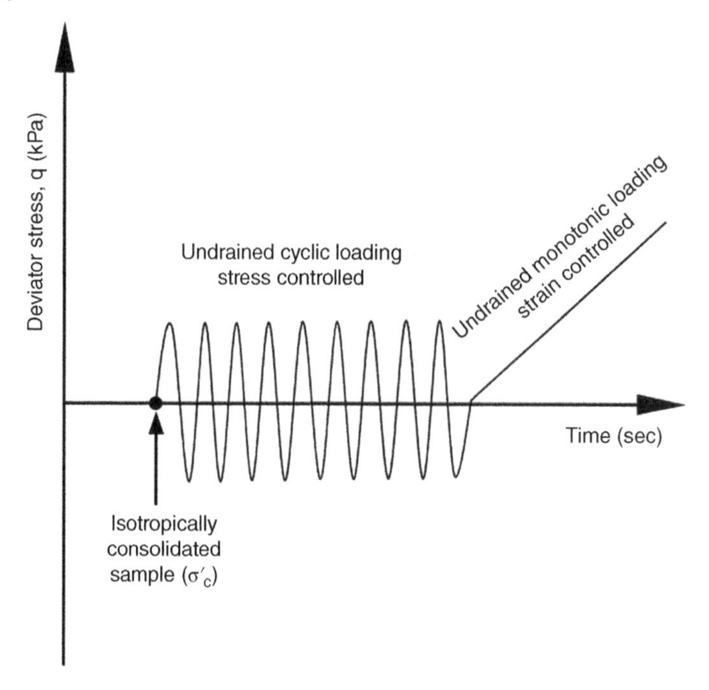

Figure 4.15 Testing scheme adopted for the multi-stage element test to study extreme loading condition due to an earthquake.

4.7.2.2 Cyclic Simple Shear Apparatus

Figure 4.17 shows a CSS apparatus together with the schematics of a sample shearing pattern taken from SAGE laboratory. The apparatus is capable of applying cyclic loads (both in vertical and horizontal direction) using two electro-mechanical dynamic actuators. The vertical and horizontal displacements can be measured using encoders within the servo motors. Different wave forms can also be applied: sinusoidal, square, triangular, saw tooth, haversine. The apparatus is also capable of applying user-defined waveforms. In the current context of offshore wind turbines, user-defined functions can be used to simulate random wind and wave loading. This apparatus has been used in Nikitas et al. (2016) to study the long-term performance of offshore wind turbine foundations.

In the study, external LVDTs (linearly varying differential transformer) were also used to record displacements and to verify the effectiveness of feedback control. Loads up to ±5kN can be applied in two directions with horizontal travel up to 25 mm and vertical travel of 15 mm. These are sufficient to study the effects of large strain levels applied to the soil. This therefore allows the study of effects of cyclic shear stress under drained and undrained conditions. The loads can be applied at frequencies of up to 5 Hz. This apparatus was used to investigate the cyclic behaviour of a silica sand (typical of North Sea) by maintaining the constant vertical consolidation stress while cycling the shear strain in horizontal direction. Typical results are presented in Figures 4.18 and 4.19.

Carefully designed and performed tests can be useful in predicting the long-term performance. Readers are referred to Chapter 5 for further details on the soil–structure

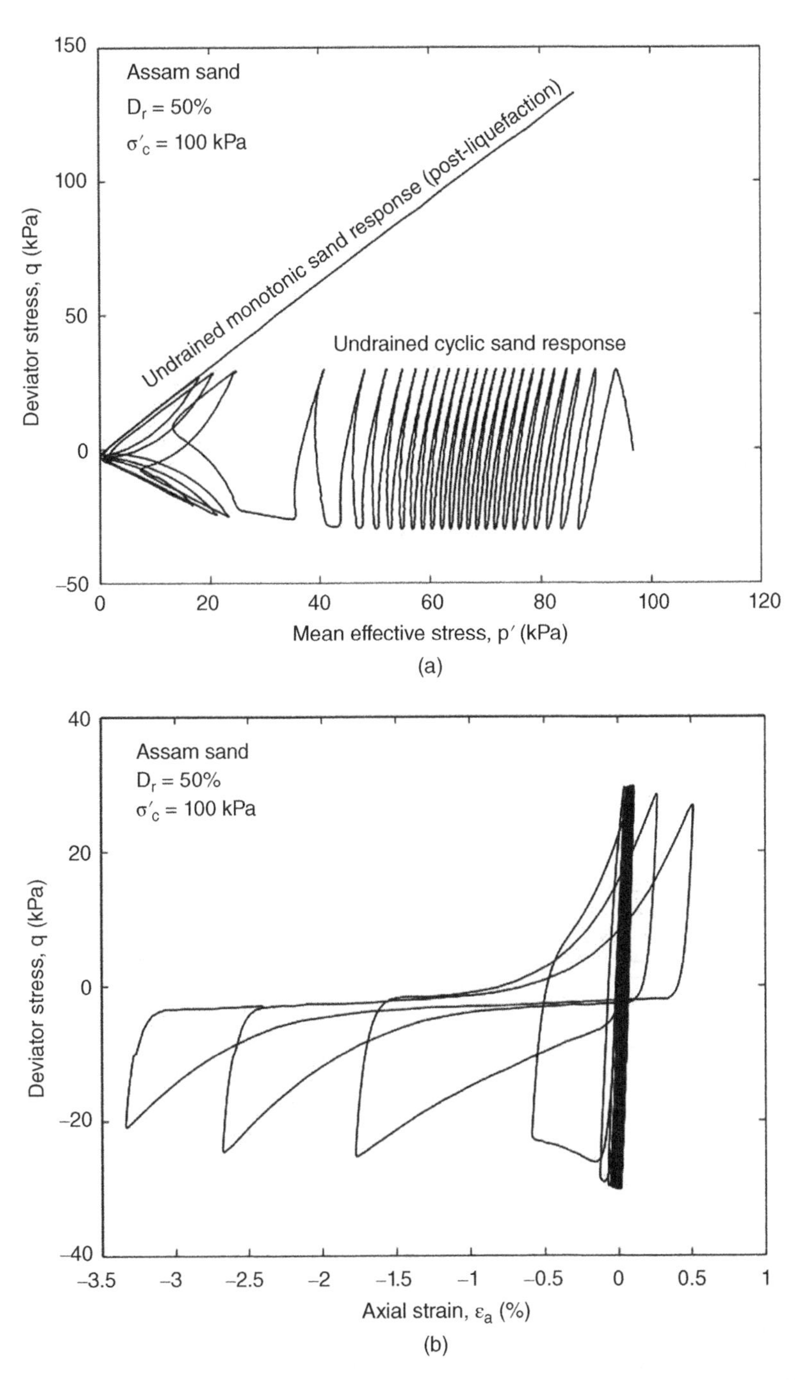

Figure 4.16 Multi-stage test using cyclic triaxial on Assam sand (a) stress path; and (b) deviator stress versus axial strain during cyclic phase.

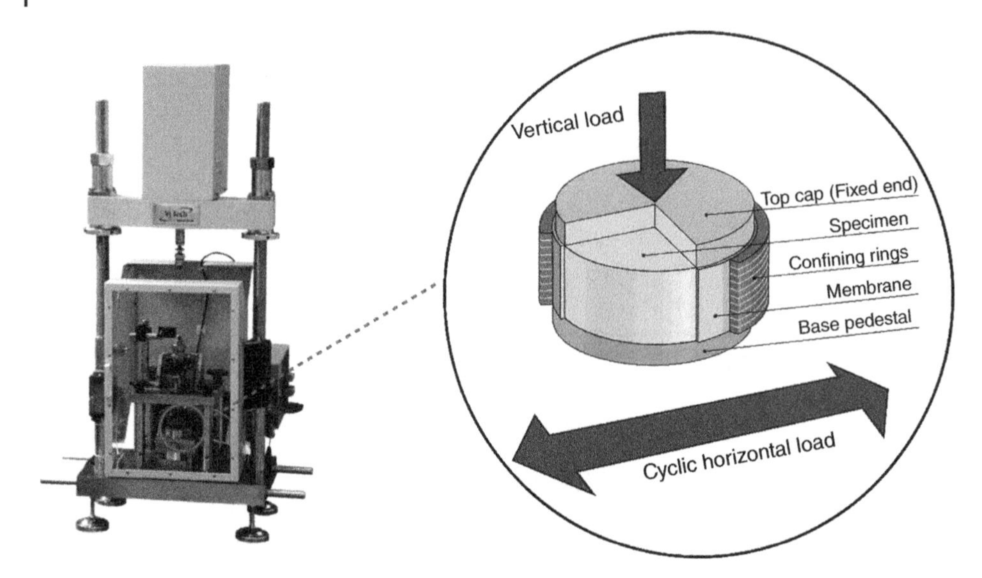

Figure 4.17 Experimental setup for cyclic simple shear test.

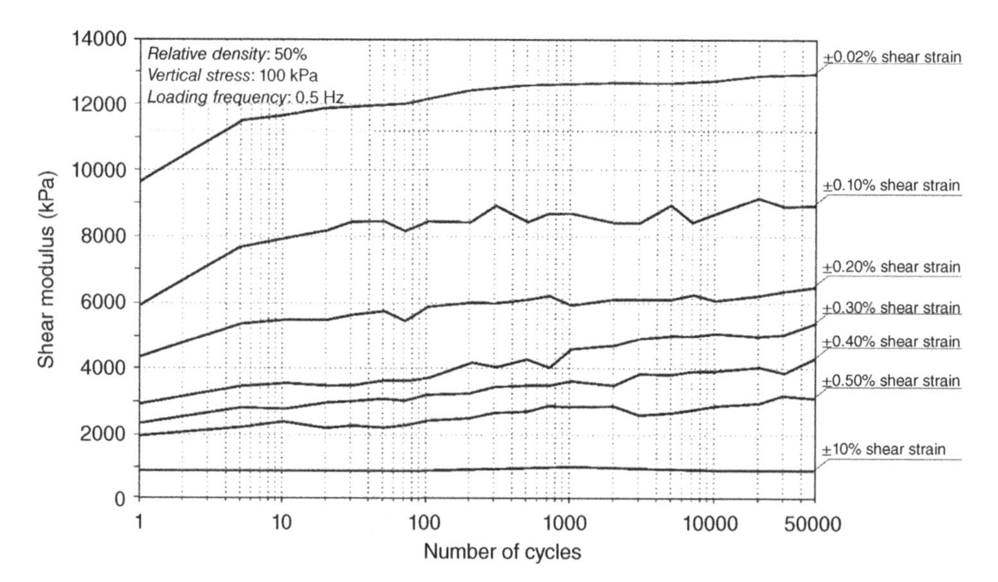

Figure 4.18 Results of red hill sand obtained from CSS apparatus.

interaction. The strain levels applied to the soil under different loading conditions from the turbine can be found in Section 5.5.

4.7.2.3 Resonant Column Tests

This apparatus can be used to map the change of stiffness for a wide range of strain levels. The basic principle involved in this test is the theory of wave propagation in prismatic rods where a cylindrical soil specimen is harmonically excited until it reaches the state of resonance (i.e. peak response). Figure 4.20 shows schematic diagram of a soil sample and

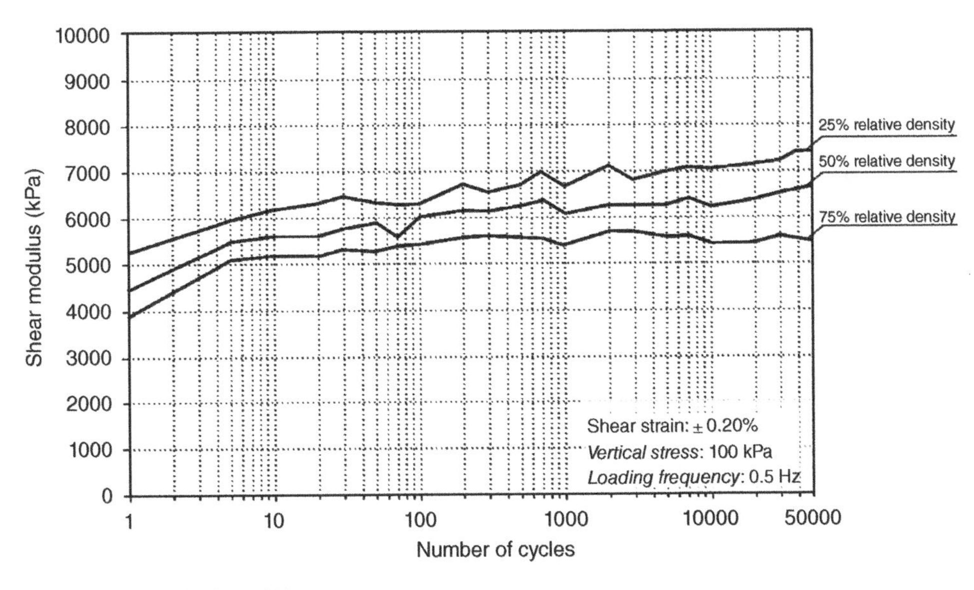

Figure 4.19 Results from CSS apparatus.

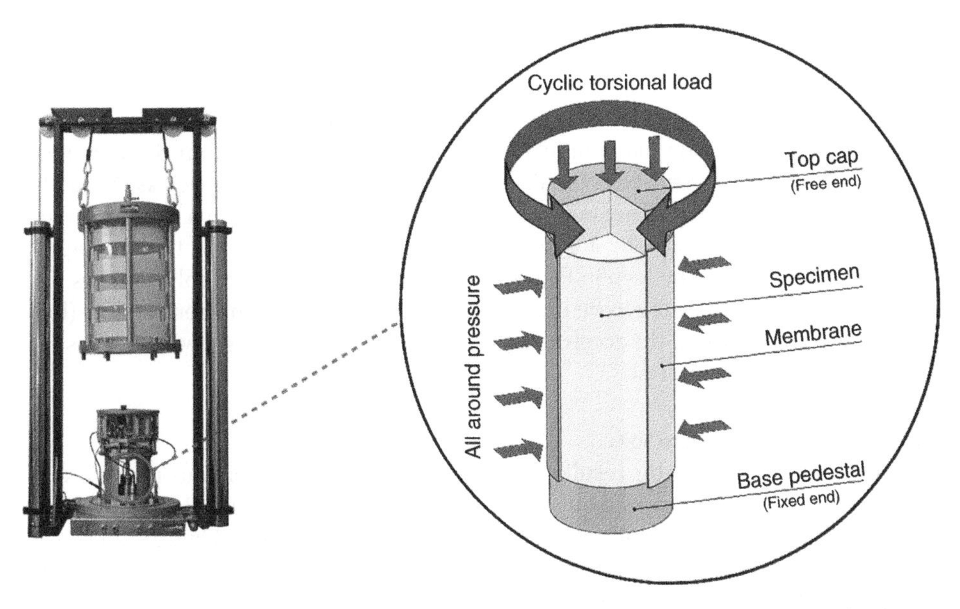

Figure 4.20 Schematic of resonant column testing layout, resonant column apparatus used at the SAGE Laboratory, University of Surrey.

the apparatus. Resonant column tests can be carried out to determine shear modulus and damping ratio of soil for a wide range of strain level.

This technique was first developed in the 1930s at the University of Tokyo by Ishimoto and Iida (1936) and Ishimoto and Iida (1937). The resonant column only saw wider use in the 1960s with the works of authors such as Hall and Richart (1963), Hardin and Richart (1963), and Drnevich et al. (1967). In the past few decades, a number of

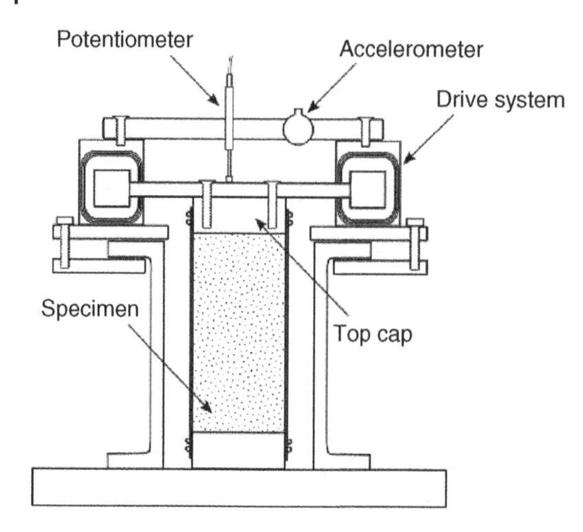

Figure 4.21 Simplified cross section schematic of one type of resonant column apparatus.

subsequent improvements and advancements have been made in resonant column testing to broaden the range of applicability. These have included the application of anisotropic stresses to the soil element (Hardin and Music 1965; Hardin and Black 1966), adaptations to the apparatus to hollow the testing of elements (Drnevich 1967), and modifications to tests specimens at large (Anderson and Stokoe 1978; Stokoe and Lodde 1978) and at high strains (Alarcon et al. 1986; Isenhower et al. 1987; Ampadu and Tatsuoka 1993).

Figure 4.20 shows a particular type of resonant column apparatus and it uses 50 mm diameter by a 100 mm long soil sample utilising a fixed-free configuration. This arrangement is the most widely utilised resonant column arrangement in research as the mathematical derivations required are fairly simple. In such an arrangement a cylindrical soil element is excited in torsion by a dynamic drive system connected to the top cap. A simplified schematic can be seen in Figure 4.21. Through the resonant column tests, the damping ratio (D) of the soil can be measured from the free vibration decay curve (logarithmic decrement) using the accelerometer. Figures 4.22 and 4.23 show typical results from the resonant column test.

4.7.2.4 Test on Intermediate Soils

Figures 4.24 and 4.25 show the results from resonant column tests on four types of soils with varying fines content carried out at Yamaguchi University (Japan). The graphs show as expected the variation of stiffness and damping with strain level. Figure 4.25 shows the plot of accumulated pore water pressure with strain level. It is interesting to note that for strain level corresponding to G_{eq}/G_{max} of 0.6–0.85, pore pressure starts to build and that is the onset of progressive collapse.

4.8 Behaviour of Soils under Cyclic Loads and Advanced Soil Testing

Behaviour of soils under cyclic and dynamic loading is still an active area of research. Soil dynamics problems can be classified based on magnitude of strain levels, frequency

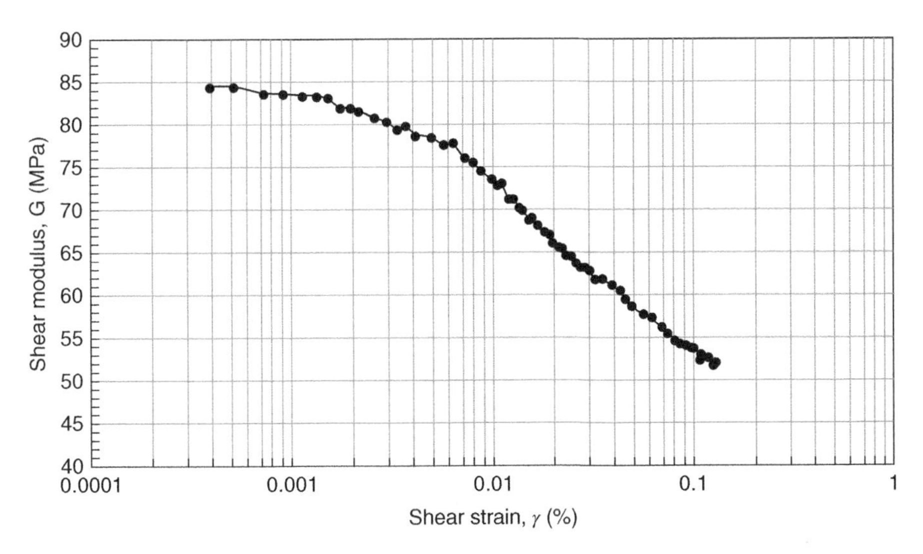

Figure 4.22 Shear strain-shear modulus graph measured by the resonant column test on Redhill-110 sand.

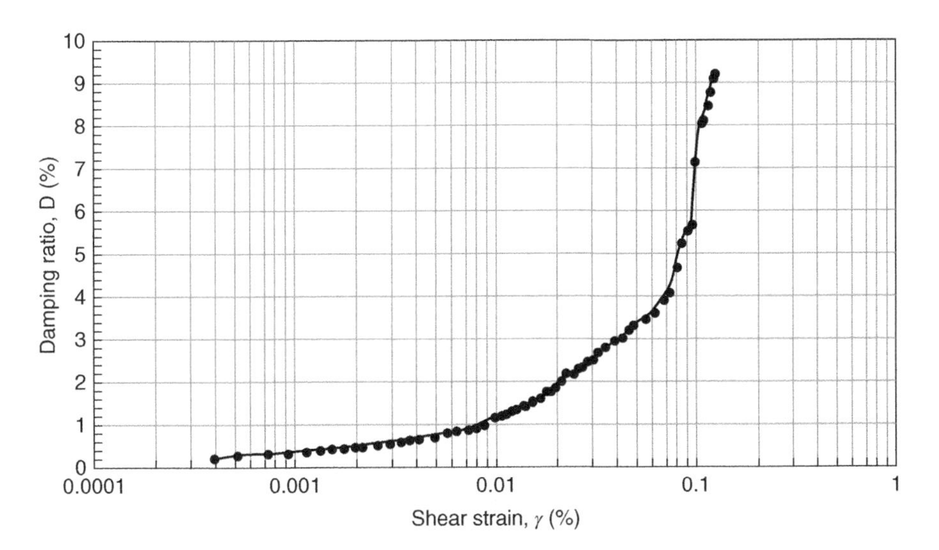

Figure 4.23 Shear strain-damping ratio graph measured by the resonant column test on Redhill-110 sand.

of loading, and number of cycles as shown in Figure 4.26. Figure 4.27 shows the different classes of soil dynamics problem and full treatment of the problem will need an entire textbook.

4.8.1 Classification of Soil Dynamics Problems

The main aim of this section is to provide a summary of the issues related to offshore wind turbine problems. Explanation of the numbers is provided in Figure 4.26:

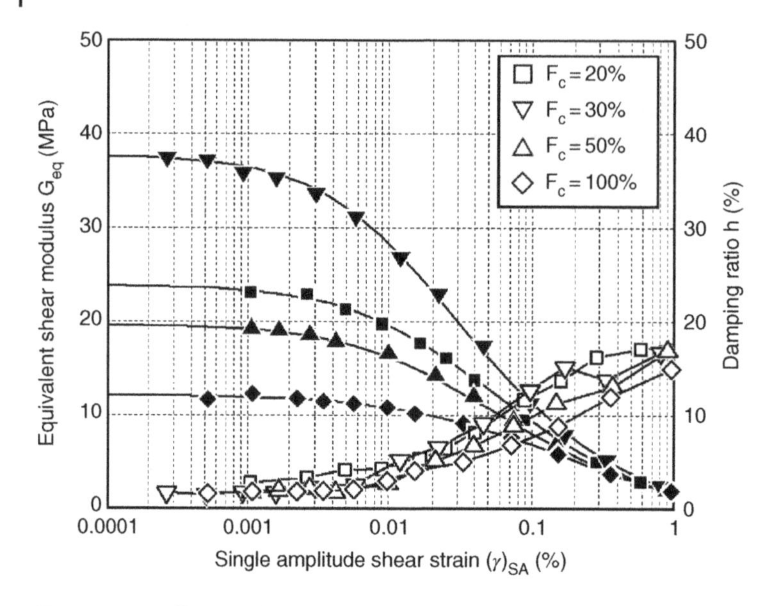

Figure 4.24 Stiffness and damping properties of soil with different strain levels.

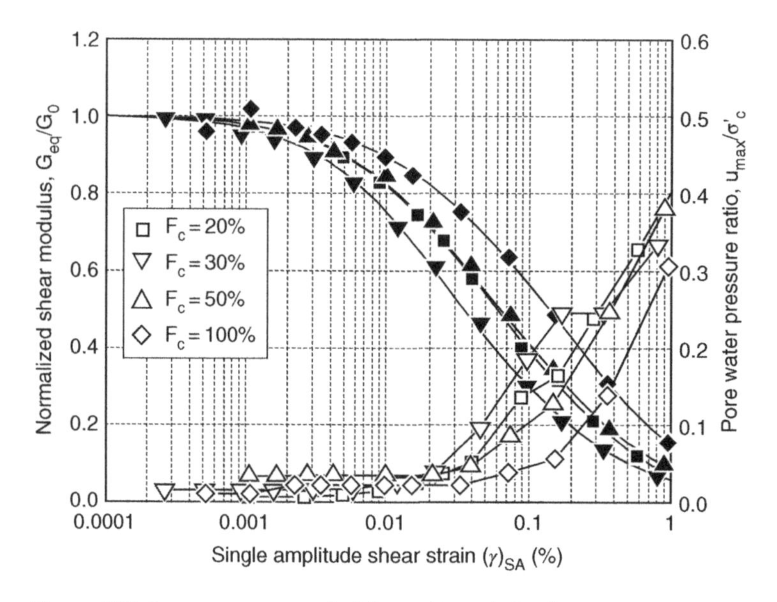

Figure 4.25 Pore water pressure build-up with strain level.

1. Earthquake loading characterised by large strain, low number of cycles, and low frequency (0.5 to 10Hz)
2. Loading on offshore structures
3. Traffic loading
4. Represents the strain in the soil due to variation of ground water table
5. Pile driving, Soil compaction

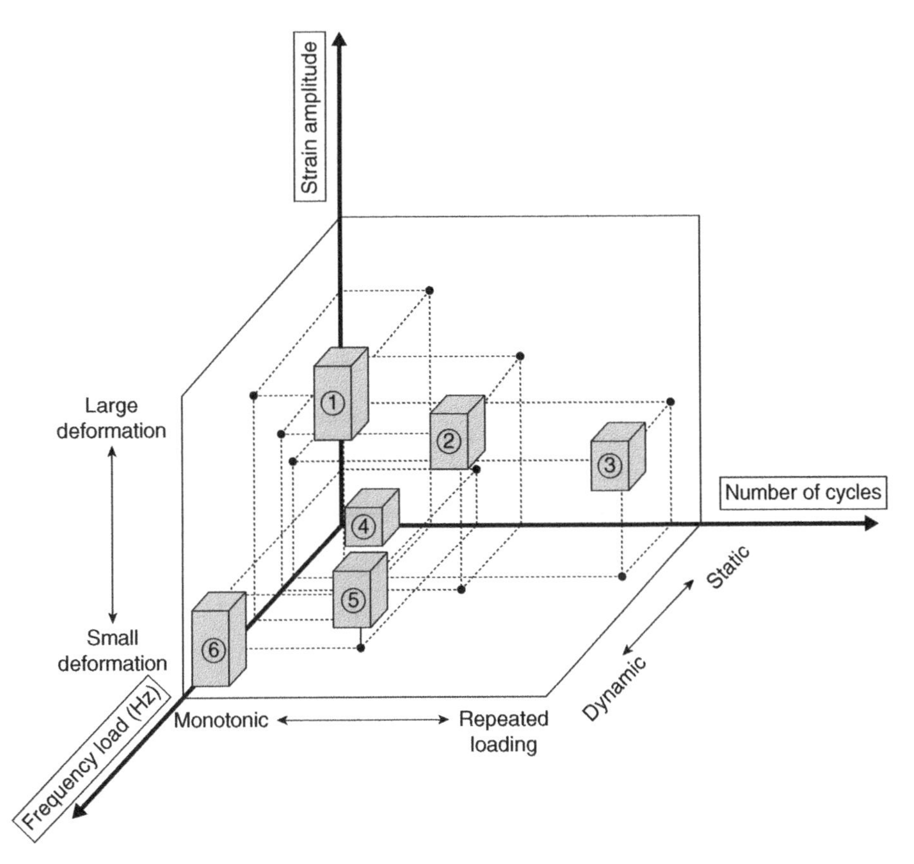

Figure 4.26 Classification of soil dynamics problem based on three parameters: frequency, number of cycles, and strain level.

6. Impulsive load (Blast) characterised by high frequency, very low number of cycles and small deformation

4.8.2 Important Characteristics of Soil Behaviour

All soils under moderate cyclic loads yield progressively, try to compact, generate positive pore water pressure, and soften. Collapsible soils such as loose sands can liquefy in the absence of drainage. The frequency of cyclic loading condition is therefore important, since the pore pressure generated by one full load cycle may only partially dissipate before the next load cycle.

For clays, the strength and stiffness degrades. The amount of stiffness degradation is a function of strain level and the number of cycles of loading. Extensive laboratory and field studies have clarified many aspects of the behaviour of clays under such loading; see, e.g. Vucetic and Dobry (1991). Stiffness degradation has been correlated with the plasticity index (PI) of the soil. For a particular cyclic strain in the soil, more plastic clays exhibit less degradation. Figure 4.28 shows a graph based on 16 sets of published experimental data from strain controlled tests on clays having different PIs and OCRs. In the graph, a PI of zero indicates sand and a PI of 200 represents highly plastic Mexico

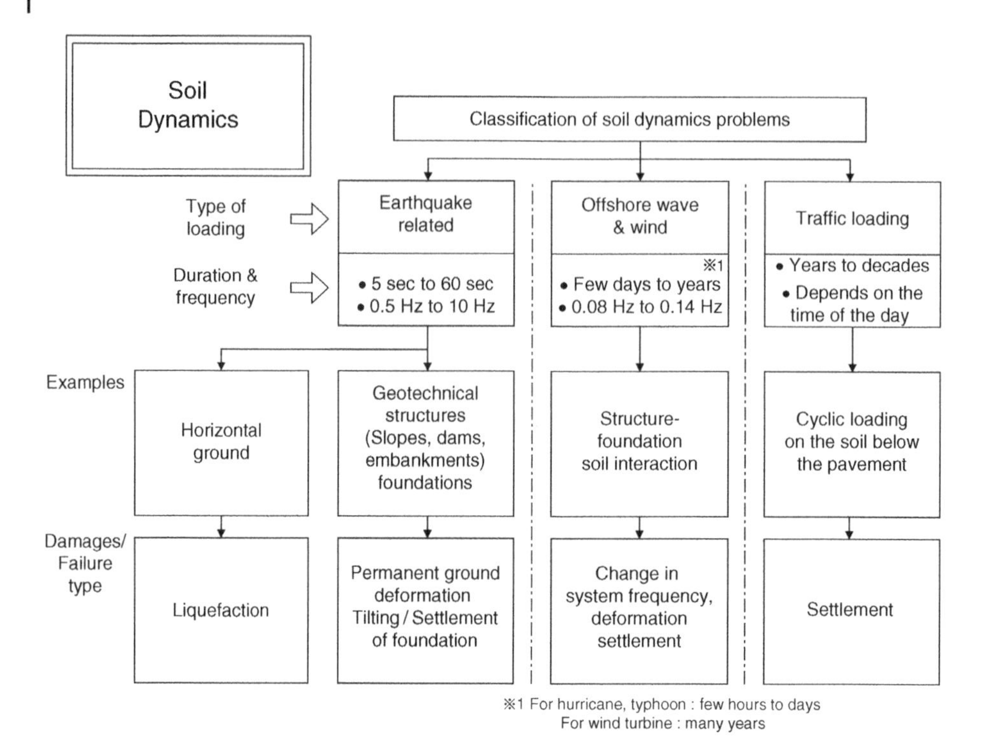

Figure 4.27 Classification of soil dynamics problems.

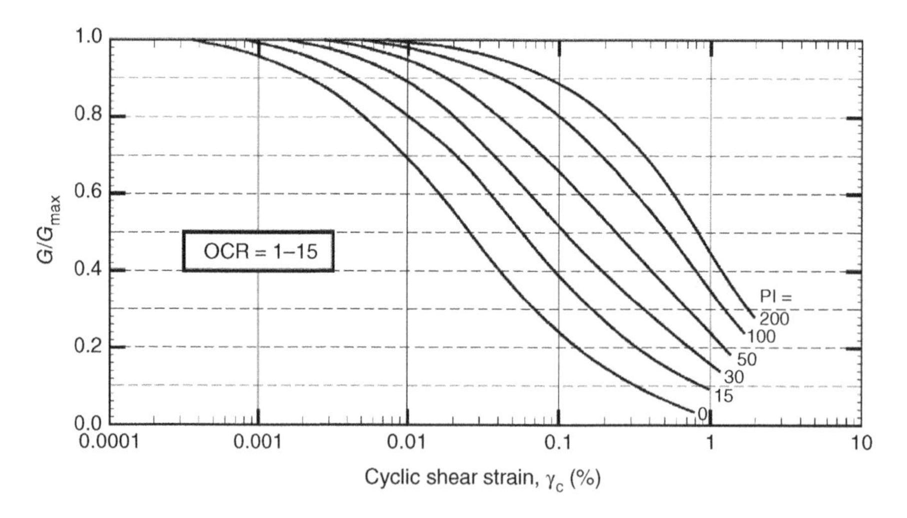

Figure 4.28 Relations between G/G_{max} versus cyclic shear strain for NC and OC clays. [Source: Vucetic and Dobry (1991)].

City clays, which show little nonlinearity up to 0.1% cyclic shear strain. G_{max} is the secant shear modulus at small strain. Figure 4.29 shows the effect of degradation on G/G_{max} for different numbers of cycles ($N = 1$, 10, 100, and 1000) for normally consolidated clays (OCR of 1).

When analysing a soil system, three important features of soil behaviour must be considered:

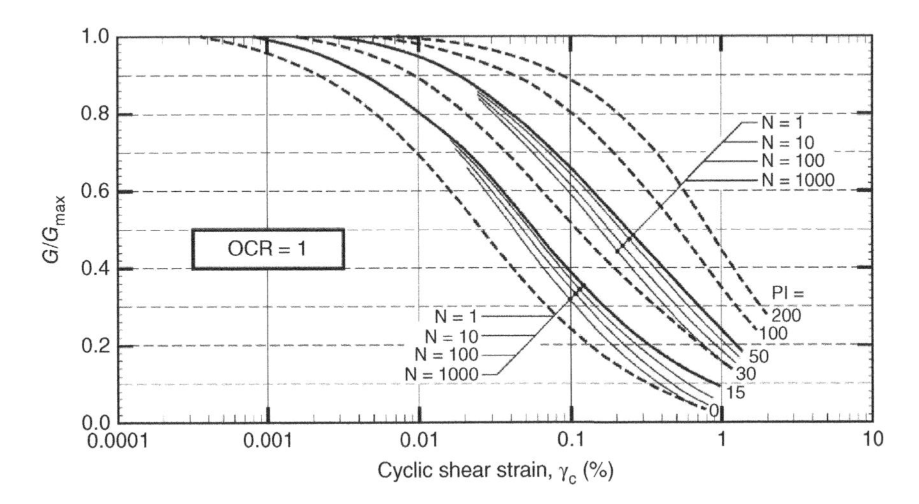

Figure 4.29 Effect of cyclic stiffness degradation on G/G_{max} versus cyclic shear strain for soils having different PI and loaded to different number of cycles. [Source: Vucetic and Dobry (1991)].

1. Soil is a nonlinear material, in which the stiffness progressively decreases with increasing shear stress, until at a sufficiently high shear stress level it deforms plastically.
2. Soil is an inelastic material. Therefore, when subject to cyclic loading, it exhibits hysteretic damping (energy loss), which increases with increasing shear strain.
3. The soil properties, including strength, stiffness, and pore water pressure, may be affected by repeated cycles of load. This is particularly relevant to saturated sands and silts that may suffer a build-up in pore water pressure with continued load application possibly resulting in liquefaction.

Five main properties (bulk density, stiffness, material damping, strength, and degradation due to cyclic loading) determine the response of a soil-structure system under dynamic load. The bulk density affects the inertia of the soil under dynamic loading. The stiffness, which is largely defined by the shear modulus, determines the natural frequency of the system and the response. Damping is a measure of energy dissipation: the higher the damping, the greater the energy loss and therefore, the smaller the response.

4.9 Typical Soil Properties for Preliminary Design

When carrying out dynamic analyses, particularly in the early stages of projects, all the soil information will not be available. Table 4.9 provides a range of standard parameters, which can be used for preliminary design if no other data are available. Readers are referred to Appendix A of Bhattacharya et al. (2019) for full details of many more correlation.

4.9.1 Stiffness of Soil from Laboratory Tests

Initial stiffness of the soil (i.e. soil stiffness at small strain) is an important parameter for design. Laboratory tests suggest that the maximum shear modulus (G_{max}) can be

Table 4.9 Typical soil properties that may be used for preliminary design.

Soil Type	Description	ϕ′ (°)	su (kPa)	G_{max} (MPa)
Gravel	Dense	40	–	230
Sand	Loose	30	–	10
	Medium Dense	32.5	–	20
		35	–	55
Silt	–	25	–	10
Clay	Soft	–	25	2
	Firm	–	50	30
	Stiff	–	100	60
Rock	Mudstone	–	–	2150
	Sandstone	–	–	3100
	Granite	–	–	15 700

Table 4.10 Variation of OCR exponent (Hardin and Drnevich 1972).

PI	0	20	40	60	80	100
k	0	0.18	0.3	0.41	0.48	0.5

expressed by Eq.

$$G_{max} = 625 \frac{(OCR)^k}{0.3 + 0.7e^2} \sqrt{p_a . \sigma'_M} \tag{4.7}$$

In another form, Eq. (4.7) can be expressed as $G_{max} = 625\ F(e)\ (OCR)^k\ p_a^{0.5}\ (\sigma_M')^{0.5}$ where $F(e)$ is a function of void ratio and k is an OCR exponent, which varies with PI, as shown in Table 4.10.

Hardin (1978) proposed that $F(e) = 1/(0.3 + 0.7e^2)$, while Jamiolkowski et al. (1991) suggested that $F(e) = 1/\mathrm{e}\ 1.3$. Please note that G_{max}, p_a and σ_M' must all be expressed in the same units. p_a is atmospheric pressure.

Ishibashi and Zhang (1993) suggest the relationship between shear modulus G and its maximum value G_{max} for fine grained soil:

$$G/G_{max} = K \cdot \sigma_m'^{m}$$

$$K = 0.5 \cdot \left\{ 1 + \tanh \left[\ln \left(\frac{0.000102 + n}{\gamma} \right)^{0.492} \right] \right\}$$

$$m = 0.272 \cdot \left\{ 1 - \tanh \left[\ln \left(\frac{0.000556}{\gamma} \right)^{0.4} \right] \right\} \cdot \exp(-0.0145 \cdot PI^{1.3})$$

$$n = \begin{Bmatrix} 0 & for\ PI = 0 \\ 3.37 \times 10^{-6} \cdot PI^{4.404} & for\ 0 < PI \leq 15 \\ 7.0 \times 10^{-7} \cdot PI^{1.976} & for\ 15 < PI \leq 70 \\ 2.7 \times 10^{-5} \cdot PI^{1.115} & for\ PI > 70 \end{Bmatrix} \tag{4.8}$$

where, σ'_m is the mean principal effective stress, γ is the shear strain, and *PI* is the *PI* of soil.

For cohesive soils, the relationship between G_0/S_U, *PI*, and *OCR* as summarised by Weiler (1988) can be used Table 4.11.

In addition to the theories proposed above, Seed and Idriss (1970) derived a simplified equation to describe the low strain stiffness for any granular material. As all formulations are strongly influenced by confining pressure and void ratio, only these factors were considered leading to the following derivation;

$$G = 1000\,K_2\,\sqrt{p'} \tag{4.9}$$

where:

p'	Mean local effective stress	$[\mathrm{kN\,m^{-2}}]$
K_2	Soil modulus coefficient	[–]

The coefficient K_2 itself has been experimentally derived from a number of different sands. It was found that under low strain conditions ($\gamma < 10^{-3}\%$) the soil modulus coefficient only depends on the void ratio (Seed et al. 1984). Typical values for K_2 are detailed in Table 4.12 taken from guidelines. The readers are also referred to updated correlations.

4.9.2 Practical Guidance for Cyclic Design for Clayey Soil

The salient features are:

1. Soil stratification and type of soil at the top layers. For a deep foundation, it is the top soil that deforms most.

Table 4.11 G_{max}/s_U values.

	OCR		
PI (%)	**1**	**2**	**5**
15–20	1100	900	600
20–25	700	600	500
34–45	450	380	300

Table 4.12 Soil modulus coefficient for a number of typical sand densities (Det Norske Veritas 2002).

Soil type	K_2
Loose sand	8
Dense sand	12
Very dense sand	16
Very dense sand and gravel	30–40

2. What is the scour depth?
3. For soils, it is required to predict under what condition, pore water pressure will start to accumulate?

Figure 4.30 shows the variation of normalised secant stiffness with strain for fully saturated cohesive soil adapted from Vucetic (1994). It is accepted that there exists a strain level, the threshold linear shear strain, γ_{tl} , for which there is no stiffness degradation and the behaviour of the soil is practically linear elastic, so that there is no permanent microstructural changes of the fabric of the soil with cyclic loading. As the strain level is increased beyond γ_{tl}, the soil behaves nonlinearly and as a result the secant shear modulus of soil will reduce. The value of γ_{tl} can be estimated from the secant shear modulus reduction curve (schematically shown in Figure 4.30) assuming that linear threshold shear strain corresponding to a ratio $G_{\text{sec}}/G_{\text{max}}$ of 0.99. There exists another value of threshold shear strain, known as volumetric threshold shear strain level (Vucetic 1994), denoted by γ_{tv}, beyond which permanent microstructural changes of the fabric do occur. In other words, beyond γ_{tv} the soil is degradable possibly due to pore pressure build-up. In the triaxial stress space, this state corresponds to the boundary between the 'recoverable' and the 'plastic zone' (Jardine 1992). γ_{tv} can be assessed from the secant shear modulus reduction curve. However, two value of $G_{\text{sec}}/G_{\text{max}}$ ratios (namely, 0.85 and 0.60) are often used, depending on the application. The value of 0.85 represents a degradation of secant shear modulus beyond which permanent deformation is expected. On the other hand, 0.6 is defined on the basis of the accumulation of excess pore pressures. These two values can represent upper and lower bound values of γ_{tv} for the problem in hand. The upper bound value will give the highest diameter that may be required. On the

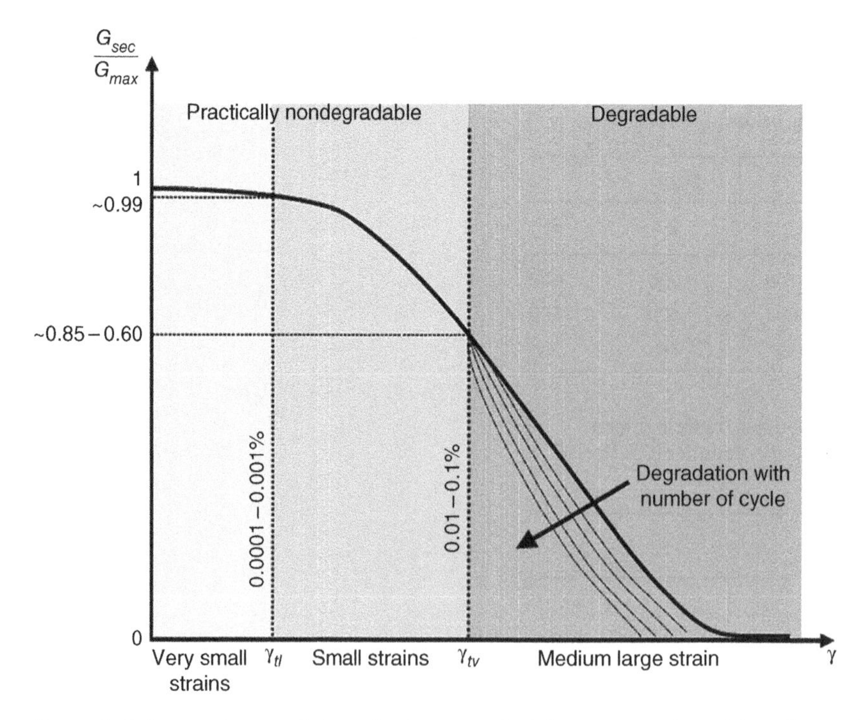

Figure 4.30 Secant shear modulus reduction curve for fully saturated clay subjected to cyclic undrained loading.

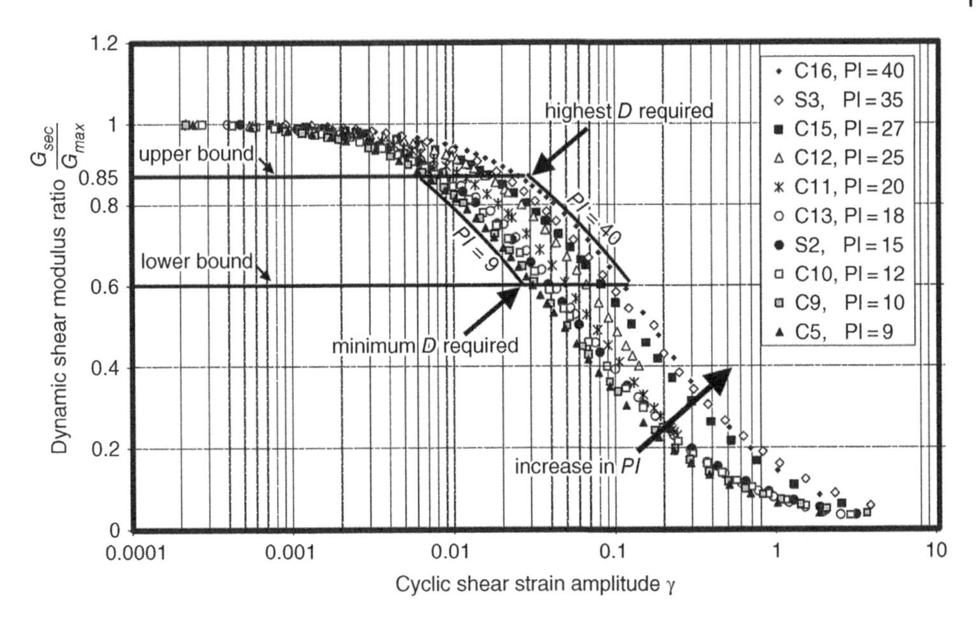

Figure 4.31 Experimental secant shear modulus reduction curve for samples of Turkish clay with different PI.

other hand, the lower bound will give the minimum diameter necessary. It is interesting to note the graphs presented in Figures 4.24 and 4.25 showing the pore water pressure build-up. While Figure 4.30 shows the schematic representation of threshold strains, Figure 4.31 plots typical experimental data obtained from cyclic triaxial tests carried out on 10 samples of Turkish clay having a PI ranging from 9 to 40 following Okur and Ansal (2007).

Figure 4.31 shows an upper and lower bound for γ_{tv} which corresponds to $G_{\text{sec}}/G_{\text{max}}$ ratio of 0.85 and 0.60, respectively. The experimental results clearly show that samples with higher PI tend to have a more linear cyclic stress–strain response at small strains and to degrade less at larger strain than soils with lower PI. Therefore, clays with higher PI are characterised by higher values of volumetric threshold shear strain. For example, considering the lower bound, γ_{tv} increases from 0.025 to 0.15 when PI increases from 9 to 40.

Figure 4.32 collates the volumetric threshold shear strain value (γ_{tv}) for 18 types of clays having PI ranging from 9 to 100. The experimental data were collected from research published by Kim & Novak (1981), Kokusho et al. (1982), Vucetic & Dobry (1988), Georgiannou et al. (1991), Lanzo et al. (1997), Stokoe et al. (1999), Massarsch (2004), Okur & Ansal (2007), Yamada et al. (2008). In the same figure, two linear correlations for evaluating γ_{tv} are also suggested for the investigated range of PI ($9 < PI < 100$). The experimental data used for deriving the plot in Figure 4.32 are given in the Table 4.13.

4.9.3 Application to Offshore Wind Turbine Foundations

The strain level in the soil next to the monopile i.e. the strain in the mobilised shear zone can be compared to the volumetric threshold strain, γ_{tv} which is a fundamental property of a soil and can be obtained from soil testing. Section 5.5 in Chapter 5 presents a method

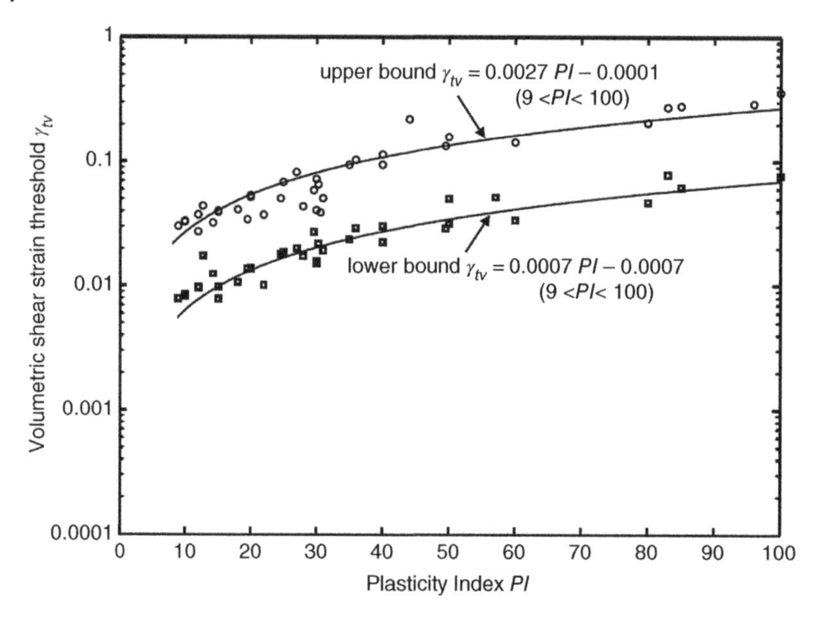

Figure 4.32 Volumetric shear strain threshold values for different PI.

to compute the soil strain and if the strain in the mobilised zone exceeds this threshold strain, degradation may be excepted. This concept is quite similar to the mobilisable strength design (MSD) concept pioneered by Osman and Bolton (2005) where the average strain in a mechanism (deformed zone or sheared zone) is linked with an element test in a soil. The above concept has been experimentally shown to be valid by Lombardi et al. (2013).

Choosing diameter of monopile for design: It will be shown in Chapter 5 that higher the diameter of the monopile, the lower is the average strain in the surrounding soil. Therefore, choosing a larger diameter pile will lower will be the degradation. The designer needs to make sure that if the strain level in the soil exceeds the γ_{tv} for the clayey soils, adequate allowance for stiffness reduction needs to be taken into account.

4.10 Case Study: Extreme Wind and Wave Loading Condition in Chinese Waters

A typhoon can lead to failure or even collapse of wind turbines, especially in the region of East and South China Sea. Typhoons can have a negative effect on wind turbine systems ranging from fracture of blades, damage to braking system, yielding of tower, and even excessive rotation foundation. However, the direct effect of typhoon is loss of power supply and emergency shutdown, which may damage the weakest part (e.g. blade and nacelle). It is therefore necessary to understand the wind condition in the Chinese waters. This section provides a brief summary:

- *Bohai Sea:* The total area of the Bohai Sea in China is approximately 77 284 km^2 and is classified in four parts: west, middle, Jinzhou, and Suizhong. Bohai Sea is the only

Table 4.13 Data pertaining to graphs in Figure 4.31.

Soil name	PI	OCR	Test apparatus	Reference
Windsor silty clay	30	2.7	Resonant column	Kim and Novak 1981
Wallaceburg silty clay	25	5.1	Resonant column	Kim and Novak 1981
Hamilton clayey silt	12	5.8	Resonant column	Kim and Novak 1981
Sarnia silty clay	14	1.8	Resonant column	Kim and Novak 1981
Alluvianal clay at Teganuma	44, 57, 58, 83, 96	NC	Cyclic triaxial	Kokusho 1982
laboratory-made clays	15, 30, 50, 100	NC	Not specified	Vucetic and Dobry 1988
Vallericca clay	31	OC	Resonant column	Georgiannou et al. 1991
Pietrafitta clay	30	OC	Resonant column	Georgiannou et al. 1991
Todi clay	28	OC	Resonant column	Georgiannou et al. 1991
laboratory-made clay	22	3	Direct simple shear	Lanzo et al. 1997
Fat clay	36	Not specified	Not specified	Stokoe et al. 1999
Not specified	10, 20, 40, 60, 80	Not specified	Not specified	Massarsch 2004
Turkish clay	9, 10, 12, 15, 18, 20, 25, 27, 35, 40	NC	Cyclic triaxial	Okur and Ansal 2007
Ariake clay	44	NC	Hollow cylinder	Yamada et al. 2008a
Onada clay	50	NC	Hollow cylinder	Yamada et al. 2008a

inner sea region of China and is a shallow sea with average water depth of about 18 m. In the winter months, due to the flow of cold air from Siberia, the average wind speed is about 6–7 m s^{-1}. On the other hand, in summer months, due to warm air flow from the Western Pacific Ocean, average wind speed is 4–5 m s^{-1}. In August, the wind speed is quite low. In this sea, wind farms are located in Jinzhou and predominant wind directions for high wind speed are southeast and southwest with mean annual wind speed of 7.21 m s^{-1} (100 m height over sea level). Typhoons and cyclones are not normal in the Bohai Sea and the wind condition is stable throughout the year. Bohai Sea has a cold climate, and in the winter period the onshore near coastline is almost frozen for three to four months.

- *Yellow Sea:* The total area of Yellow Sea 380 000 km^2 and comprises Shandong, Jiangsu, and Korean Peninsula. There are many offshore wind farms built along the coastline in Yellow Sea from north to south. In the winter, the wind is stable but in summer it varies in both magnitude and direction. Typhoons are very frequent in summer throughout Shanghai area, which is located in the south of Yellow Sea. The annual average wind speed in the north Yellow Sea is 6.80–6.88 m s^{-1} but in the south, the value is much larger and the average value is 7.45–7.55 m s^{-1}.

- *East China Sea:* The total area of East China Sea is 770 000 km^2 and comprises the continent coastline of Japan and Taiwan. There are many offshore wind farm projects located in Hangzhou Bay. The annual mean wind speed tested at 100 m in height over sea level is generally around 9.6–9.8 m s^{-1} at the exit of Hangzhou Bay. However, in terms of frequent typhoon phenomena in summer from May to September each year, wind speed remains unstable, and the value is quite large, even over 14 m s^{-1}. Cyclone effect is very important and governs the design.
- *Taiwan Strait:* The Taiwan Strait is one part of East China Sea, connected with the South China Sea in the south of China. It is common to see the strong wind (e.g. typhoon) in summer and damage the cities and structures along coastline. Usually, typhoons in the summer will come from southeast and the wind speed is quite high, even much bigger in height. The annual mean wind speed in Taiwan Strait is about 11.5–13 m s^{-1}. It will be over 20 m s^{-1} or more when a typhoon is coming. Each year, typhoons are the main natural disaster, which brings large damage to the economy and structure in Taiwan, Fujian, and Zhejiang Province, located on two sides of Taiwan Strait, respectively.
- *South China Sea:* The total area of South China Sea is about 3 500 000 km^2. It is surrounded by the continent of South China mostly, northeast to Taiwan Strait, southwest to Singapore Strait. Most of the projects are newly built or under construction in recent years. It is also common to see strong wind (e.g. typhoon) and to bring rainwater in the summer each year. The annual mean wind speed in the South China Sea is 9–11 m s^{-1}. When a typhoon comes, the wind speed is over 17 m s^{-1} usually (at 100 m in height). Table 4.14 summarises the wind conditions in the Chinese waters.

4.10.1 Typhoon-Related Damage in the Zhejiang Province

Zhejiang Province, located on the west of East China Sea, experienced typhoon 23 times since 1997 (start time of wind farm project), causing hundreds of millions of dollars of economic loss. It is of interest to discuss the effects of 10 August 2006 Typhoon Saomai, which had 68 m s^{-1} wind speed that devastated many wind farms. This is the most powerful typhoon to hit China in a half century. It killed 104 people and left at least 190

Table 4.14 Wind conditions summary in China Seas (unit: m/s).

Sea name	Wind speed range (m s^{-1})	Mean wind speed (m s^{-1})
Bohai Sea	6.4–8.2	7.45
Yellow Sea (north)	6.8–7.5	6.86
Yellow Sea (south)	7.0–8.2	7.48
East China Sea	7–11	9.68
Taiwan Strait	11.5–13	12.04
South China Sea	8.8–9.8	9.18

missing. Furthermore, it blacked out cities and smashed more than 50 000 houses in the southeast part of the country. Figure 4.33 shows the Hedingshan wind farm in Zhejiang Province before the cyclone along with the aerial photograph of the typhoon. The typhoon destroyed 28 wind turbine structures and blades, including five tower collapses (600 kW turbine), with a total a loss of about 70 million RMB (Figures 4.34–4.36).

4.10.2 Wave Conditions

Bohai Sea, Yellow Sea, East China Sea, and South China Sea are connected with each other, shaped as a bow from north to south along the China coastline. A tropical and subtropical monsoon climate also influences the China Sea waves. Thus, in Chinese offshore wind farm projects, sea wave is a main factor that influences the design. The wave height distribution in the South China Sea is variable owing to its large area of sea region. High wave height occurs northeast of South China Sea and southeast of the Indo-China Peninsula, where waves are about 2.4–2.6 m in height. The maximum significant wave

Figure 4.33 Hedingshan wind farm (before typhoon Saomai) and aerial view of the typhoon. [Photo Courtsey Prof Lizhong Wang].

Figure 4.34 Damages to blades and tower due to typhoon Saomai. [Photo Courtsey Prof Lizhong Wang].

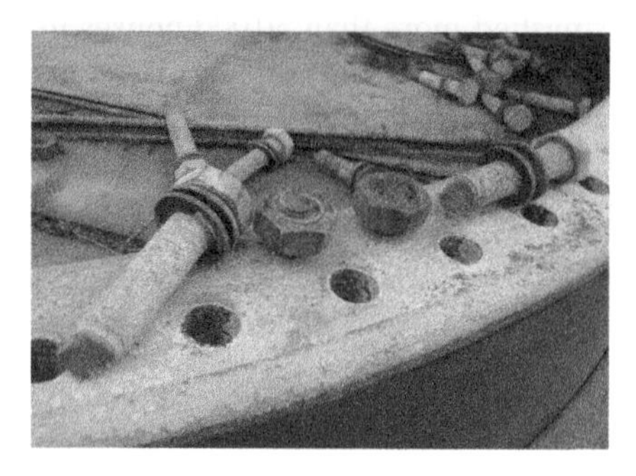

Figure 4.35 Damages to the connections.

Figure 4.36 Damages to the foundations.

Table 4.15 Wave height and period in the Chinese waters.

China Sea	Region	Wave height (m)	Wave period (s)
Bohai Sea	Bohai Strait	1.2	4.8
	Others	<1	<4.5
Yellow Sea	North	1.2	5
	Centre	1.4	5
	South	1.6	6
East China Sea	Shanghai coastline	1.6	6
	Zhejiang coastline	1.8	7
	Taiwan Strait	2.4	9
South China Sea	Luzon Strait	2.8	10
	Indo-China Peninsula	2.6	8
	Others	<2	6

height is larger than 2.8 m west of the Luzon Strait. Comparatively, the significant wave height in the north is less than 2 m. Wave period ranges from 6 to 10 seconds, with similar distribution of wave height. Table 4.15 shows the distribution of wave height and period for the Chinese waters.

Chapter Summary and Learning Points

1. Obtaining reliable engineering parameters for calculation is one of the most important tasks. Geological, geophysical, and geotechnical studies are necessary for a wind farm design. These studies can be integrated to form a ground model.
2. Soil structure interactions for offshore wind turbines are complex. It will be shown in Chapter 5 that one of the workable solutions is to estimate the strain in the soil in the mobilised/deformed zone. This chapter describes the behaviour of soils over a wide range of strains.
3. Element test of soils that can be carried out to understand the behaviour over a wide range of strain are described.

5

Soil–Structure Interaction (SSI)

Learning Objectives:

1. Soil–structure interaction (SSI) affects the overall behaviour of the wind turbine system in mainly three ways: (i) How the load is being transferred from the superstructure to the ground and under what condition the structure would collapse/fail. This is closely related to ULS; (ii) Modes of vibration of the whole system i.e. the manner through which wind turbine system loads the soil periodically (cyclically or dynamically); (iii) Long-term performance prediction, which will take into account the effects of large number of cycles of loading acting on the structure throughout its lifetime.
2. The aim of the chapter is to show that stiffness of the foundation is at the heart of many important calculations. Factors include natural frequency of the whole wind turbine, deformations (serviceability limit state [SLS]), fatigue, and long-term deformation prediction.
3. It will be shown that foundation design is dictated by stiffness calculations. Different methods to estimate foundation stiffness are therefore shown. They are classified as: simplified method (10-minute method where pocket calculators or spreadsheets can be used, based on closed-form solutions), standard method (one-hour method, where conventional code-based methods can be used), advanced method (one-week method, which needs experienced numerical modellers and high-quality soil testing data).
4. These are new structures with limited monitoring data; field records are scarce. Field records available in the public domain are also used to compare with the experimental findings. Readers are advised to consult the latest research in this subject.

SSI for offshore wind turbine (OWT) supporting structures is essentially the interaction of the foundation/foundations with the supporting soil due to the complex set of loading. The readers are referred to Chapter 2 for discussion on loading. This chapter reviews the different aspects of SSI for foundations that are being used or proposed. Due to cyclic and dynamic nature of the loading that acts on the wind turbine structure, the dominant SSI will depend to a large extent on the global modes of vibration of the overall structure. This chapter discusses the various SSI and the available calculation methods.

Design of Foundations for Offshore Wind Turbines, First Edition. Subhamoy Bhattacharya.

Companion website: www.wiley.com/go/bhattacharya/offshorewindturbines

5.1 Soil–Structure Interaction (SSI) for Offshore Wind Turbines

SSI will be affected in three main ways:

1. *Load-transfer mechanism.* The load transfers from the superstructure to the ground through the foundation, and this is the soil–structure interaction whereby stresses are generated in the soil. Monopiles will load the soil very differently than jackets. For a monopile, the main interaction is lateral pile–soil interaction (LPSI) due to the overturning moment and the lateral load. On the other hand, for a jacket, the main interaction is the axial load transfer. Therefore, the SSI depends on the choice of foundation and essentially how the soil surrounding the pile is loaded. The readers are referred to Section 2.5 in Chapter 2 for further discussion on this topic. This mainly guides the ultimate limit state (ULS) criteria for foundation design.
2. *Modes of vibration.* The modes of vibrations are dependent on the types of foundations, i.e. whether the foundation is a single shallow foundation or a summation of few shallow foundations or one deep foundation (monopile) or a combination of few deep foundations. Essentially, if the foundation is very stiff vertically – we expect sway bending modes, i.e. flexible modes of the tower. On the other hand, wind turbine generators (WTG) supported on shallow foundation will exhibit rocking modes as the fundamental modes. This will be low frequency and it is expected that there will be two closely spaced modes coinciding with the principle axes. Two closely spaced modes can create additional design issues: such as beating phenomenon, which can have an impact in FLS (fatigue limit state). Further discussions can be found in Section 3.2 in Chapter 3.
3. *Long-term performance.* Understanding SSI is important when predicting the long-term performance of these structures. Based on the discussion in Chapters 2 and 3, SSI can be cyclic as well as dynamic and will affect the following three main long-term design issues:
   (a) Determine whether the foundation will tilt progressively under the combined action of millions of cycles of loads arising from the wind, wave and 1P (rotor frequency) and 2P/3P (blade passing frequency). Figure 2.21 shows a simplified estimation of the midline bending moment acting on a monopile type foundation, and it is clear that the cyclic load is asymmetric, which depends on the site condition, i.e. relative wind and wave component. It must be mentioned that if the foundation tilts more than the allowable, it may be considered failed based on SLS criteria and may also lose the warranty from the turbine manufacturer.
   (b) It is well known from the literature and discussed in Chapter 4 that repeated cyclic or dynamic loads on a soil causes change in the properties, which, in turn, can alter the stiffness of foundation. A wind turbine structure derives its stiffness from the support stiffness (i.e. the foundation), and any change in natural frequency may lead to the shift from the design/target value and as a result the system may get closer to the forcing frequencies. This issue is particularly problematic for soft-stiff design (i.e. the natural or resonant frequency of the whole system is placed between upper bound of 1P and the lower bound of 3P) as any increase or decrease in natural frequency will impinge on the forcing frequencies

and may lead to unplanned resonance. This may lead to loss of years of service, which is to be avoided.

(c) Predict the long-term behaviour of the turbine, taking into consideration wind and wave misalignment aspects. Wind and wave loads may act in different directions. The blowing wind creates the ocean waves, and ideally they should act collinearly. However, due to operational requirements (i.e. to obtain steady power), the rotor often needs to feather away from the predominant direction (yaw action), which creates wind–wave misalignment.

5.1.1 Discussion on Wind–Wave Misalignment and the Importance of Load Directionality

Following Chapter 3, the most important design drivers for foundation design are the ultimate, fatigue, and serviceability limit state requirements. The assumption of aligned wind and waves is acceptable for ULS design. However, when the long-term performance is assessed, the directionality of wind and wave loading must be incorporated. This can be carried out probabilistically to assess the long-term performance of OWT foundations, as shown by Arany (2017). The environmental state is represented by a Bayesian network of interdependent variables representing wind and wave parameters, wind direction, and wind–wave misalignment. The number of environmental states occurring in the lifetime of the turbine is generated through simple Monte Carlo methods, and a fast-frequency domain approach is used to assess the damage for each environmental state. The complete lifetime simulation is repeated many times to establish the probability distribution of lifetime damage in different directions around the pile. It is demonstrated that the lifetime damage, taking into account load directionality, is significantly reduced compared to aligned wind and wave assumption.

This important aspect is demonstrated in Figure 5.1 through the use of fixed and moving coordinates. The notations of ξ and η directions are introduced for the along-wind (fore–aft) and cross-wind (side-to-side) directions, respectively. We further introduce another coordinate system fixed in space: north–south direction Y (with positive Y values pointing north), and the west–east direction X (with positive values towards the east). When discussing loads on the foundation, it is important to note that the ξ and η directions are not fixed in space, that is, the two coordinate systems (ξ,η) and (X,Y) have the same origin, but they do not coincide at all times. This is because the (ξ,η) coordinate system rotates with the nacelle yaw angle ϕ with the positive ξ axis being approximately normal to the rotor plane (neglecting the nacelle tilt angle) and points in the direction of the mean wind speed (neglecting nacelle yaw misalignment). This is important because when fatigue loads on the foundation have to be analysed, then the directions the designer should be concerned about are loads in the (X,Y) system, recognising that the foundation obviously does not rotate with the nacelle. It is also important to emphasise here that because of the above effect, all load directions in the (X,Y) coordinate system (that is, e.g. all sections around the circumference of the monopile) are subjected to load cycles due to both along-wind and cross-wind load processes throughout the lifetime of the turbine.

If the most severe loads (incorporating the dynamics of the structure) can be demonstrated to act in the along-wind direction and with collinear wind, waves, and currents, then it is a conservative to use this load scenario for ULS purposes. Not only

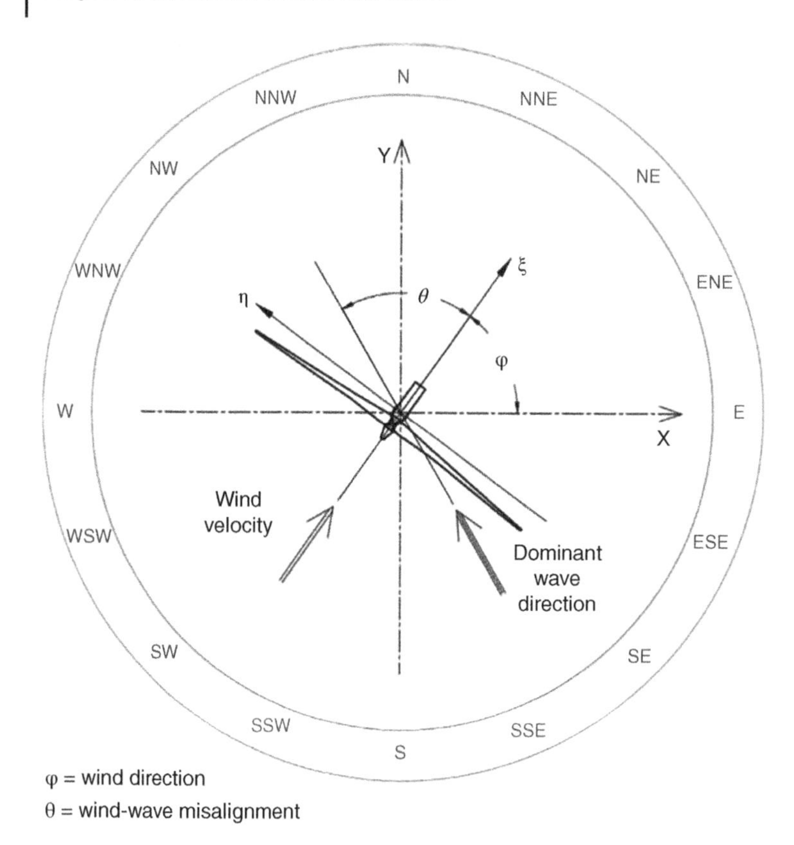

Figure 5.1 Spatially fixed (*X*,*Y*) and yaw angle-following (ξ,η) coordinate systems.

the wind–wave misalignment but also the general direction of the load can be neglected for a circularly symmetric foundation (e.g. monopiles, caissons, or round gravity-based structures), or a single-most severe load direction can be chosen for a substructure without circular symmetry (jackets, tripods, floating structures, etc.). However, when assessing the number of load cycles throughout the lifetime, then assuming that all load cycles occur in the same direction produces overly conservative estimations for the total fatigue damage suffered by the structure or indeed the total number of load cycles causing accumulated tilt or deflection.

It is therefore essential to understand the mechanisms that may cause the changes in dynamic characteristics of the structure and if it can be predicted through analysis. An effective and economic way to study the behaviour (i.e. understanding the physics behind the real problem) is by conducting carefully and thoughtfully designed scaled model tests in laboratory conditions simulating (as far as realistically possible) the application of millions of cyclic lateral loading by preserving the similitude relations. The readers are referred to Bhattacharya et al. (2018), where scaled model tests for analysis and design of OWT foundations are discussed.

Considerable amount of research has been carried out to understand various aspects of cyclic and dynamic soil–structure interaction (DSSI); see, e.g. Leblanc et al. (2009), Lombardi et al. (2013), Bhattacharya et al. (2013a,b), and Bhattacharya (2014a,b). The studies showed that to assess the SSI, it is necessary to understand not only the loading on offshore wind turbines but also the modes of vibration of the overall system.

Years of offshore engineering practice have provided clarity regarding load transfer to the foundation. The methods to estimate the capacity of foundations have been developed and are available in codes of practices. By capacity, we mean the ultimate load-carrying capacity of the foundation. Examples of commonly used foundations are given in this chapter, together with references for further reading.

Modes of vibration will dictate the interaction of the foundations with the supporting soil. Furthermore, if the foundation–soil interaction is understood, the long-term behaviour of the foundation can be predicted through a combination of high-quality cyclic element testing of soil and numerical procedure to incorporate them in different interactions.

5.2 Field Observations of SSI and Lessons from Small-Scale Laboratory Tests

Limited monitoring of offshore wind turbines indicates that the dynamic characteristics of these structures may change over time and have the potential to compromise the integrity of the structure due to fatigue and resonance phenomena. A few examples are given here.

5.2.1 Change in Natural Frequency of the Whole System

Scaled model tests carried out by Bhattacharya et al. (2013a,b), Yu et al. (2014), Guo et al. (2015), and Cox et al. (2014) indicate that the natural frequency of a wind turbine system may change owing to dynamic soil–structure interaction. Change in the natural frequency of the Hornsea Met Mast structure supported on a twisted-jacket foundation is also reported by Lowe (2012). Three months after the installation, the natural frequency dropped from its initial value of 1.28–1.32 to 1.13–1.15 Hz, shown schematically in Figure 5.2. On the other hand, Figure 5.3 shows the results from Lely wind farm following Kuhn (2000). In this figure, the design value of frequency is 0.4 Hz and the measured value after 6 years is 0.63 Hz.

5.2.2 Modes of Vibration with Two Closely Spaced Natural Frequencies

As discussed in Chapter 3, as was observed in scaled model tests, there can be multiple resonance peaks for certain types of structure. This is indicative of closely spaced natural frequencies corresponding to different modes of vibration. This was observed and noted in Brent B Condeep platform, see Figure 5.4 which was selected for an extended instrumentation program sponsored by several oil companies operating in the North Sea. Accelerations were recorded and results plotted in frequency domain following Hansteen (1980). The two first modes of vibration are 1.78 seconds (0.56 Hz) and 1.72 seconds (0.58 Hz), representing bending modes in the two horizontal directions. The third mode of 1.19 s (0.84 Hz) corresponds to torsional mode of vibration.

It is of interest to discuss some aspects of the Condeep (**Con**crete **deep** water structure) structure, shown in Figure 5.4. The foundation consists of 19 cells having an overall

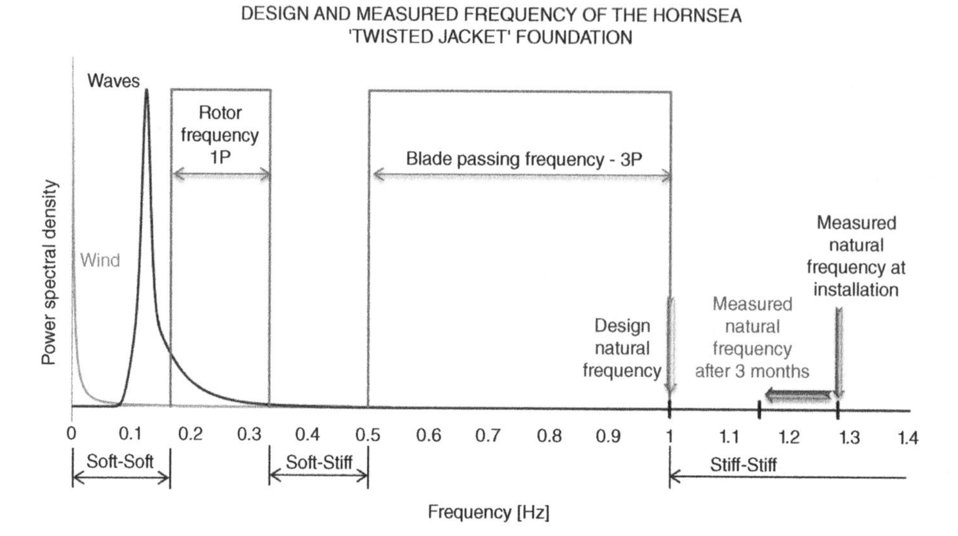

Figure 5.2 Change in natural frequency in a twisted-jacket supporting a met mast.

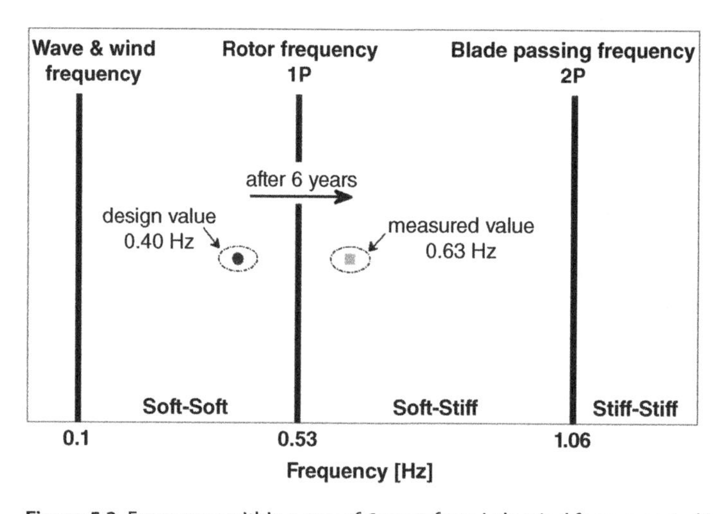

Figure 5.3 Frequency within a gap of 6 years from Lely wind farm reported in Kuhn (2000).

diameter of 100 m resting on 45 m of stiff-to-hard overconsolidated clay with some layers of dense sand. The structure is designed for 100 years wave height of 30 m.

5.2.3 Variation of Natural Frequency with Wind Speed

A schematic sketch of the measured difference between the design and actual natural frequency of the Siemens SWT-3.6-107 type 3.6 MW turbines at the Walney 1 site is shown in Figure 5.4. There are two salient features: (i) The difference of 6% between prediction and observed is reasonable, given the complexity of the structure and the uncertainty in the calculation methods mainly the ground stiffness measurements; (ii) scatter increases on natural frequency at higher wind speed until about 14 m s^{-1}. As

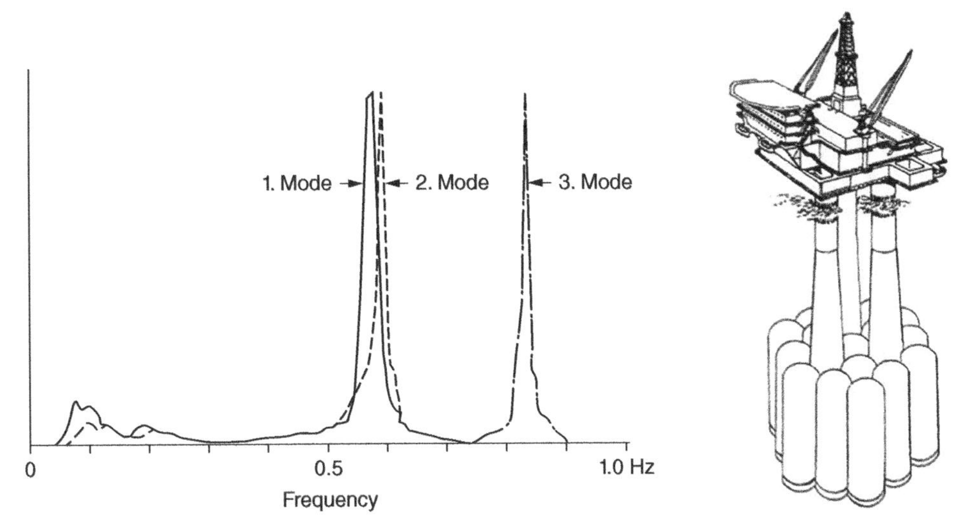

Figure 5.4 Observed multiple peaks for Brent B Condeep platform following Hansteen (1980) and image from of Condeep platform from Oide and Andersen (1984).

wind speed increases, the moment at the foundation increases, which will increase the strains in the soil. As discussed in Chapter 4, with increasing strain level, stiffness of soil decreases and damping increases. Another interesting point to note is that as the wind speed increases beyond 14 m s^{-1}, the controlling action will kick in (either pitch control or yaw control) and the load on the foundation will not increase. This may explain the reduction in the scatter (Figure 5.5).

5.2.4 Observed Resonance

Resonance has been reported in a wind turbine site in the German North Sea under operational conditions. Figure 5.6 shows the frequency domain plot based on Hu et al. (2014). The explanation of resonance is due to the proximity of the natural frequency to the 1P and 3P.

5.3 Ultimate Limit State (ULS) Calculation Methods

Based on the design considerations presented in Chapter 3, together with the discussion presented in Section 5.1, this section lists the different ULS calculations that are necessary. The ULS is of critical importance to the foundation design and is usually the first step. These are available in many textbooks and therefore are not repeated here. Some notable textbooks and monographs are Dean (2010), Randolph and Gourvenec (2011), Encyclopaedia of Maritime and Offshore Engineering (2018), and Salgado (2006).

5.3.1 ULS Calculations for Shallow Foundations for Fixed Structures

Section 1.5.1 of Chapter 1 provides different types of shallow foundations for WTG structures. They are often a good alternative to deep foundations in good ground conditions, and their stability depends on their weight. Often, jackets are temporarily

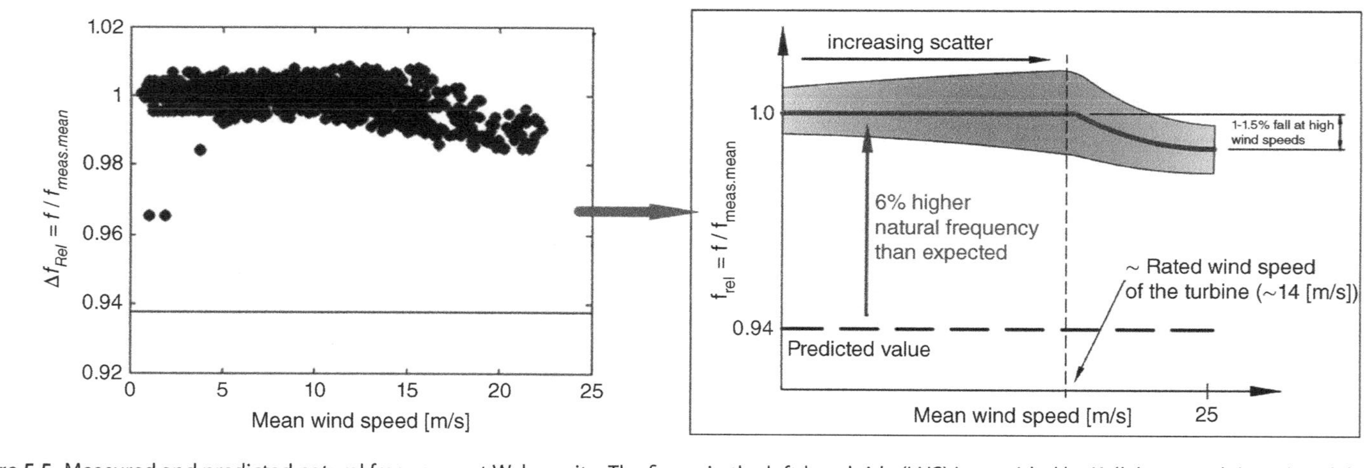

Figure 5.5 Measured and predicted natural frequency at Walney site. The figure in the left-hand side (LHS) is provided by Kallahave et al. (2012) and the interpretation is provided in the schematic sketch in right-hand side (RHS).

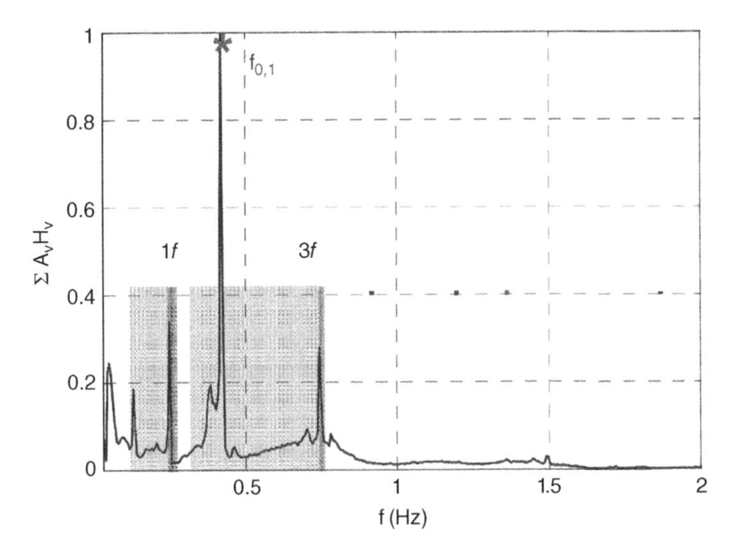

Figure 5.6 Observed resonance in German North Sea following Hu et al. (2014).

supported on shallow foundations (essentially a steel plate known as mudmats) before their main fountains are driven. The shapes of shallow foundations can be rectangular, square, or circular in plan. They can be even multicellular, and for mudmats the shape can be irregular. To improve load-bearing capacity, skirts are often attached to the foundations. The load acting on such foundations are: vertical load (V), moment (M), and horizontal load (H). The analysis of such foundations is carried out through the adaptation, modifications, and extension of Terzaghi's bearing capacity equations. The original bearing capacity equation given by Eq. (5.1) is developed for strip footing (i.e. footing under a long wall) resting on the surface of a homogenous deposit and for vertical loads (V) only. Offshore foundations are invariably subjected to lateral loads (H) and overturning moments (M) and can have different shapes and in most cases, will be embedded in the seabed. Equation (5.1) was first proposed by Terzaghi (1943) and subsequently Meyerhof (1956) advocated a slightly modified bearing capacity equation to account for any embedment of the foundation. Hansen (1970) described a number of alterations to the bearing capacity factors based on theoretical foundation behaviour. All of these methods have led to the generalised bearing capacity equation detailed is almost all codes of practices such as British Standards Institution (BSI) (2004) and Det Norske Veritas (DNV). Equation (5.2) shows a generalised form.

$$\frac{V}{A} = q_{ult} = c'N_c + \gamma z N_q + \frac{1}{2}\gamma B N_\gamma \tag{5.1}$$

$$q_{ult} = c'N_c b_c s_c i_c + \gamma z N_q b_q s_q i_q + \frac{1}{2}\gamma B N_\gamma b_\gamma s_\gamma i_\gamma \tag{5.2}$$

where:

q_{ult}	Ultimate bearing capacity	$(\mathrm{N\,m^{-2}})$
c'	Apparent soil cohesion	$(\mathrm{N\,m^{-2}})$
z	Foundation depth	(m)

B	Foundation breath	(m)
γ	Soil unit weight	$(\mathrm{N\,m^{-3}})$
N_c, N_q, N_γ	Bearing capacity factors	(–)
b_c, b_q, b_γ	Base inclination factor	(–)
s_c, s_q, s_γ	Shape factors	(–)
i_c, i_q, i_γ	Inclination factors	(–)

In this simplified approach, the generalised loading problem (V, M, H) is first transformed into a (V, H) problem by reducing the area of the foundation through an effective area approach, explained in the next section. The (V, H) problem is taken into account through the inclination factors i_c, i_q, i_γ.

5.3.1.1 Converting (V, M, H) Loading into (V, H) Loading Through Effective Area Approach

Moment (M) capacity can be considered as a function of the vertical capacity and the eccentric distance from the centreline that the resultant design loads will be applied. This can be expressed simply as:

$$e = \frac{M}{V} \tag{5.3}$$

where:

M	Moment	(Nm)
V	Vertical load	(N)
e	Eccentric loading point	(m)

When eccentric loading is considered, a reduction in the foundation area is required; this reduced area is called the effective foundation area. The effective area is defined such that its geometric centre coincides with that of the resultant load. In the case of a circular foundation, the effective area can be calculated from the following formulation:

$$A_{eff} = 2\left[\frac{D^2}{4} arccos\left(\frac{2e}{D}\right) - e\sqrt{\frac{D^2}{4} - e^2}\right] \tag{5.4}$$

where:

A_{eff}	Effective foundation area	$(\mathrm{m^2})$
D	Caisson diameter	(m)
e	Eccentric loading point	(m)

An example of how this reduction would be applied is shown in Figure 5.7. This is recommended in American Petroleum Institute (API) and DNV regulations.

5.3.1.2 Yield Surface Approach for Bearing Capacity

In the *modified bearing capacity factor* approach described in the earlier section, each loading condition (V, M, H) is considered separately. However, it is obvious that the capacity of one loading orientation is intrinsically linked to that of the others, as it is

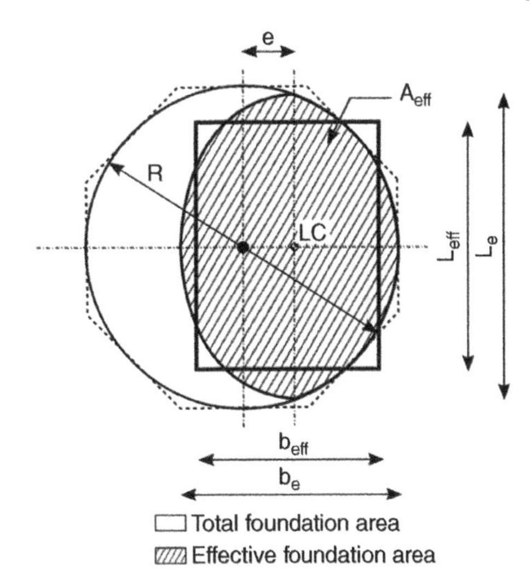

Figure 5.7 Effective foundation area method following Det Norske Veritas (2013).

the same soil element surrounding the foundation that supports all the three types of loads. Houlsby and Martin (1992) developed an integrated approach for V, M, H loading through the use of plasticity theory. The method is similar to the structural engineering approach (combined axial and bending or the combination of stresses in a material), where a yield surface for an offshore foundation is developed. Research shows that the shape of the yield surface is a rugby ball and is shown in Figure 5.8. Any point in the surface of (V, M, H) space is a yield point, i.e. beyond which plastic deformation will occur.

5.3.1.3 Hyper Plasticity Models

The behaviour within the yield surface can be characterised. However, as the load increases on the foundation the yield surface is expected to expand accordingly, and this growth will occur in a plastic manner. To describe this expansion, it is necessary to consider an appropriate strain-hardening law. Such a theory is known as hyper-plasticity. To provide a theoretical basis for strain hardening, Gottardi and Houlsby (1995) considered the strain hardening that occurred during a simple load penetration test. These are advanced methods of analysis.

5.3.2 ULS Calculations for Suction Caisson Foundation

Different variants of suction caissons are considered as suitable foundation and are therefore discussed in more details. The caisson can be considered as a solid embedded object, and under the application of a moment load the caisson will rotate about a single point as a rigid body with or without the bending of the walls/skirts. Apart from moment, the caisson may settle under the vertical load or translate due to lateral load. For preliminary design, often simple analytical methods suffice, and Figure 5.9 shows a generalised sign convention following Butterfield et al. (1997). For mono-caisson, moment-carrying capacity is critical and the point of rotation is associated with such calculations. It is often assumed that the point of rotation lies above the foundation level, i.e. within the confines of the caisson skirt. Under the application of a moment

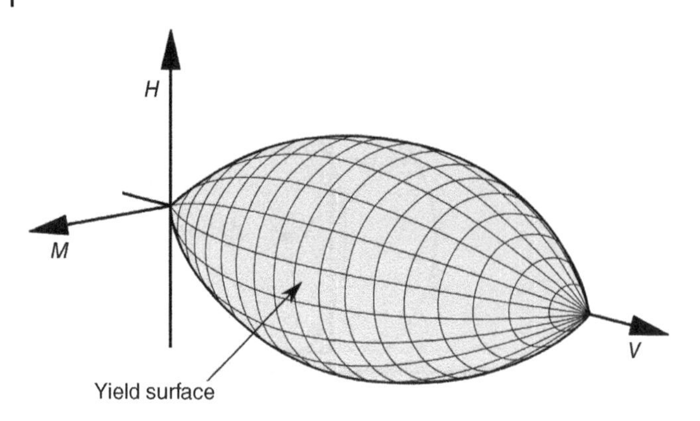

Figure 5.8 Rugby ball failure surface (Source: Villalobos et al. (2009)).

load, the deflected shape is described by the rotation of two parallel walls representing the caisson skirt. The rotation of the two parallel walls is such that it is the same as that of the original caisson, as shown in Figure 5.10. The general description of the capacity of a suction caisson has been considered by a number of research groups – Byrne (2000), Villalobos (2006), Schakenda et al. (2011), Cox and Bhattacharya (2017), and Cox (2015) – with the aim of providing a simple method for predicting the ultimate capacity of such a foundation under a series of different loading conditions. Reliable solutions may also be obtained through the use of finite element methods.

5.3.2.1 Vertical Capacity of Suction Caisson Foundations

For a typical case, the general bearing capacity equation can be simplified and reduced. This leads to the following equation for a caisson on sand:

$$V' = \frac{1}{4}\pi D^2 \left(\gamma' z N_q + \frac{1}{2}\gamma' D N_\gamma\right) \tag{5.5}$$

where

V'	Vertical foundation capacity	(N)
D	Caisson diameter	(m)

It is generally accepted that the general bearing capacity equation (with the reasonable correction factors) models the foundation capacity well. It may be noted that the mobilised angle of friction should be used due to the heaving of the soil plug within the caisson that occurs during installation. Using peak friction angle may lead to an overestimation of capacity. Villalobos (2006) derived empirical calculation for the vertical capacity of a caisson foundation. This is based on the summation of the end-bearing capacity and skin friction resisting any vertical movement of the caisson:

$$V' = \frac{\pi D^2}{4}\sigma'_v + D\gamma' z(k_p - k_a)(2z_m - z)\tan\delta \tag{5.6}$$

where:

V'	Vertical foundation capacity	(N)
σ'_v	Effective vertical earth pressure	$(\mathrm{N\,m^{-2}})$
z	Foundation depth	(m)

k_p	Coefficient of passive soil resistance	(–)
k_a	Coefficient of active soil resistance	(–)
z_m	Depth to point of caisson rotation	(–)
δ	Soil–structure interface angle of friction	(°)

5.3.2.2 Tensile Capacity of Suction Caissons

The vertical tensile capacity also needs to be considered if caissons are used in multi-pod arrangements or as anchors. The tensile capacity of a caisson can be attributed to one of two separate load-bearing mechanisms. Under a high frequency loading, negative pore pressures and suction may be generated around and within the caisson skirt and is dependent on the speed of loading and the permeability of the surrounding ground. Under low-frequency loading situations, the tensile capacity will simply be the summation of the frictional components acting on the caisson skirt and the reverse bearing capacity of the foundation. This is further discussed in Example 6.11 of Chapter 6, for example.

5.3.2.3 Horizontal Capacity of Suction Caissons

Generally, the derivation of horizontal capacity of a surface foundation is well agreed upon. The resistance to loading is simply the frictional force generated between the foundation and the supporting soil. This generalised sliding resistance is given by a number of authors in addition to being detailed in DNV regulations (Det Norske Veritas 2013) as follows:

$$H < A_{eff} \cdot c' + V \cdot \tan\phi \tag{5.7}$$

where:

H	Horizontal capacity	(N)
A_{eff}	Effective foundation area	(m^2)
ϕ	Angle of internal friction	(°)

In the case of an embedded foundation it is reasonable to assume that the additional resistive capacity of the passive soil resistance acting upon the caisson skirt can be considered in addition the frictional resistance already considered. Initially, such a formulation was derived by Byrne (2000) for a suction caisson subjected to a pure horizontal loading, and the capacity of a caisson under such conditions was defined as follows:

$$H = V\ \tan\phi' + \frac{\gamma' D z^2}{2}(k_p - k_a) \tag{5.8}$$

where:

H	Horizontal foundation capacity	(N)
V	Vertical foundation capacity	(N)
D	Caisson diameter	(m)
z	Foundation depth	(m)

k_p Coefficient of passive soil resistance (–)

k_a Coefficient of active soil resistance (–)

This formulation was revised by Villalobos (2006) considering the rotation of the foundation at a known depth and is given below.

$$H = V \tan\phi' + \frac{\gamma' D}{2}(k_p - k_a)(2\alpha^2 - z^2) \tag{5.9}$$

where:

α Caisson rotational depth (detailed later) (m)

For all such cases, the vertical weight is the sum of the buoyant structure and the buoyant weight of the soil contained within the caisson skirt.

$$V = V' + \frac{\gamma' h \pi D^2}{4} \tag{5.10}$$

From analysis, it is evident that a greater horizontal capacity can be achieved with an increase in the vertical dead weight of the structure.

5.3.2.4 Moment Capacity of Suction Caissons

Unlike the derivation of both the vertical and horizontal foundation capacity, there has been a significant amount of debate and analysis over the true definition of the moment capacity of a caisson foundation. All the proposed methods are slightly more complex than its predecessor. Despite the increasing complexity, the difference between the solutions is minimal, and all are suitable for a preliminary estimate. The load-carrying capacities are estimated based on the lateral earth pressure distribution, skin friction along the walls, eccentric vertical load, and the friction from the underlying slip surface. Using the principle of superposition, the influence of each individual component can be combined to provide a global estimate as to the capacity of the foundation. Figure 5.11 shows the formulation based on Vaitkunaite et al. (2012).

Considering the two walls separately, the coulomb earth pressure theory can be applied. Using equilibrium, the following calculation concerning the moment capacity of a suction caisson rotation about a point at its base:

$$M = R_d \cdot e + E_d \frac{D}{2} \tan\phi + E_1(z - z_1) - Hz - E_2(z - z_2) \tag{5.11}$$

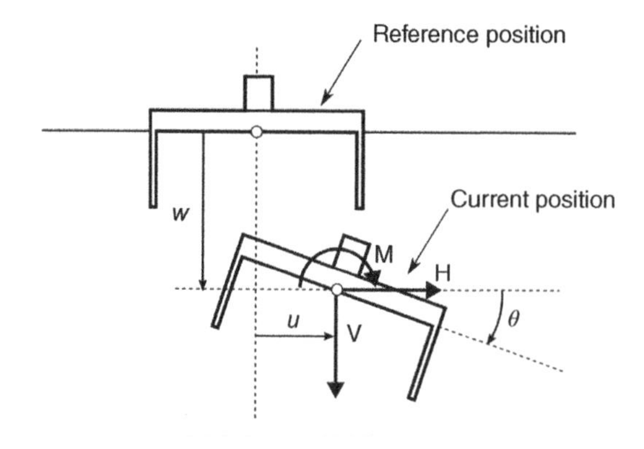

Figure 5.9 Standardised sign convention after Butterfield et al. (1997).

or this can be expressed in the form

$$M = V \cdot e + \frac{\gamma' z^2 \overline{D} D}{4}(k_p + k_a)\tan\phi + \frac{\gamma' z^3 \overline{D}}{6}(k_p - k_a) - H \cdot z \tag{5.12}$$

where:

M	Moment wise foundation capacity	(Nm)
V, R_d	Instantaneous vertical foundation capacity	(m)
z	Foundation depth	(m)
$\overline{D}$	Effective foundation diameter	(m)
k_p	Coulomb coefficient of passive soil resistance	(–)
k_a	Coulomb coefficient of active soil resistance	(–)
z_1	Depth to passive point of action	(m)
z_2	Depth to active point of action	(m)

This formulation can be evaluated considering a differing depth of rotation. The full derivation and design methodology proposed by Ibsen and MBD offshore power A/S can be found in the patent application filed by Schakenda et al. (2011).

In a similar manner, Villalobos (2006) also considered the effects of a shifting centre of rotation on the moment-resisting capacity of a suction caisson foundation. Utilising an analysis method similar to that of Ibsen, an alternative formulation for the rotational strength was obtained; see Figure 5.10. Unlike Ibsen, Villalobos included the frictional resistance generated between the soil plug and the surrounding soil mass, see Figure 5.12.

Considering the superposition of each separate variable, the capacity of a suction caisson foundation can be found.

$$M = \frac{D\gamma'(k_p - k_a)}{3}(z^3 - 2z_m^3) + \frac{1}{2}D^2\gamma'(k_p + k_a)[(z - z_m)^2 + z_m^2]\tan\delta + \frac{z^2\gamma'\pi D^2}{4}\tan\phi' \tag{5.13}$$

Cox (2014) improved the solution of Villalobos (2006), and the model is presented in Figure 5.13. In the calculations of Villalobos the distribution of vertical earth pressure

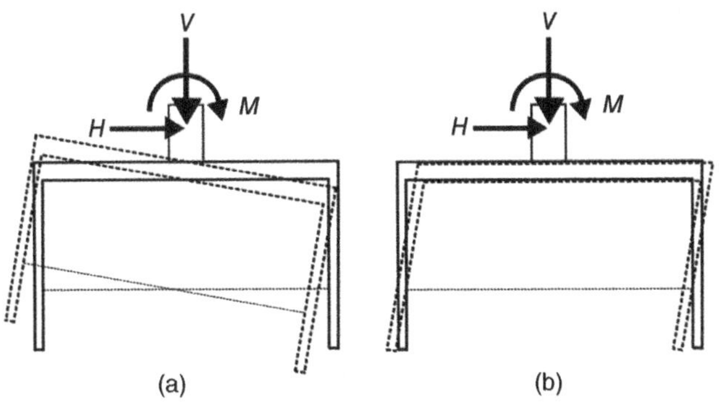

Figure 5.10 Caisson moment loading. (a) true deflection of the caisson; (b) assumed deflection for the purposes of calculation after Schakenda et al. (2011).

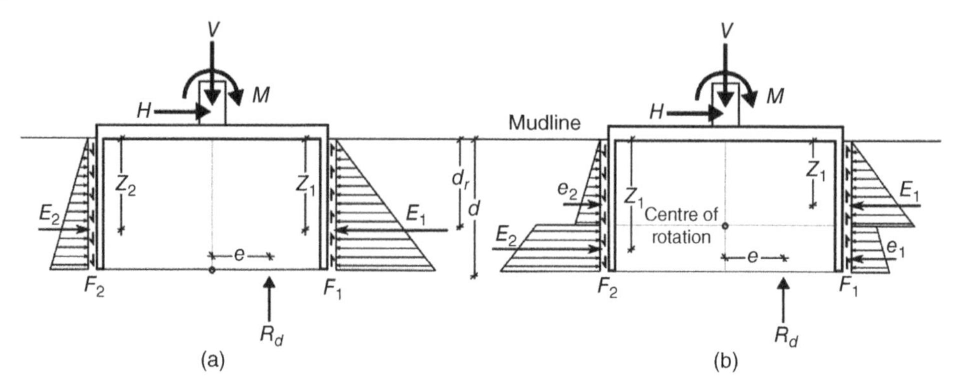

Figure 5.11 Caisson moment capacity as described by (Vaitkunaite et al. 2012). (a) rotation at bottom; (b) rotation at a midpoint.

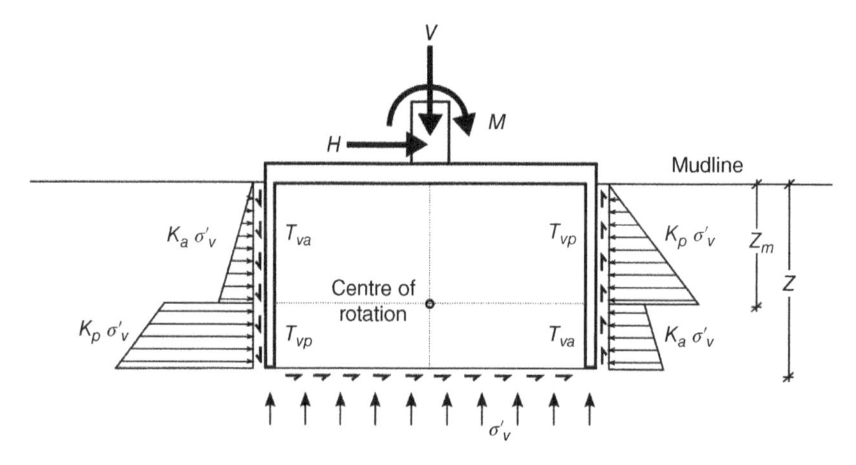

Figure 5.12 Caisson moment capacity as described by Villalobos (2006).

on the bottom of the caisson was assumed to be even over its area. By considering an eccentricity to the loading profile, it is possible to consider the additional load-carrying capacity that may be achieved. The frictional effect of shear along the base is still considered and an average load is taken to calculate the friction generated:

$$M = \frac{\gamma' z^2 D^2 \tan\delta}{2\pi}(k_p + k_a) + V[e + (z - z_m)\tan(\phi)] + \frac{D\gamma'}{6}[z_m{}^3 + (z - z_m)^2(2z + z_m)](k_p - k_a) - Hz_m \tag{5.14}$$

5.3.2.5 Centre of Rotation

Considering the three types of moment formulations proposed, it is clear that a change in the rotation depth can influence the foundation capacity. The effect of a shifting depth of rotation on the capacity of a caisson is considered in Figure 5.14 by taking an example from Cox (2014). The corresponding moment capacity with changing depth of rotation may be noted. For practical purposes, the depth of rotation corresponding to the minimum caisson capacity will be the depth at which the foundation will rotate. This assumes, however, that the rotation will be small and that the active and passive zones are fully developed.

Using the above assumption, Cox (2014) developed a formulation to calculate the depth of rotation for given geometrical and soil parameters. This is essentially based on differentiating the moment equation and finding the minima. The centre of rotation can be calculated using the following equation:

$$\alpha_{min} = \sqrt{\frac{D\gamma' z^2(k_p - k_a) + 2H + (2V')\tan\phi'}{2D\gamma'(k_p - k_a)}} \tag{5.15}$$

where:

V'	Vertical foundation capacity	(N)
α_{min}	Caisson rotational depth at minimum capacity	(m)

This general calculation indicates that the depth of rotation for a shallow caisson (with a Z/D ratio of 0.5) will be approximately at the base of the caisson and for caissons with a greater aspect ratio the centre of rotation will move proportionally upwards.

5.3.2.6 Caisson Wall Thickness

The thickness of a suction caisson also needs to be designed to resist buckling both during installation and operation. All forces such as the vertical, horizontal, moment, and pressure loads acting on the caisson structure need to be evaluated and appropriately resisted by the caisson structure. Design guidelines are provided in DNV-OS-C202 (Det Norske Veritas 2010a) to specify that an appropriate caisson thickness is used to ensure the caisson doesn't buckle.

This standard is generally used industry-wide for the design of thin-walled vessels; there has, however, been some discussion on the applicability of these regulations to the design of suction caissons by LeBlanc (2009). This uncertainty is based on the failure of the Wilhelmshaven suction caisson OWT. In this instance, the caisson buckled during installation after it was struck by one of the installation vessels. On subsequent analysis by LeBlanc (2009), it was discovered that minor imperfections in the circularity of a thin-walled vessel such as a suction caisson can cause the buckling load to be significantly reduced.

5.3.3 ULS Calculations for Pile Design

The ULS calculations for pile foundations needs typically three types of calculations:

1. Axial pile capacity (geotechnical): Here the ultimate load corresponds to the failure of the pile whereby the shaft resistance and the end-bearing resistance is fully mobilised and the pile fails in excessive settlement. Typically, if the axial pile-head displacement is more than 10% of the pile diameter, it may be considered that the ultimate capacity is reached. The parameters that are needed are soil parameters (typically strength parameters) and the geometry of the pile (length, diameter, and wall thickness).
2. Axial pile capacity (structural): If the pile is laterally unsupported due to excessive scour in the upper depths or soil layers momentarily liquefying due to earthquake, it may buckle (Euler type global buckling). This must be considered in seismic areas. Long, slender thin-walled piles are particularly vulnerable.

3. Structural capacity of the section (plastic moment capacity, M_P as well as moment at first yield, M_Y): This shows the maximum moment that the pile section can carry before the material of the pile yields. This is a function of the pile geometry (diameter, wall thickness) and material properties (yield strength). An example problem is shown in Chapter 6 (Example 6.11).
4. Moment-carrying capacity of the pile (geotechnical). This is important for monopile design, and in this condition, the soils surrounding the pile fails.

5.3.3.1 Axial Pile Capacity (Geotechnical)

The failure load or ultimate load of a pile is the summation of shaft resistance (also known as skin friction or side friction) along the embedded length of a pile and end-bearing at the pile toe. Mathematically, the side friction is the integral of the shear stress over the cylindrical surface of the pile, and the end bearing is the integral of the normal stress on the pile tip. Most, if not all, offshore piles are thin-walled steel section, and they may be closed-ended (also known as plugged) or open-ended (also known as unplugged). For closed-ended pile, the pile drives as if it is solid section, and therefore there are two terms: end bearing of the whole area of the pile and side friction of the outer pile surface. On the other hand, for open-ended conditions, the pile cuts around a plug of soil and the plug remains almost static in the ground. In this case, there are three terms: end bearing of the annulus of the section, side friction of the outer pile surface, and side friction in the inner pile surface. Figure 5.15 shows one particular case.

It is very difficult to ascertain whether a pile will behave plugged or unplugged (Figure 5.15). Therefore, a pragmatic approach is taken in most codes of practices whereby axial capacity is considered for two cases and the lower of the two is taken.

Axial capacity: open-ended pipe pile and the capacity is given by Eq. (5.16):

$$Q_{ult} = \sum_{1}^{N} f_S \cdot A_S + \sum_{1}^{N} f_i \cdot A_i + q_P \cdot A_w \tag{5.16}$$

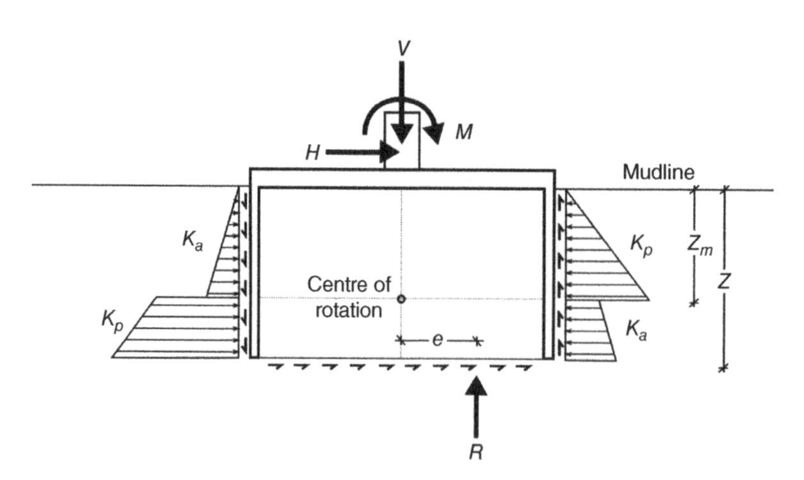

Figure 5.13 Caisson moment capacity as described by Cox (2014).

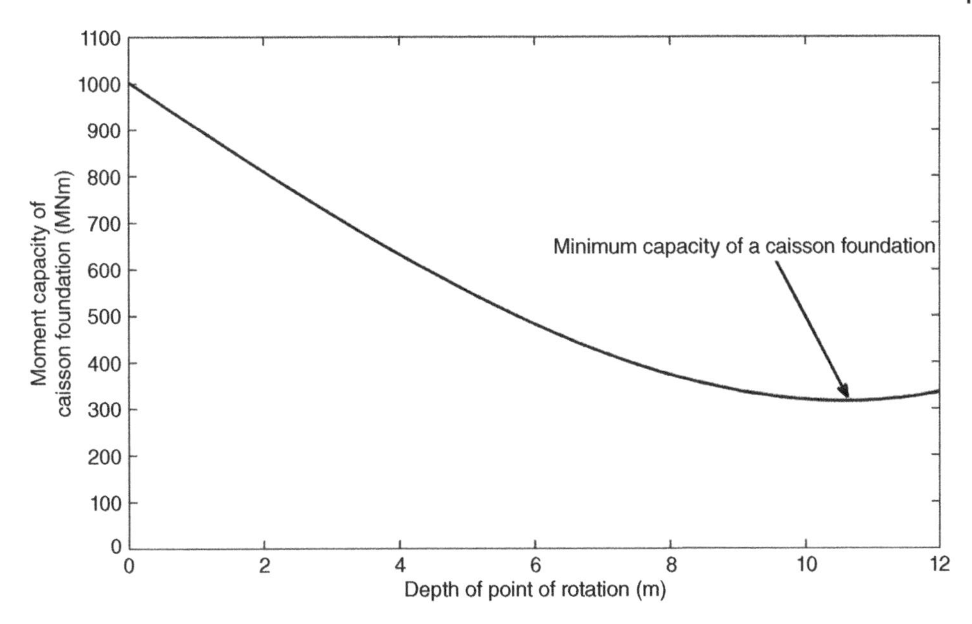

Figure 5.14 Changing moment capacity of theoretical suction caisson based on the moment capacity formulation described by Cox (2014) ($D = 12$ m, $Z = 12$ m). Source: Reproduced with permission of PHD Thesis.

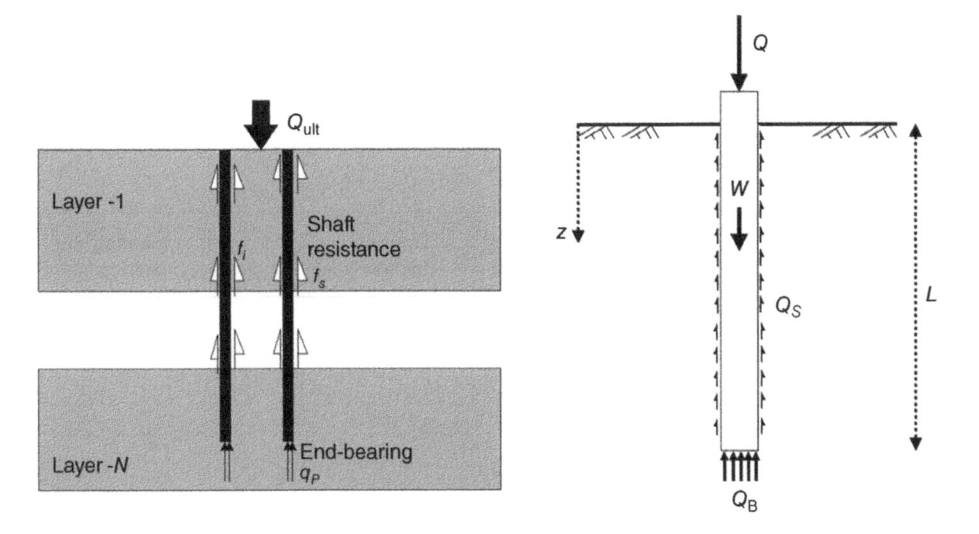

Figure 5.15 Static capacity of a pile (unplugged condition) and plugged condition.

where:

Q_{ult} = Ultimate static capacity

f_s = Unit outside shaft friction

A_s = Outside shaft area of pile

f_i = Unit inside shaft friction

A_i = Inside shaft area of pile

q_p = Unit end bearing capacity

A_w = Cross-sectional area of steel wall at toe of pile

N = Number of layers

The capacity of the closed-ended pile is given by Eq. (5.17):

$$Q_{ult} = \sum_{1}^{N} f_S \cdot A_S + q_P \cdot A_P \tag{5.17}$$

where:

A_p = Gross end-bearing area

There are different methods to estimate the capacity and the notable ones are: API, NGI, Fugro, UWA (University of Western Australia) and ICP (Imperial College Pile) method.

API method is widely used for small-diameter piles and the parameters can be obtained from soil testing and site investigation. The following are from the twenty-first edition of the *API-RP2A – WSD 2000* method (RP stands for Recommended Practice, WSD stands for working stress design). The readers are referred to the latest codes of practice or best practice guide for such purpose. However, the fundamentals of the calculations remains the same.

Parameters for sandy soil (Coarse grained soil) – API approach (Table 5.1)

Unit shaft friction

$$f = K \cdot \sigma_v' \cdot \tan\delta \tag{5.18}$$

Table 5.1 Design parameters for coarse-grained soil[a] from API.

Density	Soil type	Soil-pile friction angle (Degrees)	Limiting friction (kPa)	Nq	Limiting end bearing (MPa)
Very Loose	Sand	15	47.8	8	1.9
Loose	Sand-Silt[b]				
Medium	Silt				
Loose	Sand	20	67.0	12	2.9
Medium	Sand-Silt[b]				
Dense	Silt				
Medium	Sand	25	81.3	20	4.8
Dense	Sand-Silt[b]				
Dense	Sand	30	95.7	40	9.6
Very Dense	Sand-Silt[b]				
Dense	Gravel	35	114.8	50	12.0
Very Dense	Sand				

a) Where detailed information such as *in-situ* cone tests is available, values and limits may be modified.
b) Sand-Silt fractions – strength values generally increase with increasing sand fraction.

where:

K = Coefficient of lateral earth pressure

σ'_v = Effective overburden pressure at the point in question

δ = Angle of friction between soil and pile wall

Unit end bearing

$$q_u = \sigma'_v \cdot N_q \quad (5.19)$$

where:

N_q = Bearing capacity factor for deep foundation [See later on]

σ'_v = Effective overburden pressure at the pile tip

Parameters for clayey soil (fine grained soil) – API approach

Unit shaft friction

$$f = \alpha \cdot S_u \quad (5.20)$$

where:

α = A dimensionless factor, derived as outlined below

S_u = Undrained shear strength of the soil at the point in question

Unit end bearing

$$q_u = 9 \cdot S_u \quad (5.21)$$

API (2000) method, suggests the following:

Case 1:

For $S_u < \sigma'_v$

$$\alpha = \frac{1}{2}\left(\frac{S_u}{\sigma'_v}\right)^{-\frac{1}{2}} \quad (5.22)$$

Case 2:

For $S_u > \sigma'_v$

$$\alpha = \frac{1}{2}\left(\frac{S_u}{\sigma'_v}\right)^{-\frac{1}{4}} \quad (5.23)$$

For simplified calculations, it may be assumed that equal shaft resistance acts at the inside and outside of open-ended piles. There are excellent text books where the axial capacity of a pile is dealt with in great detail. One such is Randolph and Gourvenec (2011).

5.3.3.2 Axial Capacity of the Pile (Structural)

The allowable axial load in terms of buckling instability depends on the following:

- Length of the pile likely to be unsupported due to scour and liquefaction
- Boundary condition of the pile below and above the unsupported zone
- Bending stiffness of the pile (EI)

Basic Euler equation may be used. The readers are referred to the paper by Bhattacharya et al. (2005).

$$P_{all} = \frac{\pi^2 EI}{L_{eff}^2}$$

Where L_{eff} is the normalised unsupported length (L) of the pile. L_{eff} is L for fixed headed pile (i.e. typical jacket pile). On the other hand, for free-headed pile (i.e. monopile) L_{eff} is $2L$.

The unsupported length of the pile (L) can be estimated by the adding three terms: Scour depth + Liquefiable zone depth + Depth of fixity at the bottom of the unsupported length of the pile. In the absence of detailed investigation, this can be taken as three times the diameter of the pile. The readers are referred to Bhattacharya and Goda (2013) for details of the calculations.

5.3.3.3 Structural Sections of the Pile

Methods to obtain plastic moment capacity (M_P) and moment at first yield (M_Y) are provided in Example 6.10 of Chapter 6. Furthermore, the pile section has to be checked for general and local buckling, and this section provides guidelines in this respect. Based on the methods described in Chapter 2, one can calculate the total bending moment by assuming collinear wind and waves. This is by summing the bending moments due to each, and applying an environmental load factor $\gamma_L = 1.35$ as prescribed in (DNV 2010a,b).

$$M_\gamma = (M_U + M_W) \cdot \gamma_L$$

The ULS depends on the length of the pile, as it determines whether the monopile or the soil will fail first. A simplified methodology proposed in (Poulos and Davis 1980) recommends that for practical parameter sets, the pile fails first through yielding. The ultimate overturning moment the pile can carry is then calculated simply by:

$$\mathrm{M_y} = \frac{2f_{yk}I_P}{D_P}$$

where $f_y = 355$ MPa is the pile material's yield strength. With this the pile can take the maximum load if $M_\gamma < M_y$.

General and Local Buckling For bar buckling under compression and bending moment, the Germanischer Lloyd (2005) code prescribes the following formula:

$$\frac{N_d}{\kappa N_p} + \frac{\beta_m M_d}{M_p} + \Delta n \leq 1.0$$

where N_d is the design axial force, κ is the reduction factor for bar buckling, N_p is the plastic compression resistance, β_m is the moment coefficient for bar buckling, M_d is the design bending moment, $M_p = W_p f_y / \gamma_M$ is the plastic moment resistance with W_p being the plastic section modulus and $\gamma_M = 1.1$ the material safety factor. The following equations can be used to calculate each term. (For the meaning of symbols, please refer to the nomenclature as well as the code.)

$$\kappa = \frac{1}{\Phi + \sqrt{\Phi^2 - \overline{\lambda}^2}} \leq 1 \text{ for reduced slenderness } \overline{\lambda} > 0.2, \text{ and } \kappa = 1 \text{ for } \overline{\lambda} \leq 0.2 \tag{5.24}$$

where:

$$\Phi = 0.5 \cdot [1 + \alpha \cdot (\overline{\lambda} - 0.2) + \overline{\lambda}^2] \quad \overline{\lambda} = \frac{L_{PT}/r_P}{\sqrt{E_P/f_{yk}}} \quad \alpha = 0.21 \text{ (pile imperfection factor)}$$

$$W_p = \frac{D_P^3 - (D_P - 2t_P)^3}{6} \qquad \Delta n = 0.25\kappa\overline{\lambda}^2 \leq 0.1$$

Local (shell) buckling needs to be checked as well, and according to Germanischer Lloyd (2005) the following formula can be used

$$\left(\frac{\sigma_x}{\sigma_{xu}}\right)^{1.25} + \left(\frac{\sigma_\varphi}{\sigma_{\varphi u}}\right)^{1.25} \leq 1.0 \tag{5.25}$$

where σ_φ is the design circumferential (hoop) stress, $\sigma_{\varphi u}$ is the ultimate circumferential stress for shell buckling, σ_x is the design axial stress, and σ_{xu} is the ultimate axial stress for buckling. The following equations can be used to calculate each term. The ultimate shell buckling stress for axial stress is

$$\sigma_{xu} = \frac{\kappa_2 f_{yk}}{\gamma_m} \tag{5.26}$$

where $\kappa_2 = 1.0$ for $\overline{\lambda}_{sx} \leq 0.25$, $\kappa_2 = 1.233 - 0.933\overline{\lambda}_{sx}$ for $0.25 < \overline{\lambda}_{sx} \leq 1.0$, $\kappa_2 = 0.3/\overline{\lambda}_{sx}^{3}$ for $1.0 < \overline{\lambda}_{sx} \leq 1.5$, and $\kappa_2 = 0.2/\overline{\lambda}_{sx}^{2}$ for $\overline{\lambda}_{sx} \geq 1.5$, with $\overline{\lambda}_{sx} = \sqrt{f_{yk}/\sigma_{xi}}$ and σ_{xi} is the ideal buckling stress for axial stress:

$$\sigma_{xi} = 0.605 \cdot C_X E \frac{t_P}{R_P} \tag{5.27}$$

with

$$C_X = 1.0\frac{\sigma_{M_t}}{\sigma_x} + C_{XN}\frac{\sigma_N}{\sigma_x},$$

$$C_{XN} = 1 + 1.5\left(\frac{R_P}{L_P}\right)^2 \frac{t_P}{R_P} \quad \text{for} \quad \frac{L_P}{R_P} \leq 0.5\sqrt{\frac{R_P}{t_P}},$$

$$C_{XN} = 1 - \frac{0.4\frac{L_P}{R_P}\sqrt{\frac{t_P}{R_P}} - 0.2}{\eta} \geq 0.6 \quad \text{for} \quad \frac{L_P}{R_P} > 0.5\sqrt{\frac{R_P}{t_P}}$$

$\eta = 1$ for both ends simply supported, $\eta = 3$ for one end simply supported, one end clamped, $\eta = 6$ for both ends clamped. The safety factor can be calculated according to the following:

$$\gamma_m = 1.1 \quad \text{for} \quad \overline{\lambda}_{sx} \leq 0.25,$$

$$\gamma_m = 1.1\left(1 + 0.318\frac{\overline{\lambda}_{sx} - 0.25}{1.75}\right) \quad \text{for} \quad 0.25 < \overline{\lambda}_{sx} < 2.0$$

$$\gamma_m = 1.45 \quad \text{for} \quad \overline{\lambda}_{sx} \geq 2.0$$

The ultimate shell-buckling stress for circumferential stress is

$$\sigma_{\varphi u} = \frac{\kappa_1 f_{yk}}{\gamma_M} \tag{5.28}$$

where $\kappa_1 = 1.0$ for $\overline{\lambda}_{s\varphi} \leq 0.4$, $\kappa_1 = 1.274 - 0.686\overline{\lambda}_{s\varphi}$ for $0.4 < \overline{\lambda}_{s\varphi} < 1.2$, $\kappa_1 = 0.65/\overline{\lambda}_{s\varphi}^2$ for $\overline{\lambda}_{s\varphi} \geq 1.2$ with $\overline{\lambda}_{s\varphi} = \sqrt{f_{yk}/\sigma_{\varphi i}}$. The ideal buckling stress for circumferential stress is

$$\sigma_{\varphi i} = 0.92 \cdot C_\varphi \sqrt{\frac{R_P}{t_P}} \text{ for short to medium length piles } \frac{L_P}{R_P} \leq 1.63 C_\varphi \sqrt{\frac{R_P}{t_P}}, \text{ and}$$

$$\sigma_{\varphi i} = E_P \left(\frac{t_P}{R_P}\right)^2 \left[0.275 + 2.03 \left(\frac{C_\varphi}{\frac{L_P}{R_P}\sqrt{\frac{t_P}{R_P}}}\right)^4\right] \text{ for long piles } \frac{L_P}{R_P} \leq 6.63 C_\varphi \sqrt{\frac{R_P}{t_P}} \tag{5.29}$$

where $C_\varphi = 1.5$ for both ends clamped, $C_\varphi = 1.25$ for one end simply supported, one end clamped, $C_\varphi = 1.0$ for both ends simply supported, $C_\varphi = 0.6$ for one end clamped, one end free. The material safety factor is $\gamma_M = 1.1$.

5.3.3.4 Lateral Pile Capacity

For long-term deformation production, moment carrying capacity needs to be estimated. There are many methods to carry out such calculations: simplified method (hand-based method), standard method (Beam on Winkler Foundation – discussed later in this chapter) or finite element method. This section presents one simplified method based on Broms (1964) approach. For derivation and details, please refer to Chapter 7 of Poulos and Davis (1980).

Constant Soil Resistance with Depth In ground conditions where the soil resistance is assumed to be constant with depth (OCR soils), and where soil fails first (i.e. the pile does not fail through a plastic hinge formation), the ultimate capacity can be calculated using the following formulae.

$$g = 2(L_P + D_P + e) - \frac{\sqrt{16(L_P + D_P + e)^2 - 4(4L_P e - 2D_P e + 4D_P L_P - 2.5D_P^2 + 2L_P^2)}}{2} \tag{5.30}$$

$$f = \frac{F_R}{9 s_u D_P} \tag{5.31}$$

$$M_R = F_R(e + 1.5D_P + 0.5f) = 2.25 D_P g^2 s_u \tag{5.32}$$

$$L_P = 1.5D_P + f + g \tag{5.33}$$

where s_u is the undrained shear strength, e is the eccentricity of loading, M_R is the moment capacity of the pile, F_R is the horizontal load carrying capacity of the pile, D_P and L_P are the diameter and embedded length of the pile, respectively (Figure 5.16).

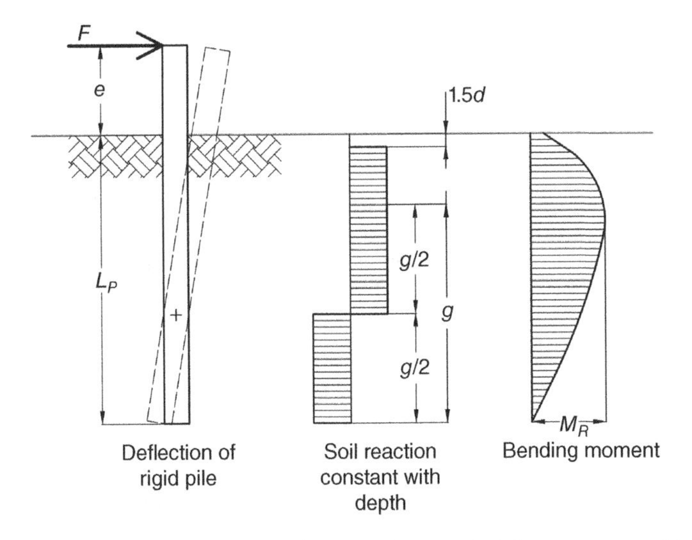

Figure 5.16 Lateral capacity for pile with ground stiffness constant with depth.

Linear Soil Resistance with Depth In ground conditions, where the soil resistance is assumed to increase linearly with depth (e.g. some cohesionless soils and lightly overconsolidated clay), the horizontal load and moment capacity of a piled foundation, assuming that the soil fails first (no plastic hinge is formed in the pile), is expressed using the following equations (Figure 5.17):

$$F_R = \frac{0.5\gamma D_P L_P^3 K_P}{e + L_P} = \frac{3}{2}\gamma' D_P K_P f^2 \tag{5.34}$$

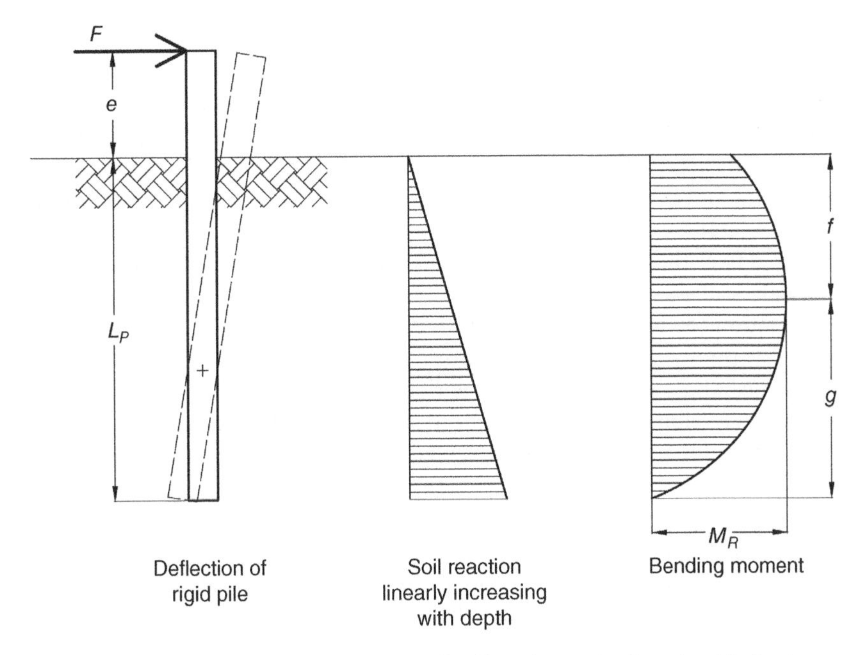

Figure 5.17 Lateral pile capacity with ground stiffness increasing linearly with depth.

$$K_P = \frac{1 + \sin \phi'}{1 - \sin \phi'} \tag{5.35}$$

$$M_R = F_R \left(e + \frac{2}{3} f \right) \tag{5.36}$$

$$f = 0.82 \sqrt{\frac{F_R}{D_P K_P \gamma}} \tag{5.37}$$

where γ' is the submerged unit weight of the cohesionless soil (assumed constant with depth); D_P and L_P are the pile diameter and embedded length, respectively; e is the load eccentricity ($e = M/F$); ϕ' is the effective angle of internal friction; and M_R and F_R are the moment and the horizontal load carrying capacity of the foundation, respectively. For derivation and details, refer to Chapter 7 of Poulos and Davis (1980).

5.4 Methods of Analysis for SLS, Natural Frequency Estimate, and FLS

There are different approaches to incorporate the effects of SSI. In the context of OWT design, this can be classified into three categories as shown in Table 5.2.

5.4.1 Simplified Method of Analysis

In a simplified method of analysis, the following steps may be adopted:

1. To find the loads on the foundation for various load combinations. For a monopile or a mono-caisson type foundation, this would mean obtaining vertical load, overturning moment (M_y), and horizontal load (F_x).
2. The second step is to choose a foundation and obtain the stiffness values, denoted by K_L, K_R, and K_{LR} in Figure 5.7. The stiffness can be obtained in a variety of ways. The simplest is based on closed-form solutions that need only a few parameters: pile dimensions, Young's modulus of soil at a depth of one pile diameter, and variation of soil stiffness with depth. Tables 5.3 and 5.4 provide closed-form solution for stiffness for rigid piles and flexible piles. Other methods such as Winkler-type solution (i.e. referred to as standard method and discussed in Section 5.4.3) or finite element method (referred to as the advanced method and discussed in Section 5.4.4) may also be used to obtain stiffness values. The foundation stiffness is required for two calculations: deformation (deflection ρ and rotation θ at mudline) and natural frequency estimation. A few points may be noted regarding these springs:
   (a) The properties and shape of the springs (load-deformation characteristics, i.e. lateral load-deflection or moment-rotation) should be such that the deformation is acceptable under the working load scenarios expected in the lifetime of the turbine.
   (b) The initial values of the springs (stiffness of the foundation) are necessary to compute the natural period of the structure using linear Eigen value analysis.
   (c) The values of the springs will also dictate the overall dynamic stability of the system due to its nonlinear nature. It must be mentioned that these springs are not

Table 5.2 Methods of SSI analysis.

	Simplified	Standard	Advanced
Details of the method	In this method, the foundation is replaced by a set of springs: K_L, K_R, and K_{LR}. Therefore, the model is similar to a beam supported on a set of non-linear springs. Figure 5.18 explains the method. If spreadsheets are developed, this method takes about 10 min to carry out the calculations.	This method is based on Beam on nonlinear Winkler foundations. The pile–soil interaction or the foundation soil interaction is modelled as a series of nonlinear springs. This method is not expensive in terms of computation and does not need specialist geotechnical knowledge.	This method is based on advanced numerical analysis and the possible methods are finite element, finite difference, discrete element. Finite element is most commonly used in practice and expertise is necessary. The readers are referred to Potts (2013) Rankine Lecture
What are the ground parameters required?	Two parameters are required: (a) ground profile i.e. the variation of the ground stiffness along the depth. (b) Soil stiffness at a depth of one diameter.	The parameters needed are stress–strain of soils (i.e. some form of soil shear tests preferably triaxial tests). Some formulations may need Subgrade modulus data and relative density of the soil. These are routine soil tests.	Soil parameters depending on soil models.
What this method can do?	This method can predict the natural frequency of the system using the formulation presented in Arany et al. (2016). The method can also predict the deformation of the pile head.	This method can be used to obtain foundation stiffness which can be then be used for structural/dynamic analysis. Bending moment and deflection profile along the length of the pile can also be obtained.	FE model can be used to obtain foundation stiffness, deflection, and moment in the pile. The method can also estimate the strain field in the soil in the deformed/mobilised zone.
Limitation of the method	Cannot produce a bending moment and deflection profile of the pile	The pile–soil interactions are modelled as a series of discrete springs. However, in reality, the springs are not independent.	This is very expensive and needs trained personnel.

Table 5.3 Stiffness formulae by different researchers for slender piles in various soil profiles.

Lateral stiffness K_L	Cross-coupling stiffness K_{LR}	Rotational stiffness K_R
Randolph (1981), slender piles, both for homogeneous and linear inhomogeneous soils		
$\frac{1.67E_{S0}D_P}{f_{(v_S)}}\left(\frac{E_{eq}}{E_{S0}}\right)^{0.14}$	$-\frac{0.3475E_{S0}D_P^2}{f_{(v_S)}}\left(\frac{E_{eq}}{E_{S0}}\right)^{0.42}$	$\frac{0.1975E_{S0}D_P^3}{f_{(v_S)}}\left(\frac{E_{eq}}{E_{S0}}\right)^{0.7}$
Pender (1993), slender piles, homogeneous soil		
$1.285E_{S0}D_P\left(\frac{E_{eq}}{E_{S0}}\right)^{0.188}$	$-0.3075E_{S0}D_P^2\left(\frac{E_{eq}}{E_{S0}}\right)^{0.47}$	$0.18125E_{S0}D_P^3\left(\frac{E_{eq}}{E_{S0}}\right)^{0.738}$
Pender (1993), slender piles, linear inhomogeneous soil		
$0.85E_{S0}D_P\left(\frac{E_{eq}}{E_{S0}}\right)^{0.29}$	$-0.24E_{S0}D_P^2\left(\frac{E_{eq}}{E_{S0}}\right)^{0.53}$	$0.15E_{S0}D_P^3\left(\frac{E_{eq}}{E_{S0}}\right)^{0.77}$
Pender (1993), slender piles, parabolic inhomogeneous soil		
$0.735E_{S0}D_P\left(\frac{E_{eq}}{E_{S0}}\right)^{0.33}$	$-0.27E_{S0}D_P^2\left(\frac{E_{eq}}{E_{S0}}\right)^{0.55}$	$0.1725E_{S0}D_P^3\left(\frac{E_{eq}}{E_{S0}}\right)^{0.776}$
Poulos and Davis (1980) following Barber (1953), slender pile, homogeneous soil		
$\frac{k_h D_P}{\beta}$	$-\frac{k_h D_P}{\beta^2}$	$\frac{k_h D_P}{2\beta^3}$
Poulos and Davis (1980) following Barber (1953), slender pile, linear inhomogeneous soil		
$1.074n_h^{\frac{3}{5}}(E_P I_P)^{\frac{2}{5}}$	$-0.99n_h^{\frac{2}{5}}(E_P I_P)^{\frac{3}{5}}$	$1.48n_h^{\frac{1}{5}}(E_P I_P)^{\frac{4}{5}}$
Gazetas (1984) and Eurocode 8 Part 5 (2003), slender pile, homogeneous soil		
$1.08D_P E_{S0}\left(\frac{E_{eq}}{E_{S0}}\right)^{0.21}$	$-0.22D_P^2 E_{S0}\left(\frac{E_{eq}}{E_{S0}}\right)^{0.50}$	$0.16D_P^3 E_{S0}\left(\frac{E_{eq}}{E_{S0}}\right)^{0.75}$

Gazetas (1984) and Eurocode 8 Part 5 (2003), slender pile, linear inhomogeneous soil		
$0.60D_P E_{S0}\left(\frac{E_{eq}}{E_{S0}}\right)^{0.35}$	$-0.17D_P^2 E_{S0}\left(\frac{E_{eq}}{E_{S0}}\right)^{0.60}$	$0.14D_P^3 E_{S0}\left(\frac{E_{eq}}{E_{S0}}\right)^{0.80}$
Gazetas (1984) and Eurocode 8 Part 5 (2003), slender pile, parabolic inhomogeneous soil		
$0.79D_P E_{S0}\left(\frac{E_{eq}}{E_{S0}}\right)^{0.28}$	$-0.24D_P^2 E_{S0}\left(\frac{E_{eq}}{E_{S0}}\right)^{0.53}$	$0.15D_P^3 E_{S0}\left(\frac{E_{eq}}{E_{S0}}\right)^{0.77}$
Shadlou and Bhattacharya (2016), slender pile, homogeneous soil		
$\frac{1.45E_{S0}D_P}{f_{(v_s)}}\left(\frac{E_{eq}}{E_{s0}}\right)^{0.186}$	$-\frac{0.30E_{S0}D_P^2}{f_{(v_s)}}\left(\frac{E_{eq}}{E_{s0}}\right)^{0.50}$	$\frac{0.18E_{S0}D_P^3}{f_{(v_s)}}\left(\frac{E_{eq}}{E_{S0}}\right)^{0.73}$
Shadlou and Bhattacharya (2016), slender pile, linear inhomogeneous soil		
$\frac{0.79E_{S0}D_P}{f_{(v_s)}}\left(\frac{E_{eq}}{E_{s0}}\right)^{0.34}$	$-\frac{0.26E_{S0}D_P^2}{f_{(v_s)}}\left(\frac{E_{eq}}{E_{S0}}\right)^{0.567}$	$\frac{0.17E_{S0}D_P^3}{f_{(v_s)}}\left(\frac{E_{eq}}{E_{S0}}\right)^{0.78}$
Shadlou and Bhattacharya (2016), slender pile, parabolic inhomogeneous soil		
$\frac{1.02E_{S0}D_P}{f_{(v_s)}}\left(\frac{E_{eq}}{E_{S0}}\right)^{0.27}$	$-\frac{0.29E_{S0}D_P^2}{f_{(v_s)}}\left(\frac{E_{eq}}{E_{S0}}\right)^{0.52}$	$\frac{0.17E_{S0}D_P^3}{f_{(v_s)}}\left(\frac{E_{eq}}{E_{S0}}\right)^{0.76}$

Parameter definitions:

$$E_{eq} = \frac{E_P I_P}{\frac{D_P^4 \pi}{64}}$$

$$f_{(v_S)} = \frac{1 + v_S}{1 + 0.75v_S}$$ for Randolph (1981) and

$$f_{(v_s)} = 1 + |v_s - 0.25|$$ for Shadlou and Bhattacharya (2016)

$$\beta = \sqrt[4]{\frac{k_h D_P}{E_P I_P}}$$

Table 5.4 Stiffness formulae by different researchers for rigid piles in various soil profiles.

K_L	K_{LR}	K_R
Poulos and Davis (1980) following Barber (1953), rigid pile, homogeneous soil ($n = 0$)		
$k_h D_P L_P$	$-\frac{k_h D_P L_P^2}{2}$	$\frac{k_h D_P L_P^3}{3}$
Poulos and Davis (1980) following Barber (1953), rigid pile, linear inhomogeneous soil ($n = 1$)		
$\frac{1}{2} L_P^2 n_h$	$-\frac{1}{3} L_P^3 n_h$	$\frac{1}{4} L_P^4 n_h$
Carter and Kulhawy (1992), rigid pile, rock		
$\frac{3.15 G^* D_P^{\frac{2}{3}} L_P^{\frac{1}{3}}}{1 - 0.28\left(\frac{2L_P}{D_P}\right)^{\frac{1}{4}}}$	$-\frac{2 G^* D_P^{\frac{7}{8}} L_P^{\frac{9}{8}}}{1 - 0.28\left(\frac{2L_P}{D_P}\right)^{\frac{1}{4}}}$	$-\frac{4 G^* D_P^{\frac{4}{3}} L_P^{\frac{5}{3}}}{1 - 0.28\left(\frac{2L_P}{D_P}\right)^{\frac{1}{4}}}$
Shadlou and Bhattacharya (2016), rigid pile, homogeneous soil ($n = 0$)		
$\frac{3.2 E_{SO} D_P}{f_{(v_s)}} \left(\frac{L_P}{D_P}\right)^{0.62}$	$-\frac{1.7 E_{SO} D_P^2}{f_{(v_s)}} \left(\frac{L_P}{D_P}\right)^{1.56}$	$\frac{1.65 E_{SO} D_P^3}{f_{(v_s)}} \left(\frac{L_P}{D_P}\right)^{2.5}$
Shadlou and Bhattacharya (2016), rigid pile, linear inhomogeneous soil ($n = 1$)		
$\frac{2.35 E_{SO} D_P}{f_{(v_s)}} \left(\frac{L_P}{D_P}\right)^{1.53}$	$-\frac{1.775 E_{SO} D_P^2}{f_{(v_s)}} \left(\frac{L_P}{D_P}\right)^{2.5}$	$\frac{1.58 E_{SO} D_P^3}{f_{(v_s)}} \left(\frac{L_P}{D_P}\right)^{3.45}$
Shadlou and Bhattacharya (2016), rigid pile, parabolic inhomogeneous soil ($n = 1/2$)		
$\frac{2.66 E_{SO} D_P}{f_{(v_s)}} \left(\frac{L_P}{D_P}\right)^{1.07}$	$-\frac{1.8 E_{SO} D_P^2}{f_{(v_s)}} \left(\frac{L_P}{D_P}\right)^{2.0}$	$\frac{1.63 E_{SO} D_P^3}{f_{(v_s)}} \left(\frac{L_P}{D_P}\right)^{3.0}$
Parameter definitions:		
$E_{eq} = \frac{E_P I_P}{\frac{D_P^4 \pi}{64}}$	$f_{(v_s)} = 1 + \lvert v_s - 0.25 \rvert$	$\beta = \sqrt[4]{\frac{k_h D_P}{E_P I_P}}$

only frequency dependent but also change with cycles of loading due to dynamic soil–structure interaction.

3. Once the stiffness values K_L, K_R, and K_{LR} are estimated, deformations in the foundation can be estimated using Eq. (5.38) assuming linearity in load deformation relationship:

$$\begin{bmatrix} F_x \\ M_y \end{bmatrix} = \begin{bmatrix} K_L & K_{LR} \\ K_{LR} & K_R \end{bmatrix} \begin{bmatrix} \rho \\ \theta \end{bmatrix} \tag{5.38}$$

where F_x is the lateral force in the direction of the x-axis as defined in Figure 5.18, M_y is the fore–aft overturning moment (around the y-axis), K_L is the lateral spring, K_R is the rotational spring, K_{LR} is the cross-coupling spring, ρ is the displacement in the x direction, and $\theta = \partial\rho/\partial z$ is the slope of the deflection (tilt or rotation).

The deformations can then be easily expressed using (Figure 5.18)

$$\rho = \frac{K_R}{K_L K_R - K_{LR}^2} F_x - \frac{K_{LR}}{K_L K_R - K_{LR}^2} M_y \tag{5.39}$$

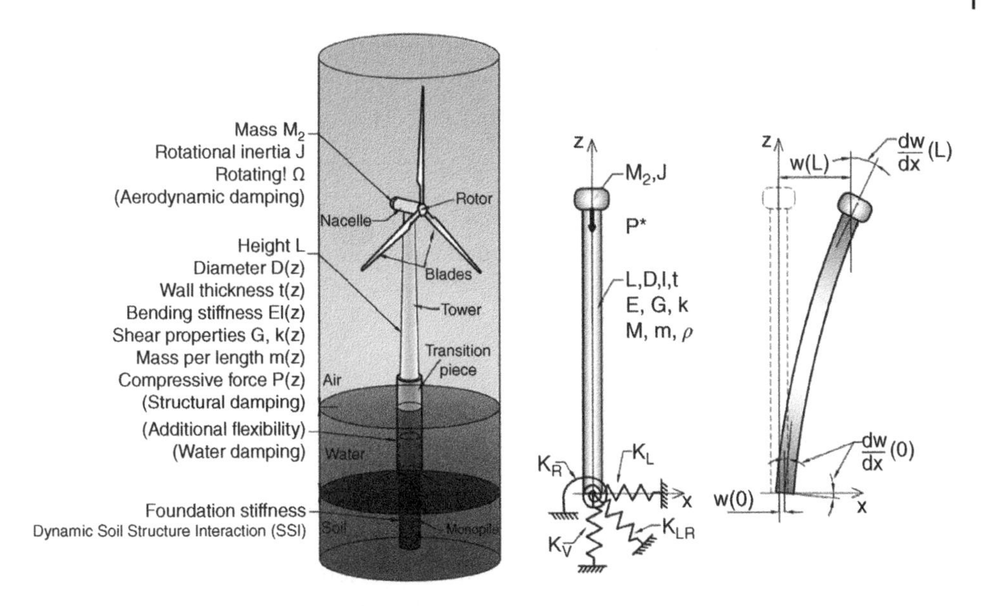

Figure 5.18 Simplified model for SSI analysis.

$$\theta = -\frac{K_{LR}}{K_L K_R - K_{LR}^2} F_x + \frac{K_L}{K_L K_R - K_{LR}^2} M_y \tag{5.40}$$

4. The natural frequency of the system shown in Figure 5.18 can be estimated following method developed in Arany et al. (2015b, 2017). This simplified methodology builds on the simple cantilever beam formula to estimate the natural frequency of the tower, and then applies modifying coefficients to take into account the flexibility of the foundation and the substructure. This is expressed as:

$$f_0 = C_L C_R C_S f_{FB} \tag{5.41}$$

where C_L and C_R are the lateral and rotational foundation flexibility coefficients, C_S is the substructure flexibility coefficient and f_{FB} is the fixed-base (cantilever) natural frequency of the tower. The readers are referred to Appendices.

The fixed-base natural frequency of the tower is expressed simply with the equivalent stiffness k_0 and equivalent mass m_0 of the first mode of vibration as

$$f_{FB} = \frac{1}{2\pi}\sqrt{\frac{k_0}{m_0}} = \frac{1}{2\pi}\sqrt{\frac{3E_T I_T}{L_T^3\left(m_{RNA} + \frac{33}{140} m_T\right)}} \tag{5.42}$$

where E_T is the Young's modulus of the tower material, I_T is the average area moment of inertia of the tower, m_T is the mass of the tower, m_{RNA} is the mass of the rotor-nacelle assembly and L_T is the length of the tower. The average area moment of inertia is calculated as

$$I_T = \frac{1}{16} t_T \pi (D_b^3 + D_t^3) \tag{5.43}$$

where D_b is the tower bottom diameter, D_t is the tower top diameter. The average wall thickness and the average tower diameter are given by Eq. (5.44).

$$t_T = \frac{m_T}{\rho_T L_T D_T \pi} D_T = \frac{D_b + D_t}{2} \tag{5.44}$$

where ρ_T is the density of the tower material (steel). The coefficients C_L and C_R are expressed in terms of the nondimensional foundation stiffness values:

$$\eta_L = \frac{K_L L_T^3}{EI_\eta} \eta_{LR} = \frac{K_{LR} L_T^2}{EI_\eta} \eta_R = \frac{K_R L_T}{EI_\eta} \tag{5.45}$$

where K_L, K_{LR}, K_R are the stiffness parameters, EI_η is the equivalent bending stiffness of the tower calculated as

$$EI_\eta = E_T I_T \cdot f(q) \quad where \quad q = \frac{D_b}{D_t} f(q) = \frac{1}{3} \cdot \frac{2q^2(q-1)^3}{2q^2 \ln q - 3q^2 + 4q - 1} \tag{5.46}$$

The derivation of the above equations are provided in Appendices A to C.

Using the calculated nondimensional stiffness values, the foundation flexibility coefficients are given as

$$C_R(\eta_L, \eta_R, \eta_{LR}) = 1 - \frac{1}{1 + a\left(\eta_R - \frac{\eta_{LR}^2}{\eta_L}\right)} C_L(\eta_L, \eta_R, \eta_{LR}) = 1 - \frac{1}{1 + b\left(\eta_L - \frac{\eta_{LR}^2}{\eta_R}\right)} \tag{5.47}$$

where $a = 0.5$ and $b = 0.6$ are empirical coefficients (Arany et al. 2016, 2017).

The substructure flexibility coefficient is calculated by assuming that the monopile goes up to the bottom of the tower. The distance between the mudline and the bottom of the tower is L_S, and $E_P I_P$ is the bending stiffness of the monopile. The foundation flexibility is expressed in terms of two dimensionless parameters, the bending stiffness ratio χ and the length ratio ψ

$$\chi = \frac{E_T I_T}{E_P I_P} \quad \psi = \frac{L_S}{L_T} \text{ and } C_S = \sqrt{\frac{1}{1 + (1+\psi)^3 \chi - \chi}} \tag{5.48}$$

A spreadsheet can be easily used to carry out the calculations. A solved example is carried out in Chapter 6. This method is calibrated for 10 offshore wind turbines and can be found in Arany et al. (2016).

ASIDE: Note on the Natural Frequency Estimates

It is of key importance to predict the natural frequency of the offshore wind turbine–support structure–foundation system because both under- and overprediction of the natural frequency may be unconservative. This is because the structure is excited in a wide frequency band from wind turbulence, waves, aerodynamic, and mass imbalance loads at the rotational frequency range (1P) and blade passage and rotational sampling loads at the blade passing frequency (2P or 3P).

5.4.2 Methodology for Fatigue Life Estimation

The analysis of fatigue life of the substructure has to be carried out, which is typically done following DNV-RP-C203 – 'Fatigue design of offshore steel structures' (DNV 2005). This section is aimed at providing a simple methodology for the conceptual design of monopiles, and therefore fatigue life issues related to other components of the substructure (e.g. transition piece, grouted connection, J-tubes, etc.) are naturally omitted. In terms of fatigue analysis of the structural steel of the pile wall under bending moment, one has to calculate the stress levels caused by various load cases. The material factor $\gamma_M = 1.1$ and load factor $\gamma_L = 1.0$ can be used.

With these, the maximum stress levels σ_m caused by the load cases can be calculated as

$$\sigma_m = \gamma_L M_{max} \frac{D_P}{2I_P} \tag{5.49}$$

where M_{max} is the maximum bending moment that occurs in the given load case, D_P and I_P are the pile diameter and area moment of inertia, respectively. The maximum cyclic stress amplitude is given as

$$\sigma_c = \gamma_L \frac{(M_{max} - M_{min})}{2} \frac{D_P}{2I_P} \tag{5.50}$$

where M_{min} is the lowest bending moment occurring in each of the load cases.

In typical practical cases, the fatigue analysis of the structural steel of a monopile results in sufficient fatigue life with a high margin. However, the welds of flush ground monopiles are more prone to fatigue-type failure, as fatigue crack initiation typically occurs around the welds before it would occur in the structural steel. The fatigue analysis of welds of flush ground monopiles is carried out using the C1 and D classes defined in DNV (2005). A thickness correction factor has to be applied as monopile welds are almost always thicker than 25 mm. These curves build on tests carried out specifically for the requirements of the offshore oil and gas (O&G) industry. Currently, research and testing are ongoing in the SLIC Joint Industry Project (Brennan and Tavares 2014) to develop S–N curves representative of the load regime, geometry, materials, environmental conditions, and manufacturing procedures of the offshore wind industry. More detailed fatigue analyses through, e.g. finite element analysis (FEA) may need to be carried out once a more detailed design is available, as fatigue-type failure is expected to occur in weak points in the structure (e.g. holes, welds, and joints) where stress concentration is expected and crack initiation is more likely. Furthermore, a crack propagation approach is generally more suitable for detailed fatigue design and simple S–N curve fatigue analyses are often not satisfactory to predict the fatigue life of certain structural details.

5.4.3 Closed-Form Solution for Obtaining Foundation Stiffness of Monopiles and Caissons

In a simplified three-springs approach (see Figure 5.8), foundation stiffness needs to be calculated. They are: K_V (vertical stiffness), K_L (lateral stiffness), K_R (rocking stiffness), and K_{LR} (cross-coupling). It is required to note that the vertical stiffness is not required for simplified calculations as the structure is very stiff vertically. However, values of K_V

are required for rocking modes of vibration. In a simplified approach or a closed-form solution approach, the input required to obtain K_L, K_R, and K_R are:

1. Pile dimensions.
2. Ground profile, i.e. soil stiffness variation with depth. In line with Eurocode practice (EC8, Part 5), three types of profiles are considered. They are: (i) constant stiffness with depth which is typical of overconsolidated clay profile; (ii) linearly varying stiffness with depth which is typical of normally consolidated (NC) clay; and (iii) soil stiffness varying with square root of depth, which is typical of sandy soil. Figure 5.8 plots the variation.
3. Soil stiffness at a depth of one pile diameter.

Alternatively, some formulations define the soil with the modulus of subgrade reaction k_h or the coefficient of subgrade reaction n_h (the rate of increase of k_h with depth).

Note: Terminologies such as subgrade modulus, coefficient can be confusing. Codes of practices, textbooks, and software require different types of inputs. It is therefore recommended to always write units.

5.4.3.1 Closed-Form Solution for Piles (Rigid Piles or Monopiles)

The first step in the calculation procedure of the pile-head stiffness is the classification of pile behaviour, i.e. whether the monopile will behave as a long flexible pile or a short rigid pile, and then using the appropriate relations to obtain K_L, K_R, and K_{LR}.

Rigid piles are short enough to undergo rigid body rotation in the soil under operational loads, instead of deflecting like a clamped beam. Slender piles, on the other hand, undergo deflection under operating loads and fail typically through the formation of a plastic hinge, and the pile toe does not 'feel' the effects of the loading at the mudline and the pile can be considered *infinitely long*. Formulae for determining whether a pile can be considered slender or rigid are available in the literature. Calibrated method for natural frequency estimate by Arany et al. (2016) suggests the following.

Flexible Behaviour or Rigid? Based on the elastic continuum approach proposed by Randolph (1981), the critical pile length can be expressed through the necessary ratio of pile length L_P to pile diameter D_P in terms of the modified shear modulus G^* of the soil and the equivalent Young's modulus of the pile (E_{eq}). With this the pile length is calculated from the diameter as

$$L_P \geq D_P \left(\frac{E_{eq}}{G^*} \right)^{\frac{2}{7}} \tag{5.51}$$

where $E_{eq} = E_P I_P / \left(\frac{D_P^4 \pi}{64} \right)$, $G^* = G_S \left(1 + \frac{3}{4} \nu_S \right)$ with G_S being the shear modulus of the soil averaged between the mudline and the pile embedment length, $E_P I_P$ is the pile's bending stiffness.

Carter and Kulhawy (1992) present an expression to determine whether the pile can be considered rigid using a similar approach to that of Randolph (1981) whereby the pile is rigid if

$$L_P \leq 0.05 D_P \left(\frac{E_{eq}}{G^*} \right)^{\frac{1}{2}} \tag{5.52}$$

Another approach is shown in Poulos and Davis (1980) following Barber (1953) using the soil's modulus of subgrade reaction k_h. In cohesive soils (which applies to the over-consolidated clayey ground profile), the modulus of subgrade reaction k_h can be considered constant with depth. The pile can be considered slender (infinitely long) if

$$L_P > 2.5\left(\frac{E_P I_P}{k_h D_P}\right)^{\frac{1}{4}} \tag{5.53}$$

and the pile can be considered rigid if

$$L_P < 1.5\left(\frac{E_P I_P}{k_h D_P}\right)^{\frac{1}{4}} \tag{5.54}$$

In normally consolidated clay or cohesionless soils (sand), the modulus of subgrade reaction approximately increases linearly with depth, according to $k_h = n_h(z/D_P)$. In such soils the pile can be considered slender if

$$L_P > 4.0\left(\frac{E_P I_P}{n_h}\right)^{\frac{1}{5}} \tag{5.55}$$

and the pile can be considered rigid if

$$L_P < 2.0\left(\frac{E_P I_P}{n_h}\right)^{\frac{1}{5}} \tag{5.56}$$

These formulae can be used to obtain the necessary length as a function of pile diameter and soil stiffness.

Other methods to find critical length can be found in Aissa et al. (2017).

In the simplified procedure to obtain foundation stiffness, two parameters are required to define the ground (soil stiffness at $1D_P$ below mudline denoted by E_{S0} and the stiffness profile, i.e. variation with depth). The stiffness profile is expressed mathematically as

$$E_S(z) = E_{S0}\left(\frac{|z|}{D_P}\right)^n \tag{5.57}$$

where homogeneous, linear inhomogeneous, and square root inhomogeneous profiles are given by $n = 0$, $n = 1$, and $n = 1/2$, respectively (see Figure 5.19).

Analytical solutions are rarely available from a subgrade approach for general cases, but simplified expressions are available for rigid and slender piles and available in (Poulos and Davis 1980). Various approaches have been developed to correlate foundation loads (horizontal load F_x and bending moment M_y) to pile-head deflection ρ and rotation θ. These expressions can be easily transformed into a matrix form of the load response in terms of three springs (K_L, K_{LR}, K_R). Some of the most common methods are found in Poulos and Davis (1980) following Barber (1953) for both rigid and slender piles; Gazetas (1984) also featured in Eurocode 8 Part 5 (European Committee for Standardization 2003) developed for slender piles; Randolph (1981) developed for slender piles in both homogeneous and linear inhomogeneous soils; Pender (1993) developed for slender piles; Carter and Kulhawy (1992) for rigid piles in rock; Higgins and Basu (2011) for rigid piles; and Shadlou and Bhattacharya (2016) for both rigid and slender piles. The

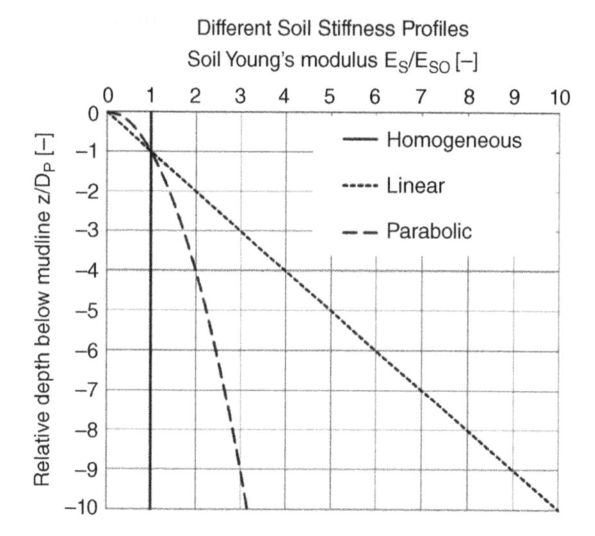

Figure 5.19 Homogeneous, linear, and parabolic soil stiffness profiles.

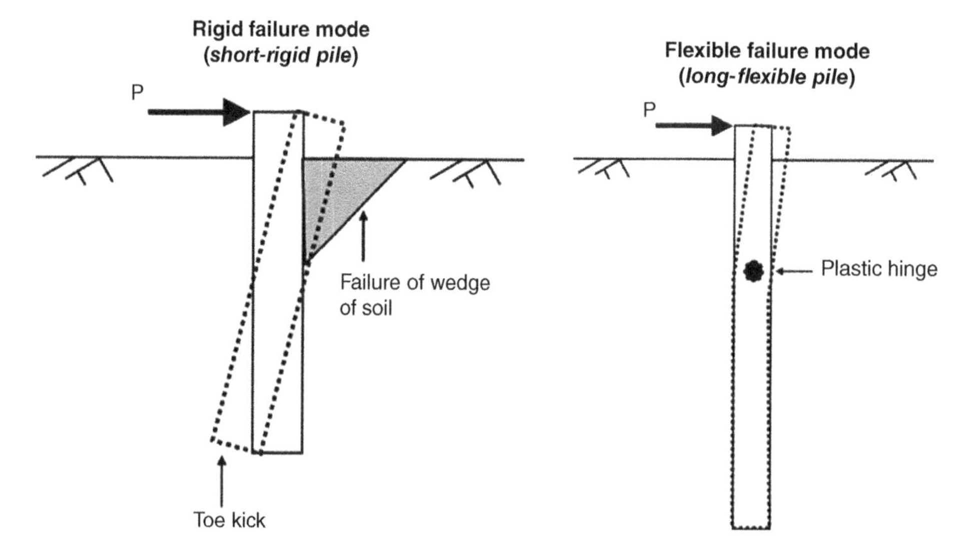

Figure 5.20 Distinguishing failure between short-rigid pile and long flexible pile.

formulae for the foundation stiffness are summarised in Table 5.3 for slender piles and Table 5.4 for rigid piles (Figure 5.20).

ASIDE

The main difference in the formulation for rigid and flexible are as follows. For rigid piles, the formulation has the term (L/D), i.e. slenderness ratio/aspect and soil stiffness (E_{SO}). On the other hand, for flexible piles, the formulation depends on pile–soil relative stiffness (E_{eq}/E_s).

Table 5.5 Impedance functions for section caissons exhibiting rigid behaviour L/D > 2. See Figure 5.21.

Ground profile See Figure 5.19 for definition	$\frac{K_L}{DE_{SO}f(v_s)}$	$\frac{K_{LR}}{D^2E_{SO}f(v_s)}$	$\frac{K_R}{D^3E_{SO}f(v_s)}$
Homogeneous	$3.2\left(\frac{L}{D}\right)^{0.62}$	$-1.8\left(\frac{L}{D}\right)^{1.56}$	$1.65\left(\frac{L}{D}\right)^{2.5}$
Parabolic	$2.65\left(\frac{L}{D}\right)^{1.07}$	$-1.8\left(\frac{L}{D}\right)^{2}$	$1.63\left(\frac{L}{D}\right)^{3}$
Linear	$2.35\left(\frac{L}{D}\right)^{1.53}$	$-1.8\left(\frac{L}{D}\right)^{2..5}$	$1.59\left(\frac{L}{D}\right)^{3.45}$

5.4.3.2 Closed-Form Solutions for Suction Caissons

Shadlou and Bhattacharya (2016) presented impedance functions for rigid deep foundations in homogeneous, parabolic, and linear ground profiles, keeping in mind the application for offshore wind turbines. Table 5.5 presents solutions that are applicable for L/D greater than 2. The definition of L and D are given in Figure 5.21. Jabli et al. (2018) presented solutions for stiffness of suction caissons having rigid skirted for $0.5 < L/D < 2$ in three types of ground profiles for Table 5.5, $(L/D > 2)$ see Table 5.6 for Table 5.6 for $0.5 < L/D < 2$. An example problem is carried out in Chapter 6 to show the application. The variation of Poisson's ratio may be noted for two types of aspect ratio (Figure 5.21).

$$f(v_s) = 1 + 0.6|0.25 - v_s|$$

$$f(v_s) = 1.1 \times \left(0.096\left(\frac{L}{D}\right) + 0.6\right) v_s^2 - 0.7v_s + 1.06$$

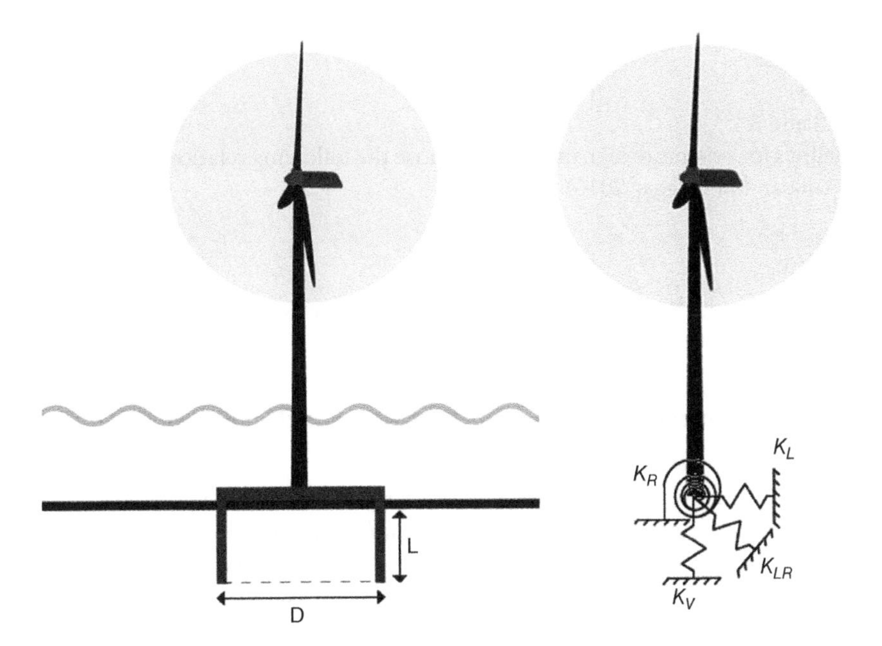

Figure 5.21 Definition of the suction caissons.

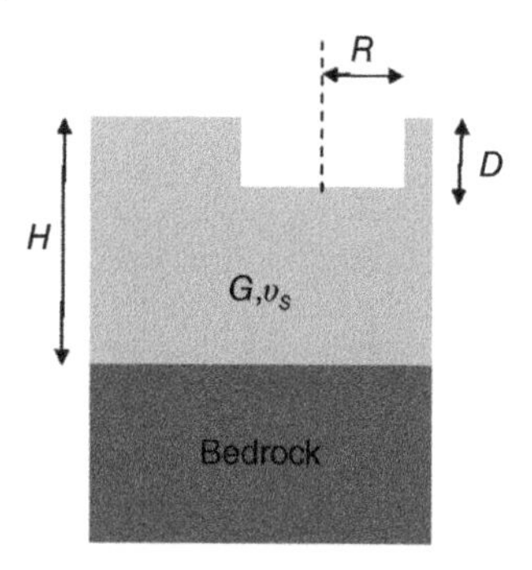

Figure 5.22 Figure defining the terms in Eq. (5.58).

Note: It must be mentioned that these solutions can be used for preliminary sizing of caisson at feasibility and tender design stage. Once the size is optimised and the project is finalised, further optimization and detailed analysis should be carried out using conventional methods.

5.4.3.3 Vertical Stiffness of Foundations (K_V)

Rigid Circular Embedded Footings The (DNV 2002) provides guidance for rigid embedded shallow foundations over a bedrock layer and may be used as a preliminary estimate for suction caissons (Figure 5.22).

$$k_v = \frac{4G_S R}{1-\upsilon_s}\left(1+1.28\frac{R}{H}\right)\left(1+\frac{D}{2R}\right)\times\left[1+\left(0.85-\frac{0.28D}{R}\right)\frac{D/H}{1-D/H}\right] \quad (5.58)$$

Vertical Stiffness of Piles Fleming et al. (1992) suggested the following for embedded piles:
Shaft friction only**

$$k_v = \frac{2\pi L_p G_s}{\zeta} \quad (5.59)$$

ζ is between 3 and 5

LRFD guidelines for seismic design of bridge propose the following relation for vertical stiffness (Sharma and El Naggar 2015):

$$k_v = 1.25\frac{E_P A}{L_P} \quad (5.60)$$

Note: In practice, t-z (axial load transfer analysis) type of analysis or calibrated FEA can be carried out to obtain the axial stiffness of the piles.

5.4.4 Standard Method of Analysis (Beam on Nonlinear Winkler Foundation) or *p-y* Method

To model laterally loaded piles, practicing engineers often use a simplified method normally referred to as *Beam-on-Nonlinear-Winkler-Foundation* BNWF) following Winkler (1867a,b) and Hetényi (1946). This is often known as *p-y method* and the main hypothesis is that the soil reaction *p*, exerted by the soil at a certain elevation on the pile shaft, is proportional to the relative pile-soil deflection, *y*. In offshore O&G industry, *p-y* method is employed to find out pile-head deformations (deflection and rotation) and foundation stiffness. The approach can be found in API (2005) and also suggested in DNV (2014). Originally, it was developed by Matlock (1970a,b); Reese et al. (1975); O'Neill and Murchinson (1983). The basis of this methodology is the Winkler approach (Winkler 1867a,b) whereby the pile–soil interaction is modelled as independent springs along the length of the pile. The well-known limitation of this method is the independent nature of these springs. However, it has been successfully used in O&G industry for more than 40 years.

According to the BNWF method, the pile is modelled by means of consecutive beam-column elements, whereas the LPSI (Lateral Pile Soil Interaction) is modelled through nonlinear springs attached to nodal points between two consecutive elements; see Figure 5.23. Each spring is defined by means of a nonlinear relationship between soil reaction per unit length of the pile p and corresponding relative soil–pile horizontal displacement, y. The coefficient of proportionality between p and y is the modulus of subgrade reaction k, with dimension of pressure divided by length. This relationship is normally referred to as *p-y* curve or soil-reaction curve. Intuitively, *p-y* curves depend on the soil and the pile diameter as it is effectively bearing capacity problem in the lateral direction whereby the pile section pushes the soil. Figure 5.23 shows a BNWF model for two types of *p-y* curves (two extremes), and it is important to highlight the importance of the shape of these curves.

1. The first case (5.23a) is representative of strain-softening behaviour. This type of behaviour is typical of most soils, i.e. sand or clay.

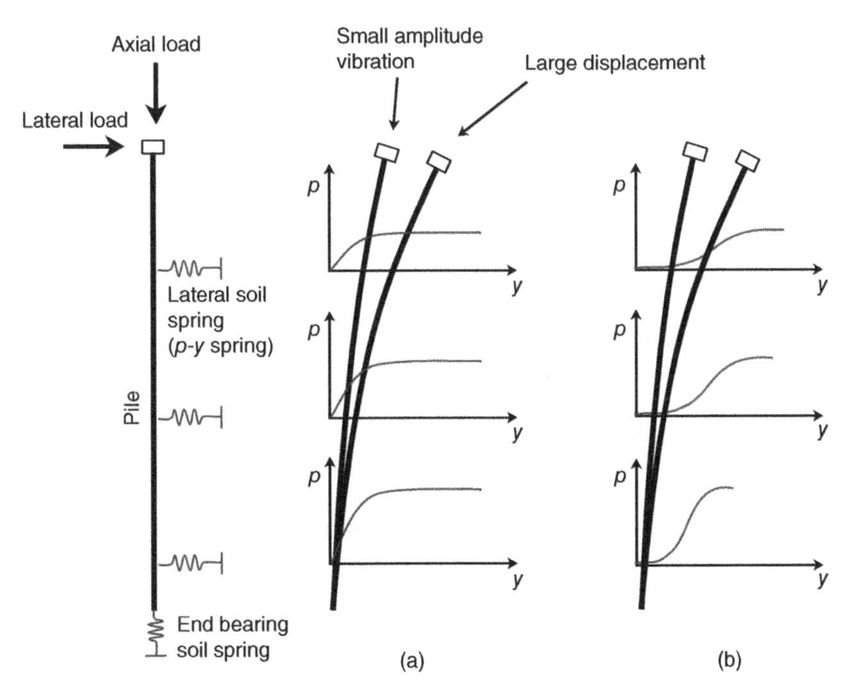

Figure 5.23 Winkler model showing pile response resulting from two types of *p-y* curves: (a) strain-softening response soils; (b) strain-hardening response.

2. The second case is representative of strain-hardening behaviour exhibited by post-liquefied soils. Once soil liquefies (i.e. effective stress is zero), the strength and stiffness reduce to zero and the *p-y* curve is aligned along the *y*-axis – the micromechanics being the soil grain particles are not in contact. However, with pile–soil relative displacement, the grains start to interlock and we note a strain-hardening behaviour.

This directly obtained results of the *p-y* model or its derivative is used for two types of calculations for offshore wind turbines:

1. Natural frequency estimate (i.e. eigen value problem which by definition is linear) requires stiffness of the pile for small amplitude vibrations. In this case, the behaviour that is of interest is the stiffness of the pile at very small pile displacement, which will lead to stiffness of the soil at very small strains. This is depicted in Figure 5.23a,b. Therefore, small strain stiffness measurement is very important and the readers are referred to the discussion in Chapter 4.
2. On the other hand, if the behaviour of the pile is required to be modelled for storm loading or ultimate behaviour, the behaviour of the soil at large strain is of interest.

Figure 5.23 schematically illustrates the effects of different shapes of *p-y* curve. For concave-downward strain-softening *p-y* curve illustrated in Figure 5.23a, it can be noted that when the relative soil–pile displacement is small, the resistance experienced by the pile depends on the initial stiffness of the soil and corresponding value of deflection. For large displacement, however, the resistance offered by the adjacent soil is governed by the ultimate strength of the soil.

In contrast, if the shape of the *p-y* curve is concave-upward, i.e. strain-hardening, the pile response is much more complex and may be significantly different from that described above. In seismic zones, if the top layer liquefies, owing to the practically zero stiffness mobilised at small displacements, the soil may offer minimal opposition to any lateral movement of the pile. This will result in enhanced P-delta moment and in extreme scenarios, buckling mode of failure of the pile. On the other hand, the higher stiffness and strength mobilised at larger displacements may prevent a complete collapse of the structure.

5.4.4.1 Advantage of *p-y* Method, and Why This Method Works

Despite its limitations of discrete nature, i.e. the springs work independently, the BNWF method is extensively used in practice because of its mathematical convenience and ability to incorporate nonlinearity of the soil and ground stratification. For example, in the closed-form method or solutions presented in Section 5.4.3 and Tables 5.5 and 5.6, only three types of profile could be modelled. On the other hand, using the BNWF method, any type of stratification could be modelled. The validity of BNWF approach is based on the assumed similarity between two mechanical system responses:

1. Load-deformation response of the pile, which takes into account the overall macro behaviour of the soil–pile system.
2. Stress–strain response of the adjacent soil being sheared as the pile moves laterally. This is related to the micro behaviour of the deforming material.

Table 5.6 Impedance functions for shallow-skirted foundations exhibiting rigid behaviour 0.5 < L/D < 2.

Ground profile	$\frac{K_L}{DE_{SO}f(v_s)}$	$\frac{K_{LR}}{D^2E_{SO}f(v_s)}$	$\frac{K_R}{D^3E_{SO}f(v_s)}$
Homogeneous	$2.91\left(\frac{L}{D}\right)^{0.56}$	$-1.87\left(\frac{L}{D}\right)^{1.47}$	$2.7\left(\frac{L}{D}\right)^{1.92}$
Parabolic	$2.7\left(\frac{L}{D}\right)^{0.96}$	$-1.99\left(\frac{L}{D}\right)^{1.89}$	$2.54\left(\frac{L}{D}\right)^{2.44}$
Linear	$2.53\left(\frac{L}{D}\right)^{1.33}$	$-2.02\left(\frac{L}{D}\right)^{2.29}$	$2.46\left(\frac{L}{D}\right)^{2.9}$

The value of $f(v_s)$ is given by the equation in Section 5.4.3.2.

In theory, the transformation from micro (stress–strain of the soil) to macro (*p-y* curves) can be made by applying appropriate scaling factors, whereby stress is converted into equivalent soil reaction, *p*; and strain is converted into equivalent relative pile-soil displacement (*y*). Dash (2010), Bouzid et al. (2013) demonstrated that appropriate scaling factors can be derived from the so-called mobilisable strength design (MSD) method. Essentially, any stress–strain curve of a soil can be used to construct a *p-y* curve through scaling and this method is gaining popularity. Lombardi et al. (2017) and Dash et al. (2017) developed *p-y* curves of liquefied soil from the stress–strain data of liquefied soil based on the scaling method developed by Bouzid et al. (2013). This method is discussed later in this chapter and is known as *scaling method* (Figure 5.24).

5.4.4.2 API Recommended *p-y* Curves for Standard Soils

The *p-y* curves are constructed by means of empirical relationships and were originally developed in the 1970–1980s from a relatively limited number of full-scale tests carried out on flexible steel piles (Matlock 1970a,b; Reese et al. 1974; Reese et al. 1975; O'Neill and Murchison 1983). Table 5.7 provides details of the reference of the research on which API *p-y* curves are based.

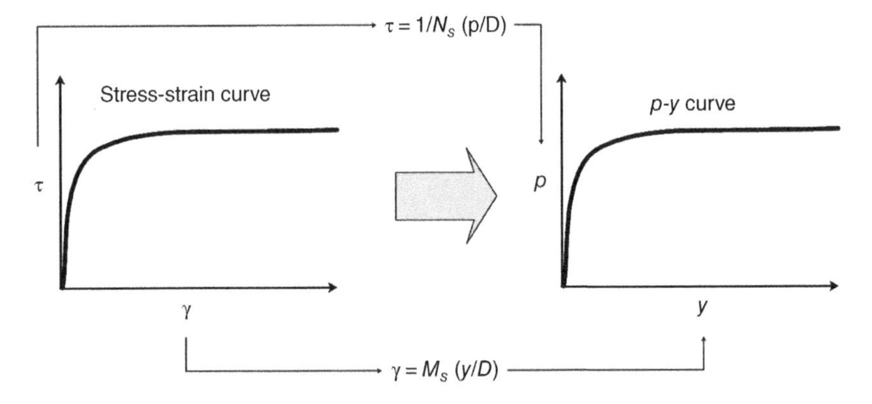

Figure 5.24 Scaling method for obtaining *p-y* curve from stress–strain behaviour. Note M_S and N_S are the scaling coefficients to convert the stress–strain to *p-y*.

Table 5.7 *p-y* curves.

Type of soil	*p-y* approach	Remarks
Soft normally consolidated (NC) marine clay	Matlock (1970a,b)	Static and cyclic
Stiff clay	Reese et al. (1975)	Static and cyclic
Sand	Reese et al. (1974), O'Neill and Murchison (1983)	Static and cyclic

5.4.4.3 *p-y* Curves for Sand Based on API

For sand: The *p-y* curve for a pile in sand is generated using the following derivation;

$$P = A\,P_u \tanh\left(\frac{K\,x}{A\,P_u}\,y\right) \tag{5.61}$$

where:

A	Cyclic loading parameter	(–)
P_u	Ultimate lateral resistance	(kN m^{-1})
k	Initial modulus of subgrade reaction	(kN m^{-3})

The ultimate lateral resistance is defined by a number of coefficients based on a series of observations. The ultimate lateral resistance is detailed as such:

$$P_u = \begin{cases} (C_1 x + C_2 D)\,\gamma' x & for\, 0 < x \le x_r \\ C_3 D\,\gamma' x & for \quad x \ge x_r \end{cases} \tag{5.62}$$

where:

$C_1, C_2, \& C_3$	Coefficient of lateral resistance	(–)
D	Average pile diameter	(m)
γ'	Effective soil unit weight	(kN m^{-3})
x_r	Transition depth	(m)

These coefficients are given in design charts as a function of the angle of internal friction as shown in (5.25). As well as characterising the ultimate lateral resistance, the initial modulus of subgrade reaction can also be obtained as shown in Figures 5.25 and 5.26.

5.4.4.4 *p-y* Curves for Clay

The *p-y* curves for clay are generated based on the following parameters:

p = Lateral soil resistance

p_u = Ultimate lateral soil resistance

y = Lateral pile deflection

y_c = $2.5\varepsilon_{50}.d$

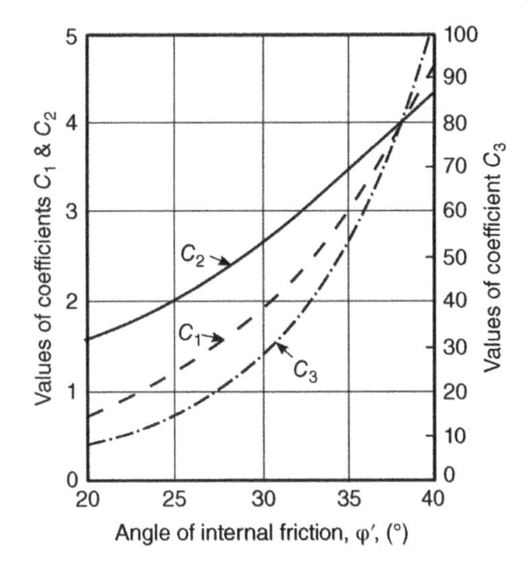

Figure 5.25 Lateral resistance coefficients as a function of internal angle of friction after American Petroleum Institute (1993).

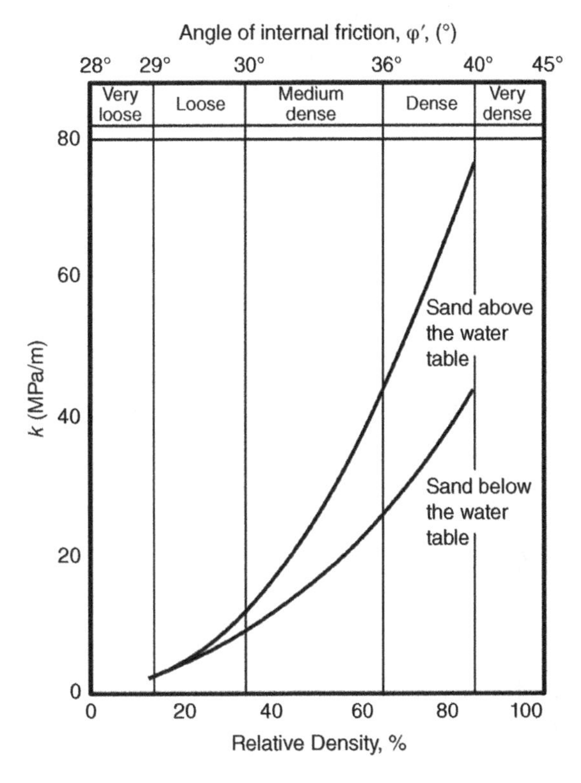

Figure 5.26 Initial modulus of subgrade reaction after Det Norske Veritas (2013) and API.

ε_{50} = Characteristic strain at 50% of failure stress in undrained tests
d = Pile diameter

The ultimate soil resistance is determined by Eq. (5.63):

$$p_u = N_P \cdot S_u \cdot d \tag{5.63}$$

Table 5.8 *p-y* coordinates for a static loading.

y (Deflection)	p (Soil reaction)
0	0
y_c	$0.5\,p_u$
$2\,y_c$	$0.63\,p_u$
$4\,y_c$	$0.8\,p_u$
$6\,y_c$	$0.9\,p_u$
$8\,y_c$	p_u
infinite	p_u

where:

S_u = Undrained shear strength of the soil

N_P = Ultimate lateral soil resistance coefficient

It is often assumed that N_P is 3 at mudline ($X = 0$) and increases to 9 at depths equal to or greater than X_r, which is given by Eq. (5.64). The depth calculated is often termed as *transition depth.*

$$X_r = \frac{6d}{\frac{\gamma' d}{S_u} + J} \tag{5.64}$$

where:

γ' = Effective unit weight of the soil

J = A constant

For static loading, a *p-y* curve can be defined using the coordinates given in Table 5.8.

Figure 5.27 shows a typical *p-y* curve for static loading based on Matlock (1970a,b). The curve is drawn by fitting a cubic parabola given by Eq. (5.65).

$$\frac{p}{p_u} = 0.5\left(\frac{y}{y_c}\right)^{\frac{1}{3}} \tag{5.65}$$

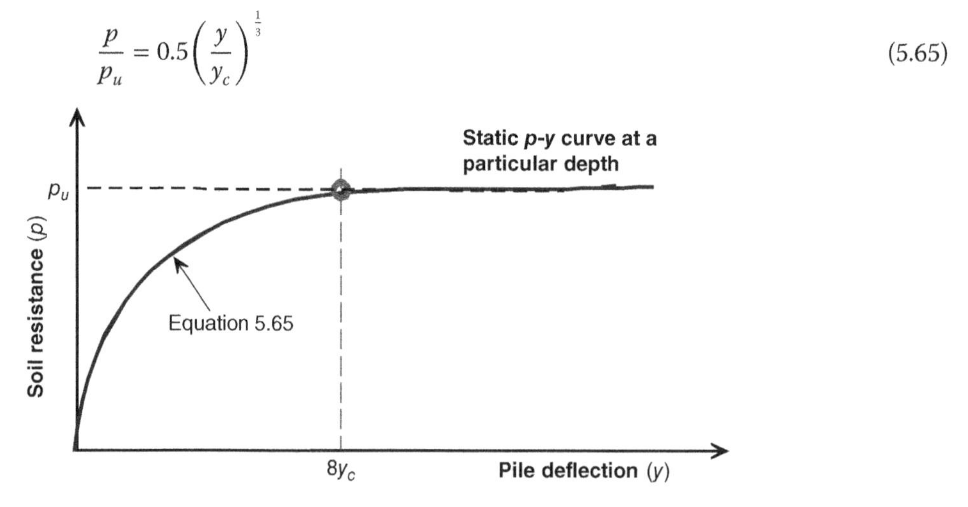

Figure 5.27 Static *p-y* curves, Matlock (1970a,b).

The static p-y curves are modified to express the deterioration due to cyclic loading.

Bhattacharya et al. (2006) recommended that for free-headed piles, consideration should be given to incorporate the shear resistance across the base of the pile. This may be incorporated into a special p-y curve at the tip of the pile, which has an ultimate resistance equal to the lateral shear of the soil across the full base of the pile. An appropriate displacement to mobilise the base shear can be idealised from elastic methods, or alternatively, from a full elasto-plastic FEA. Monopiles or anchor piles for floating systems fall under such category.

5.4.4.5 Cyclic *p-y* Curves for Soft Clay

The construction of cyclic p-y curves for soft clay is empirical and was formulated to fit the observed pile load test at the Lake Austin and Sabine sites. The maximum cyclic resistance is limited to $0.72\,p_u$. It is also assumed that complete loss in resistance occurs at the soil surface when deflections at that point reach $15\,y_c$. The coordinates for cyclic p-y curves are given in Table 5.9.

Figure 5.28 shows a typical cyclic p-y curve. It must be remembered that the curve is constructed in order to fit the observed data. Matlock (1970a,b) cautions against three aspects of the curve that are primarily empirical:

1. The position of the cyclic deterioration threshold, i.e. the coordinate $(3\,y_c,\ 0.72\,p_u)$ along the pre-plastic portion of the static p-y curve
2. The manner in which the residual resistance $0.72\,p_u(X/X_r)$ is adjusted with depth

Table 5.9 *p-y* coordinates for cyclic loading.

y (Deflection)	p (Soil reaction)
0	0
y_c	$0.5\,p_u$
$3\,y_c$	$0.72\,p_u$
$15\,y_c$	$0.72\,p_u\,(X/X_r)$
Infinite	$0.72\,p_u\,(X/X_r)$

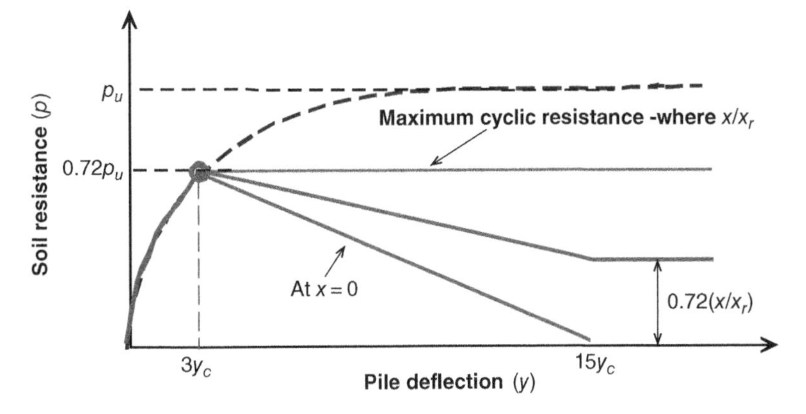

Figure 5.28 Cyclic *p-y* curves for soft clays.

3. The value of the deflection at which the residual resistance occurs

5.4.4.6 Modified Matlock Method

The *p-y* curves developed by Reese et al. (1975) for stiff clay result from test piles loaded laterally in overconsolidated clay deposits. The clays at the test site ranged in strength from 96 kPa (surface) to 290 kPa at 3.6 m and had significant secondary structure (fissures, joints, and slickensides). ε_{50} values for stiff clays recommended in the paper range from 0.7% (shear strength 50–100 kPa) to 0.4% (shear strengths 200–400 kPa). Typical overconsolidated North Sea clays, however, do not usually show significant secondary structure, nor demonstrate the same trend in variation of ε_{50} with shear strength as recommended by Reese et al. (1975).

Matlock (1979) has suggested that for stiff clays, *p-y* curves may be obtained using Matlock (1970a,b) with a modification applied above the transition depth X_r. It is in the zone above the transition depth that large deflections usually occur. Table 5.10 gives the coordinates to construct the modified *p-y* curve. Essentially, the peak point in the cyclic *p-y* curve for soft clays is omitted as shown in Figure 5.29. In this modified method, cyclic degradation starts at a deflection of y_c where the soil reaction is equal to 0.5 p_u. If soft clay cyclic curves had been used, cyclic degradation would have occurred at a deflection equal to $3y_c$ and the peak resistance would also have been approximately 50% higher.

Table 5.10 *p-y* coordinates for cyclic loading for stiff clays (modified).

y (Deflection)	p (Soil reaction)
0	0
y_c	$0.5\,p_u$
$15\,y_c$	$0.72\,p_u\,(X/X_r)$
Infinite	$0.72\,p_u\,(X/X_r)$

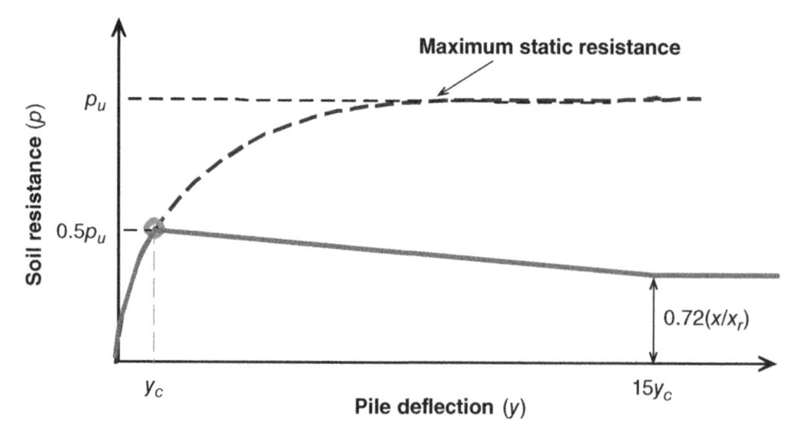

Figure 5.29 Modified Matlock cyclic *p-y* curve, based on Bhattacharya et al. (2006). This *p-y* curve is recommended in API and ISO code.

5.4.4.7 ASIDE: Note on the API Cyclic *p-y* Curves

(1) In the API (2005) design approach, cyclic deterioration has been defined on the basis of an empirical fit to a limited pile load test database from Sabine and Lake Austin. Matlock cautions against the uncertainties regarding the position of the cyclic deterioration threshold and the distribution in which the residual resistance is adjusted with depth, and the value of the deflection at which the residual resistance occurs.
(2) Centrifuge tests carried out to verify the design of piles for the URSA (Mississippi Canyon Block 809) TLP Platform showed that, for cyclic behaviour of a single pile, large lateral displacements reduce Matlock's 0.72 factor by about 20%, i.e. a factor of 0.57 (Doyle et al. 2004). This provides some justification for the factor of 0.5 used as the peak resistance for cyclic *p-y* curves in the modified Matlock approach.
(3) The data on the cyclic loading of soils indicate that soil degradation affects not only soil strength but also soil stiffness. However, the recommended cyclic *p-y* curves do not reduce the initial pile response, i.e. pile stiffness when compared with the static *p-y* curves. This is obviously at odds with observed cyclic behaviour of soils.
(4) ε_{50} is used to link the stiffness of the *p-y* curves to the soil stiffness at the site. Therefore, this is an important design parameter and must be measured through advanced soil testing. Soils with a higher OCR are expected to have higher ε_{50} values.
(5) For free-headed piles such as monopiles or anchor piles, where cyclic behaviour is very critical, it is therefore prudent to consider fundamental soil behaviour when deriving generalised cyclic *p-y* data. Incorporation of degradation or increase in stiffness due to large cyclic loads must be linked in the *p-y* curves.

5.4.4.8 Why API *p-y* Curves Are Not Strictly Applicable

(1) The *p-y* approach uses nonlinear springs and produces reliable results for the cases for which it was developed, i.e. small diameter piles and for few cycles of loading. For O&G applications, the main design requirement is ULS, i.e. avoiding plastic hinge in the pile. In other words, the ultimate strength of the soil i.e. nature of the *p-y* curves at large strain level. On the other hand, for offshore wind monopile design, SLS and natural frequency is the main design criteria and therefore initial part of the *p-y* curve is critical.
(2) The method is not validated yet for large-diameter piles. Kallehave & Thilsted (2012) reported that using API *p-y* curves underpredicts the foundation stiffness.
(3) API formulation doesn't consider the base resistance as well as shaft resistance, and this will have significant contribution for large-diameter piles. The readers are referred to Shadlou and Bhattacharya (2016) for analytical work considering the base resistance and shaft resistance.

ASIDE

Through the PISA project Byrne et al. (2015a, 2015b) large-scale tests on piles are being carried out. The test sites include Cowden, for example. The readers are referred to the test data.

Table 5.11 Reference for *p-y* curve construction for different types of soils.

Type of soils	Reference for *p-y* curves
Calcareous soils	Wesselink et al. (1988) Williams et al. (1988) Dyson (1999) Dyson and Randolph (2000)
Weak rock	Reese et al. (1997)
Weak carbonate rock	Abbs (1983)
Weak calcareous claystone	Fragio et al. (1985)

5.4.4.9 References for *p-y* Curves for Different Types of Soils

There are no code provisions or guidelines for soft rocks such as weak mudstone, sandstone, and carbonate materials. Different forms of *p-y* curves have been proposed. Table 5.11 provides references for weak rocks along with liquefied soils and partially liquefied soils.

5.4.4.10 What Are the Requirements of *p-y* Curves for Offshore Wind Turbines?

Ideally, the *p-y* curves to be used for modelling pile–soil interaction should capture the two pile–soil interaction likely to be encountered in its lifetime:

1. Effects of action of millions of cycles of loadings due to wind, wave, 1P, and 3P under operating conditions. It may be noted that the strains in the soil depend on the type of loading. As discussed in Chapter 4, sandy soils under very small strains will densify. On the other hand, sandy soils under moderate to large strains under undrained condition will partially or fully liquefy through the generation and accumulation of pore water pressure.
2. Soil stiffness and strength will change under extreme loading conditions.

5.4.4.11 Scaling Methods for Construction of *p-y* Curves

This section presents the modern way of developing *p-y* curves from stress–strain curves of the corresponding soil. The scaling method for the construction of *p-y* curves from stress–strain curves relies on the similarity between load-deflection characteristics of the pile and mechanical behaviour of the deforming soil. This involves scaling of stress and strain into compatible soil reaction p and pile deflection y, respectively. Figures 5.30 and 5.24 show a schematic of the application and the values that are necessary are M_s and N_s. It is assumed that plane strain conditions are established around the pile at any depth. As a result, soil is expected to flow around the pile from front to back. Although such an assumption is acceptable at deeper depths, the same may not be valid at the surface level when wedge type failure is likely to occur, particularly at shallow depths. In accordance with the postulated collapse mechanism, the SSI problem reduces to a series of decoupled plane strain problems as schematically illustrated in Figure 5.30. The model considered here is therefore a disc having an outer radius of R, representing the soil, and a rigid disc with outer radius r_0 that moves laterally in the deforming soil.

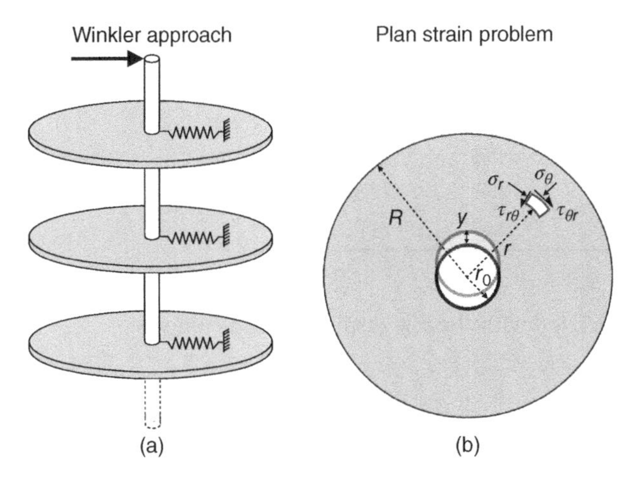

Figure 5.30 Schematic representation of the problem: (a) Winkler approach; (b) Plane strain model.

The soil is further supposed to adhere perfectly to the pile. It should be emphasised that the conceptualised problem is analogous to the plane strain problem used in plasticity theory for the evaluation of the undrained lateral capacity of a cylinder moving through an infinite medium.

This method is similar to MSD method proposed by Osman and Bolton (2005). The key feature of the MSD approach is that a single stress and a single strain are chosen to represent the behaviour of the soil mass under any given loading conditions. The representative stress is related to the applied load, and the representative strain is related to the applied displacement. If a stress–strain relationship is known for the soil, then this can be converted to an equivalent load–displacement relationship for the chosen problem. Clearly, this approach cannot be expected to describe the detailed response with great accuracy, as a number of approximations are involved, most importantly a single stress value cannot represent the stress state of the entire soil mass and a single strain state cannot represent the strains in an entire soil mass.

Nevertheless, as demonstrated by Osman and Bolton, the approach lies within the tradition of robust engineering applications for two fundamental reasons: first, the MSD can provide a realistic estimate of load-deformation behaviour, and second, it has a benefit that complex calculations are not required. The soil resistance p developed at a mobilised stress τ_{mob} is given by

$$p = N_s \tau_{mob} D \tag{5.66}$$

in which N_s is a scaling factor for stress. The mobilised shear stress τ_{mob} can be related to an average mobilised engineering shear strain $\gamma_{s\,mob}$. This can be mathematically defined as the spatial average of the shear strain γ_s in the entire volume of the deforming medium. The engineering shear strain γ_s can be defined as the difference between the major and the minor principal strains:

$$\gamma_s = \varepsilon_1 - \varepsilon_3 \tag{5.67}$$

The average shear strain mobilised in the deforming soil can be calculated from the spatial average of the shear strain in the whole volume of the deformation zone. This

Table 5.12 Summary of M_S and N_S.

Type of interface and soil model	M_S	N_S
Linear elastic soil with smooth interface	3.15	11.23
Linear elastic soil with rough interface	2.59	17.89
Rigid plastic with rough interface	2.60	9.00

mobilised shear strain can be associated with the pile displacement as follows:

$$\gamma_{smob} = \frac{\int_v \gamma_s \, dv}{\int_v dv} = M_S \frac{y}{D} \tag{5.68}$$

M_S and N_S are determined and presented in the work of Bouzid et al. (2013), Dash (2010), and Dash et al. (2017) and are summarised In (Table 5.12).

Similar scaling methods as that of Dash (2010), Bouzid et al. (2013) are also proposed by Zhang et al. (2016), Zhang and Andersen (2017), and Jeanjean (2017). Using the scaling method discussed of Bouzid et al. (2013), Lombardi et al. (2017), and Dash et al. (2017) presented *p-y* curves for liquefied soils.

5.4.4.12 *p-y* Curves for Partially Liquefied Soils

One of the first attempts to propose *p-y* curves for liquefiable soils is provided by Liu and Dobry (1995). The method, illustrated in Figure 5.31, consists of applying to the conventional *p-y* curve for nonliquefied sand a reduction factor m_p also known as *p*-multplier. m_p is based on the excess pore water pressure ratio ($r_u = \frac{u}{\sigma_v'}$) where u is the excess pore water pressure developed and σ_v' is the vertical effective stress. r_u of 1 implies full liquefaction and m_p is 0 and r_u of 0 would imply no liquefaction and m_p is 1. Linear interpolation is allowed and this model is reasonable for partially liquefied soils. This model will predict a flat curve, i.e. no springs for fully liquefied soils. While this is a valid assumption for small amplitude vibration, it is unrealistic for large amplitude. The next section shows a *p-y* curve for fully liquefied soils developed based on the scaling method.

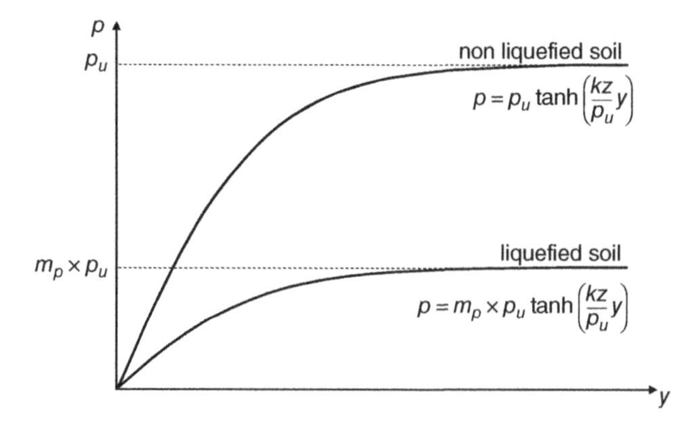

Figure 5.31 Construction of *p-y* curves for liquefiable soils according to *p*-multiplier approach.

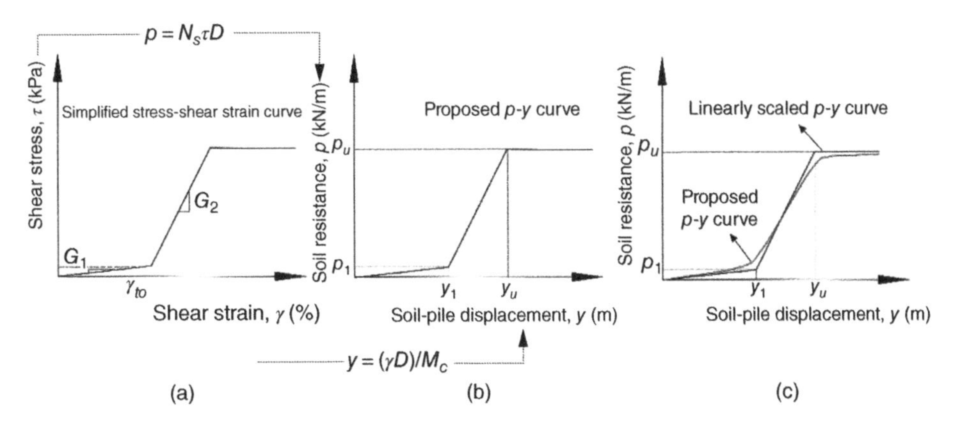

Figure 5.32 Use of scaling approach for derivation of *p-y* curves from stress–strain response of liquefied soils.

5.4.4.13 *p-y* Curves for Liquefied Soils Based on the Scaling Method

Figure 5.32 presents *p-y* curves for liquefied soils based on the work on Lombardi et al. (2017) and Dash et al. (2017). The method is schematically shown in Figure 5.32. For this method, a multistage cyclic triaxial test is required. First, the soil is cyclically loaded to full liquefaction and when the soil is fully liquefied, monotonic load is applied and the stress–strain curve is obtained. Through the advanced testing of wide range of sandy soils, Rouholamin et al. (2016) showed that stress–strain curve for liquefied soil looks like Figure 5.32a and the important parameter is the take-off strain (γ_{to}). Take-off strain is the strain at which the soil grains of the initially liquefied soil start to lock and resistance is developed. Once the simplified stress–strain model of liquefied soil is formulated, the corresponding *p-y* curve parameters are calculated by using scaling factors, M_s and N_s. Figure 5.32 schematically represents the process involved in transforming stress–strain model to the *p-y* curve of liquefied soil. The readers are referred to the paper by Dash et al. (2017) for further details and an example application.

5.4.5 Advanced Methods of Analysis

For the current application, advanced method of analysis represents finite element, discrete element, or finite difference methods. Discussion of these methods is beyond the scope of this book. Figure 5.33 shows a model of the whole problem. The advantage of this method is that complex soil behaviour can be incorporated. For example, to account for long-term performance prediction, it is necessary to employ a suitable constitutive model, capable of reproducing realistic soil behaviour when subjected to cyclic/dynamic loading conditions. Simple elastoplastic models are usually based on one yield surface without hardening (e.g. the Mohr coulomb model), with isotropic hardening (e.g. the Cam-Clay model), or on two yield surfaces (isotropic loading and deviatoric loading), e.g. Lade's. They are ideal and efficient on simulating the soil behaviour under monotonic conditions, but not suitable for model cyclic loading. Essentially, in such models, plastic deformations start to appear for a certain magnitude of loading, usually higher than the amplitude of the cycles, and hence, the simulated soil behaviour remains elastic under that limit, which is against experimental evidences, particularly for granular soils

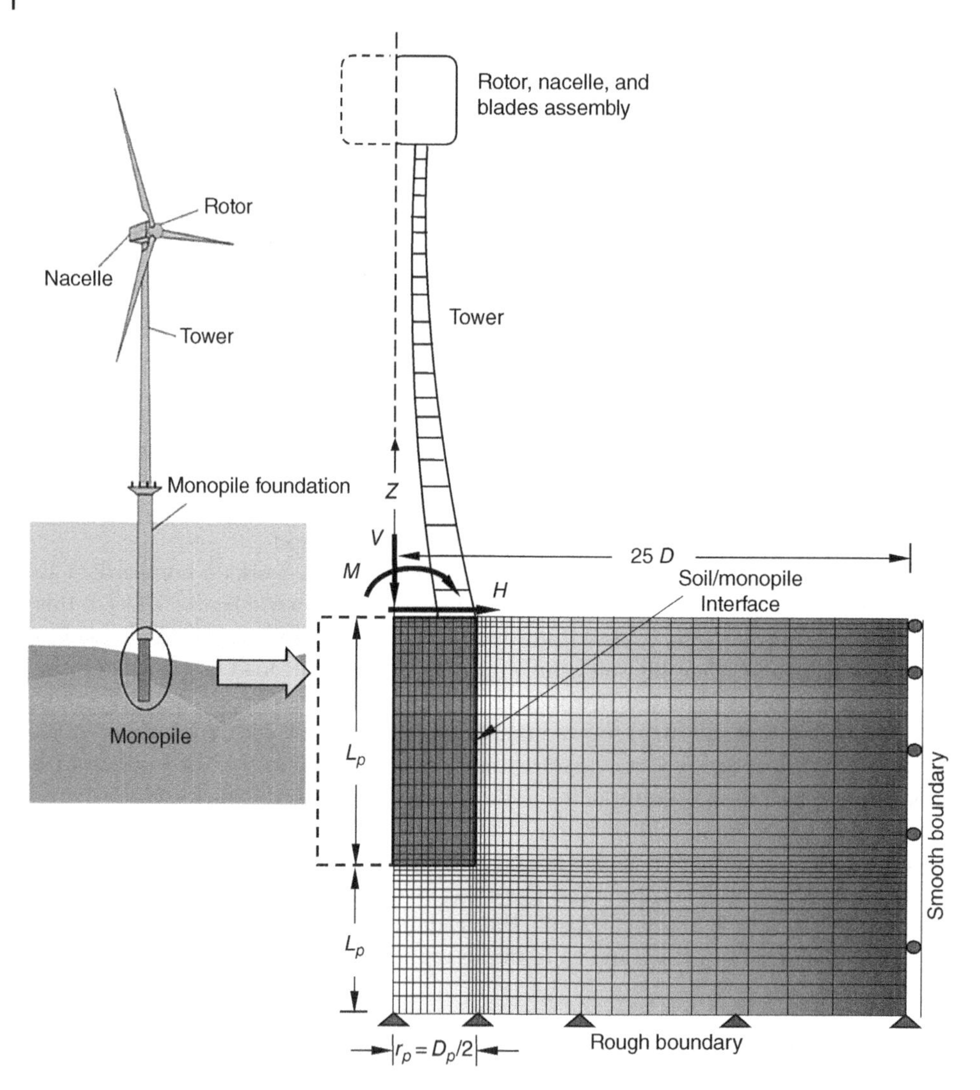

Figure 5.33 Finite element of a monopile and surrounding elastic medium.

Cambou and Hicher (2008). Figure 5.34 shows the different phenomenon that can take place in soil during cyclic loading.

Plastic phenomenon for cyclic loading can either be based on *constant stress amplitude* (so called stress controlled) or *constant strain amplitude* (strain controlled). The phenomenon called *adaptation* refers to less dissipation of energy in each cycle of loading as the number of cycles increases, until convergence to nondissipative elastic cycles. On the other hand, *accommodation* refers to change of the dissipation of energy from the beginning with an irreversible cumulative deformation and evolves towards a stabilised cycle. This is experienced in laboratory experiments for drained samples. *Ratcheting* is an irreversible strain accumulation, which keeps the same shape as that of the beginning.

Constant strain amplitude cyclic loading phenomenon results in the cyclic hardening or softening. Cyclic hardening occurs when there is an increase in the cyclic stress

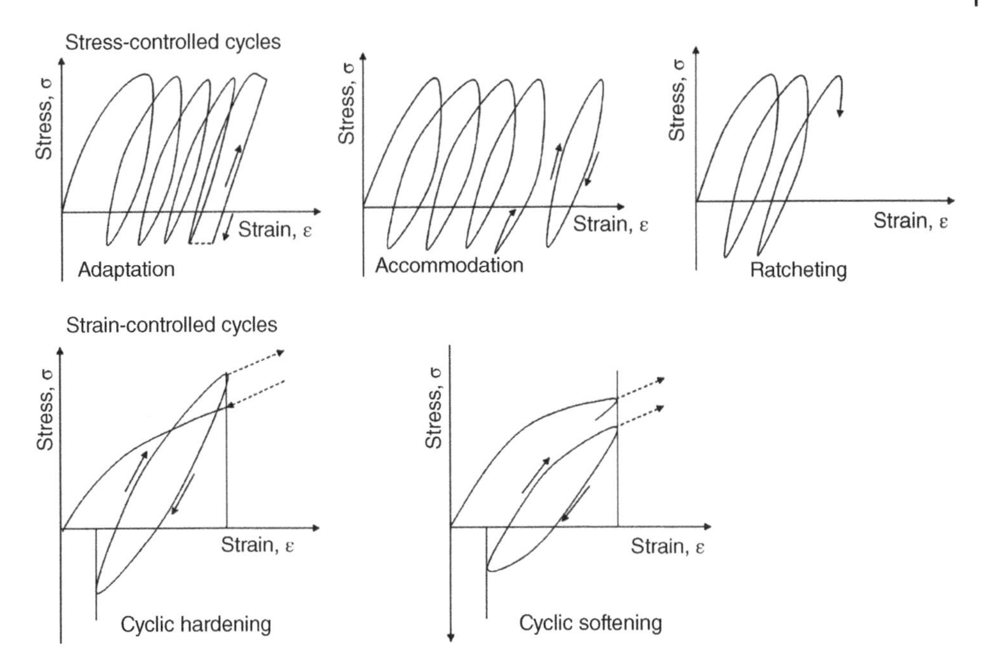

Figure 5.34 Different cyclic behaviours.

amplitude with an increase in the number of cycles, e.g. densification during testing of drained soil. On the contrary, cyclic softening occurs when there is a reduction in the cyclic stress amplitude as the number of cycles increases, e.g. an increase in pore pressure during the testing of undrained soil samples (decrease in the effective stress).

A kinematic hardening law may be employed by coupling the elastic moduli and the hardening parameter. Kinematic hardening can be used to model plastic ratcheting, which is the build-up of plastic strain during cyclic loading, as previously explained. Other hardening behaviours include changes in the shape of the yield surface in which the hardening rule affects only a local region of the yield surface, and softening behaviour in which the yield stress decreases with plastic loading. The kinematic hardening model involves only one plastic mechanism with a smooth yield surface. For nonlinear modelling, the material behaviour is characterised by an initial elastic response, followed by plastic deformation and unloading from the plastic state. The plasticity is a result of the microscopic nature of the material particles and includes shear loading that causes particles to move past one another, changes in void or fluid content that result in volumetric plasticity, and exceeding the cohesive forces between the particles or aggregates. The material is defined by model materials subject to loading beyond their elastic limit. Figure 5.35 shows a PLAXIS model for suction caisson type foundations.

The readers are referred to Lopez Querol et al. (2017) for advanced numerical model analysis of monopile.

5.4.5.1 Obtaining K_L, K_R, and K_{LR} from Finite Element Results

This section shows a method to compute the three stiffness terms (K_L, K_{LR}, and K_R) from FEA. Mathematically, two FEA analyses are required and there are *pile-head moment-rotation* (M-θ) and *pile-head horizontal load-deflection* (H-δ). Considering

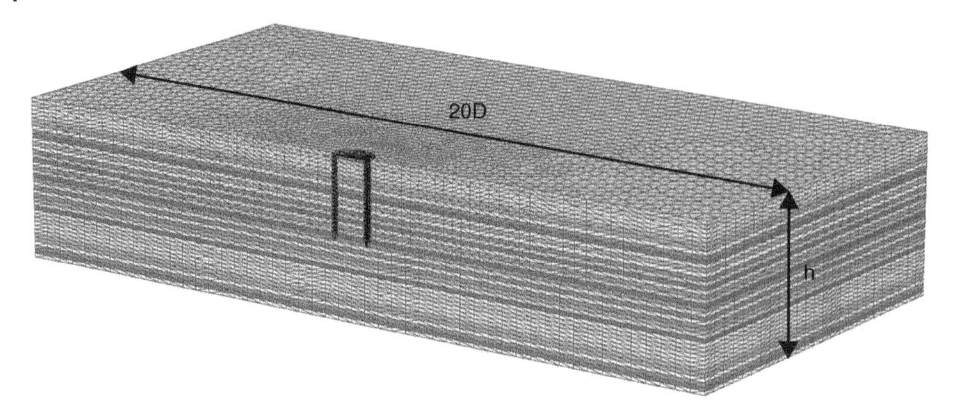

Figure 5.35 Soil model used to simulate parabolic stiffness variation in PLAXIS 3D.

the full range of the curve, i.e. until failure, pile-head load-deflection and pile-head moment-rotation curves are nonlinear, depending on the soil type. However, the initial linear range of the curves must be used to estimate pile-head rotation and deflection based on Eq. (5.69).

$$\begin{bmatrix} H \\ M \end{bmatrix} = \begin{bmatrix} K_L & K_{LR} \\ K_{LR} & K_R \end{bmatrix} \begin{bmatrix} \rho \\ \theta \end{bmatrix} \tag{5.69}$$

Eq. (5.69) can be rewritten as Eq. (5.70) through matrix operation where I (flexibility matrix) is a 2×2 matrix given by Eq. (5.71):

$$\begin{bmatrix} \rho \\ \theta \end{bmatrix} = [I] \times \begin{bmatrix} H \\ M \end{bmatrix} \tag{5.70}$$

$$I = \begin{bmatrix} I_L & I_{LR} \\ I_{RL} & I_R \end{bmatrix} \tag{5.71}$$

To obtain the stiffness components, run a numerical model for a lateral load (say $H = H_1$) with zero moment ($M = 0$) and obtain values of deflection and rotation (ρ_1 and θ_1). The results can be expressed through Eqs. (5.72)–(5.73).

$$\begin{bmatrix} \rho_1 \\ \theta_1 \end{bmatrix} = \begin{bmatrix} I_L & I_{LR} \\ I_{RL} & I_R \end{bmatrix} \times \begin{bmatrix} H_1 \\ 0 \end{bmatrix} \tag{5.72}$$

$$\rho_1 = H_1 \times I_L \Rightarrow I_L = \frac{\rho_1}{H_1}$$

$$\theta_1 = H_1 \times I_{RL} \Rightarrow I_{RL} = \frac{\theta_1}{H_1} \tag{5.73}$$

Similarly, another numerical analysis can be done for a defined moment ($M = M_1$) and zero lateral load ($H = 0$) and the results are shown in Eqs. (5.74)–(5.75).

$$\begin{bmatrix} \rho_2 \\ \theta_2 \end{bmatrix} = \begin{bmatrix} I_L & I_{LR} \\ I_{RL} & I_R \end{bmatrix} \times \begin{bmatrix} 0 \\ M_1 \end{bmatrix} \tag{5.74}$$

$$\rho_2 = M_1 \times I_{LR} \Rightarrow I_{LR} = \frac{\rho_2}{M_1}$$

$$\theta_2 = M_1 \times I_R \Rightarrow I_R = \frac{\theta_2}{M_1} \tag{5.75}$$

From the above analysis (Eqs. (5.72)–(5.75)), terms for the matrix (Eq. (5.71)) can be obtained. Eq. (5.70) can be rewritten as Eq. (5.76) through the matrix operation.

$$[I]^{-1} \times \begin{bmatrix} \rho \\ \theta \end{bmatrix} = \begin{bmatrix} H \\ M \end{bmatrix} \tag{5.76}$$

Comparing Eqs. (5.69) and (5.76), one can easily see the relation between the stiffness matrix and the inverse of flexibility matrix (I) given by Eq. (5.77). Equation (5.78) is a matrix operation that can be carried out easily to obtain K_L, K_R, and K_{LR}.

$$K = \begin{bmatrix} K_L & K_{LR} \\ K_{RL} & K_R \end{bmatrix} = I^{-1} = \begin{bmatrix} I_L & I_{LR} \\ I_{RL} & I_R \end{bmatrix}^{-1} \tag{5.77}$$

$$K = I^{-1} = \begin{bmatrix} \dfrac{\rho_1}{H_1} & \dfrac{\rho_2}{M_2} \\ \dfrac{\theta_1}{H_1} & \dfrac{\theta_2}{M_2} \end{bmatrix}^{-1} \tag{5.78}$$

It is important to note that the above methodology is only applicable in the linear range, and therefore it is advisable to obtain a load-deflection and moment-rotation curve to check the range of linearity even if a nonlinear soil model is used. If the analysis is used beyond the linear range, deflections and rotations will be underestimated. An example application is shown in Chapter 6.

The above method can also be used with the standard method of analysis. Essentially, one needs load-deflection and a moment-rotation curves.

5.5 Long-Term Performance Prediction for Monopile Foundations

Monopiles are commonly used foundations for these turbines in water depths of less than about 30 m. Figure 5.36 shows a schematic diagram of a monopile-supported wind turbine with the main cyclic and dynamic loads acting on the turbine. As discussed in Chapter 2, wind and wave provides the largest overturning moment.

Typically, in shallow to medium deep waters, the wind thrust loading at the hub will produce the highest cyclic overturning moment at the mudline. However, the frequency of this loading is extremely low and is in the order of magnitude of 100 seconds as shown in Figure 5.36. For further details the readers are referred to Chapter 2. Typical period of wind turbine structures being in the range of about 3 to 5 seconds, and therefore no resonance of structure due to wind turbulence is expected. Therefore, this is cyclic soil–structure interaction (CSSI) rather than DSSI. On the other hand, the wave loading will also apply overturning moment at the mudline and the magnitude depends on water depth, significant wave height, and peak wave period. Typical wave period will be in the order of 10 seconds (for North Sea) and will therefore have DSSI.

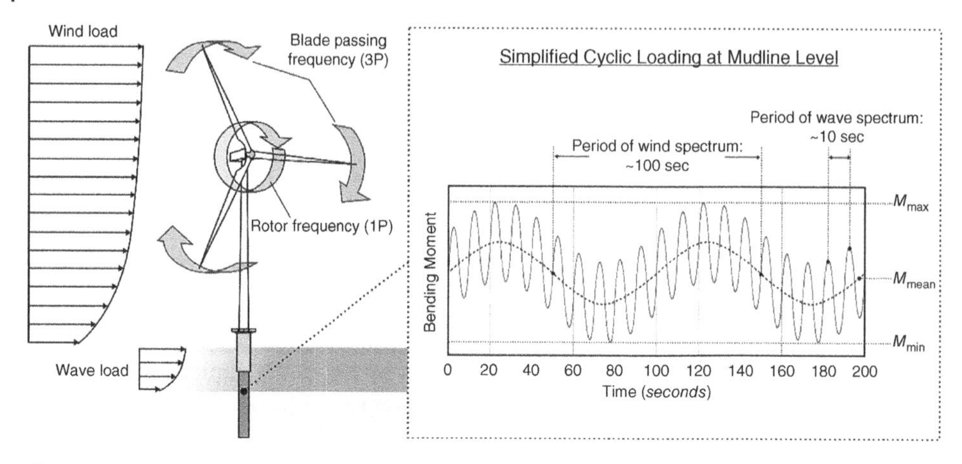

Figure 5.36 Loads acting on a typical OWT foundation and typical mudline moment.

A calculation procedure developed in Arany et al. (2017) and related examples are presented in Chapters 2 and 6, whereby overturning moments can be estimated for various load cases on wind and wave. The output of such a calculation will be relative wind and the wave loads given by M_{max}, M_{min}, and M_{mean} as shown in Figure 5.36. It is assumed in the analysis that the wind and wave are perfectly aligned, which is a fair assumption for deeper water further offshore projects (i.e. fetch distance is high). Analysis carried out in Chapter 2 showed that the loads from 1P and 3P are orders of magnitude lower than wind and wave, but they will have the highest dynamic amplifications. The effect of dynamic amplifications due to 1P and 3P will be small amplitude vibrations.

Furthermore, there are added soil–structure interactions due to many cycles of loading and the wind–wave misalignments. Typical estimates will suggest that OWT foundations are subjected to 10–100 million load cycles of varying amplitudes over their lifetime (25–30 years). The load cycle amplitudes will be random/irregular and have broadband frequencies ranging several orders of magnitudes from about 0.001 to 1 Hz. Based on the discussion above, the soil–structure interaction can be simplified into two superimposed cases and is shown in Figure 5.37.

1. Cyclic overturning moments (typical frequency of 0.01 Hz) due to lateral loading of the wind acting at the hub together with that of the wave. This will be similar to a fatigue-type problem for the soil and may lead to strain accumulation in the soil giving rise to progressive tilting. Due to wind–wave load misalignment, the problem can also be multidirectional.
2. Wave loading, on the other hand, will be dynamic to different degree (mild/moderately/high), i.e. mildly dynamic or moderately dynamic or highly dynamic. This depends on the location of the wind farm and more specifically wave spectrum and the target frequency of the turbine. An 8 MW turbine supported of monopiles having 0.22 Hz as the first eigen frequency located in the North Sea (having 0.1 Hz predominant wave frequency) will be moderately dynamic. The same turbine in the China Sea having 5 seconds predominant period will be highly dynamic.
3. Due to the proximity of the frequencies of 1P and 3P to the natural frequency of the structure, resonance in the wind turbine system is expected. Field evidence also

supports this; see Hu et al. (2014) with reference German wind farm projects. Studies presented in Chapter 2 show that these are orders of magnitude lower than that caused by the wind and the wave. This resonant dynamic bending moment will cause cyclic strain in the pile in the fore–aft direction, and this strain will be transferred to the neighbouring soil. This resonant type mechanism will involve compaction of the soil in front of and behind the pile (in the fore–aft direction). This is shown by a SDOF oscillator in Figure 5.37.

5.5.1 Estimation of Soil Strain around the Foundation

In the whole turbine structure, i.e. RNA-Tower-Transition Piece-Monopile-Soil, the element that is most likely to change with the operation is the soil. Other materials such as steel may change due to fatigue-type loading, but the change is miniscule compared to the soil. In routine engineering practice, SSI can be incorporated through the estimation of strain levels in the deformed/mobilised zone by considering the soil behaviour at that strain level. A method is developed based on the above and explained below:

Deformation of the pile under the action of the loading as described in Section 5.5 and Figure 5.37 will lead to three-dimensional soil–pile interaction, as shown schematically in Figure 5.38. Simplistically, there would be two main interactions:

1. Due to pile bending (which is cyclic in nature) and the bending strain in the pile, strain in the soil will transfer through contact friction, which will be cyclic in nature.
2. Due to lateral deflection of the pile, strain will develop in the soil around the pile.

Figure 5.38 also shows a simple methodology to estimate the levels of strains in a soil for the two types of interactions, and is given by Eqs. 5.79 and 5.80.

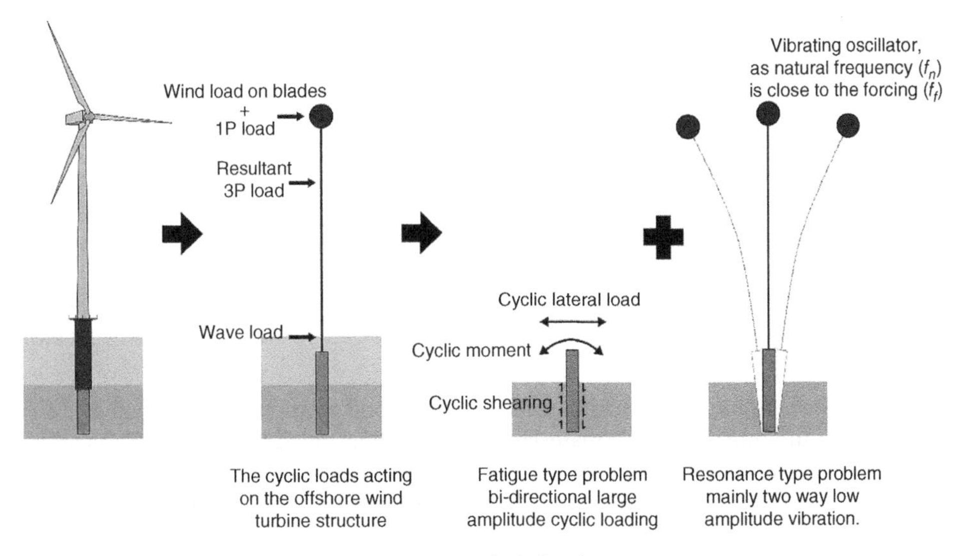

Figure 5.37 Loads acting on a monopile-supported wind turbine.

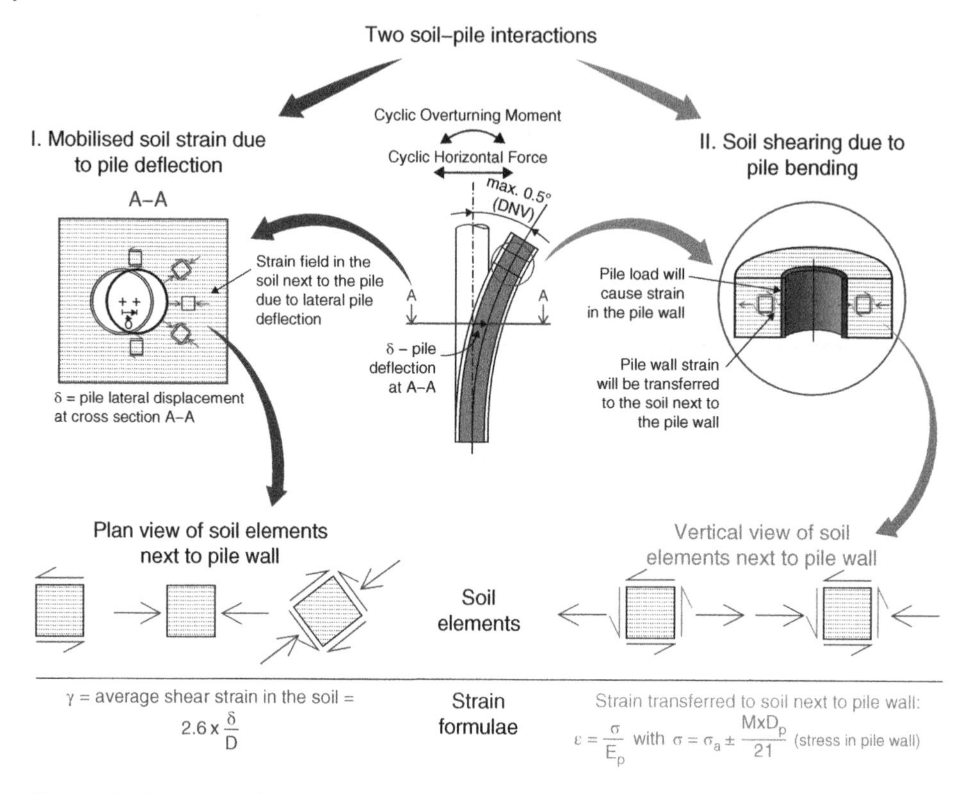

Figure 5.38 Two types of soil-pile interaction on a monopile-supported wind turbine. A simple model is proposed in Nikitas et al. (2017).

The average strain in the soil at any section in a pile due to deflection can be estimated using Bouzid et al. (2013):

$$\gamma = 2.6\frac{\delta}{D_P} \tag{5.79}$$

where δ is the pile deflection at that section (for example A-A in Figure 5.38) and D_P is the pile diameter. This is also explained in some details in Figure 5.39 and the readers are referred to Lombardi et al. (2013) for further details.

On the other hand, the shear strain in the soil next to the pile due to pile bending can be estimated using Eq. (5.80).

$$\gamma_1 = \frac{M \times D_P}{2 \times I \times E_P} \tag{5.80}$$

where M is the bending moment in the pile, I is the second moment of area of pile and E_P is the Young's modulus of the pile material. It must be mentioned that Eq. (5.80) assumes that 100% of the strain is transmitted to the soil, which is a conservative assumption and calls for further study. In practice, this will be limited to the mobilisable friction between the pile and the soil. The above methodology is used to estimate the strain levels in the soil for 15 wind turbines.

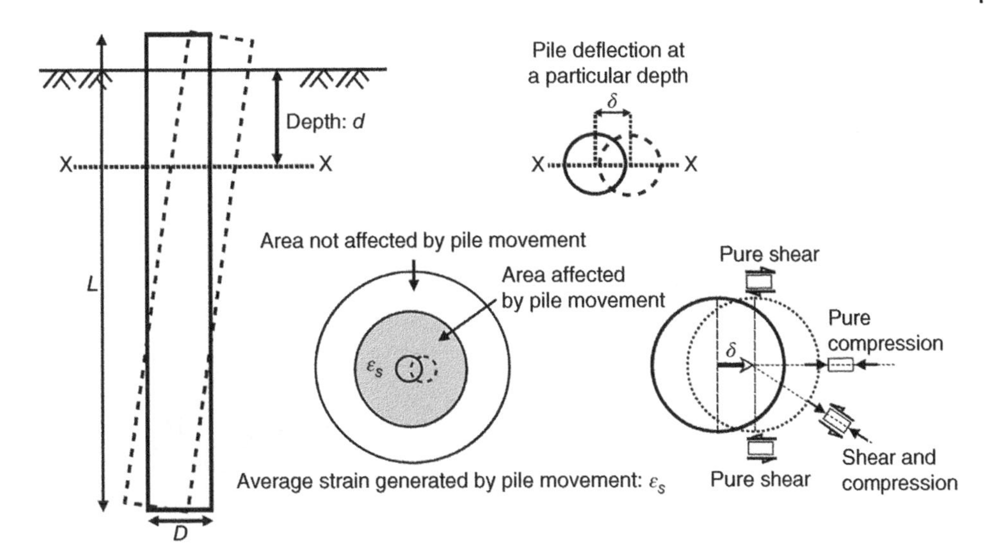

Figure 5.39 Average strain in the soil around a pile.

This strain level can then be used to carry out element tests of the soil and obtain parameters. If the strain is high for a clayey soil, residual strength may be taken for calculations. If the strain is medium, appropriate strength and stiffness degradation curves may be used. The readers are referred to Chapter 4 for further understanding of soil behaviour for a wide range of strains.

5.5.2 Numerical Example of Strains in the Soil around the Pile for 15 Wind Turbines

In this section, 15 wind turbines from 12 wind farms (Figure 5.40) are analysed to show the application of the methodology. For all 15 wind turbines, the loads are calculated using the methodology developed in Chapter 2, and an example is shown in Chapter 6. The input parameters for the load calculation are shown in Tables 5.13–5.15. For the case studies, the strains in the soil are estimated based on the methods discussed in the earlier section and depicted in Figure 5.38. They are:

1. The pile wall strain is assumed to be have been transferred fully to the soil in contact with the pile wall. This will typically extend to few soil grain diameters. Apart from these small amplitude strains (due to wind and wave), along the length of the pile small-magnitude vibrations (resonance due to the proximity of 1P and 3P loading) will cause the soil to compact. Equation (5.80) can be used for such strain calculations, and the bending moment along the pile is required for such estimate. The soil strain can be computed for different load conditions. Figures 5.41 and 5.42 plot the strain for normal working condition and extreme load condition. It may be noted that the strain in the soil due to resonance is neglected as the moments due to 1P and 3P are orders of magnitude lower than wind and wave moments.
2. Large amplitude cyclic loads (wind + wave) cause deflections along the pile length and shear strains are developed in the soil surrounding the pile. These strains, if high

Depth [m]	Lely A2	Lely A3	Irene Vorrink	Irene Vorrink	Blyth	Kentish Flats I	Barrow II	Thanet III	Belwind IV	Burbo Bank	Walney I	Gunfleet Sands	Horns rev	London Array 1	London Array 2
1	Soft clay	Soft clay	Silt	Silt	W. bedrock	Loose sand	Medium sand	Dense sand	Dense sand	Loose sand	Medium sand	Soft Clay	[illegible]	Dense sand	[illegible]
2	Soft clay	Soft clay	Silt	Silt	W. bedrock	Loose sand	Medium sand	Dense sand	Dense sand	Loose sand	Medium sand	Soft Clay	[illegible]	Dense sand	[illegible]
3	Soft clay	Soft clay	Dense sand	Dense sand	W. bedrock	Loose sand	Medium sand	Dense sand	Dense sand	Loose sand	Medium sand	Soft Clay	[illegible]	Silt (F.C.)	[illegible]
4	Soft clay	Soft clay	Soft clay	Soft clay	W. bedrock	Loose sand	Medium sand	Dense sand	Dense sand	Loose sand	Medium sand	Soft Clay	[illegible]	Silt (F.C.)	Medium sand
5	Dense sand	Dense sand	Soft clay	Soft clay	W. bedrock	Loose sand	Dense sand	Dense sand	Dense sand	Loose sand	Medium sand	Soft Clay	[illegible]	Silt (F.C.)	Medium sand
6	Dense sand	Dense sand	Soft clay	Soft clay	W. bedrock	Dense sand	Dense sand	Dense sand	Dense sand	Loose sand	Medium sand	Soft Clay	[illegible]	Silt (F.C.)	Medium sand
7	Dense sand	Dense sand	Soft clay	Soft clay	W. bedrock	Dense sand	Dense sand	Dense sand	Dense sand	Loose sand	Medium sand	Soft Clay	[illegible]	Silt (F.C.)	Medium sand
8	Dense sand	Dense sand	Dense sand	Dense sand	W. bedrock	Dense sand	Dense sand	Dense sand	Dense sand	Loose sand	Medium sand	Soft Clay	Dense sand	Stiff clay	Silt (F.C.)
9	Dense sand	Dense sand	[illegible]	[illegible]	W. bedrock	Firm clay	[illegible]	Dense sand	Dense sand	Loose sand	Medium sand	Soft Clay	Dense sand	Stiff clay	Silt (F.C.)
10	Dense sand	Dense sand	[illegible]	[illegible]	W. bedrock	Dense sand	[illegible]	Dense sand	Dense sand	Loose sand	Medium sand	Soft Clay	Dense sand	Stiff clay	Silt (F.C.)
11	Dense sand	Dense sand	Dense sand	Dense sand	W. bedrock	Dense sand	[illegible]	Dense sand	Dense sand	Loose sand	Medium sand	Firm clay	Dense sand	Stiff clay	Silt (F.C.)
12	Dense sand	Dense sand	Dense sand	Dense sand	W. bedrock	Dense sand	Stiff clay	Dense sand	Dense sand	Loose sand	Medium sand	Firm clay	Dense sand	Stiff clay	Silt (F.C.)
13	Dense sand	Dense sand	Dense sand	Dense sand	W. bedrock	Firm clay	Stiff clay	Dense sand	Dense sand	Loose sand	Medium sand	Firm clay	Dense sand	Dense sand	Silt (F.C.)
14	Dense sand	Dense sand	Dense sand	Dense sand	W. bedrock	Firm clay	Stiff clay	Dense sand	Dense sand	Loose sand	Medium sand	Firm clay	Dense sand	Dense sand	Silt (F.C.)
15	Dense sand	Dense sand	Dense sand	Dense sand	W. bedrock	Firm clay	Stiff clay	Dense sand	Dense sand	Medium sand	Medium sand	Firm clay	Dense sand	Dense sand	Silt (F.C.)
16	Dense sand	Dense sand	Dense sand	Dense sand	W. bedrock	Firm clay	Stiff clay	Stiff clay	Stiff clay	Medium sand	Medium sand	Firm clay	Dense sand	Dense sand	Medium sand
17	Dense sand	Dense sand	Dense sand	Dense sand	W. bedrock	Firm clay	Stiff clay	Stiff clay	Stiff clay	Medium sand	Medium sand	Firm clay	Dense sand	Dense sand	Medium sand
18	Dense sand	Dense sand	Dense sand	Dense sand	W. bedrock	Firm clay	Stiff clay	Stiff clay	Dense sand	Medium sand	Medium sand	Firm clay	Dense sand	Dense sand	Medium sand
19	Dense sand	Dense sand	Dense sand	Dense sand	W. bedrock	Firm clay	Stiff clay	Stiff clay	Dense sand	Medium sand	Medium sand	Firm clay	Dense sand	Dense sand	Silt (F.C.)
20	Stiff clay	Stiff clay	Dense sand	Dense sand	W. bedrock	Firm clay	Stiff clay	Stiff clay	Dense sand	Medium sand	Medium sand	Firm clay	Dense sand	Dense sand	Silt (F.C.)
21	Stiff clay	Stiff clay	Dense sand	Dense sand	W. bedrock	Firm clay	V. stiff clay	Stiff clay	Dense sand	Medium sand	Medium sand	Hard clay	Dense sand	Dense sand	Silt (F.C.)
22	Stiff clay	Stiff clay	Dense sand	Dense sand	W. bedrock	Firm clay	V. stiff clay	Stiff clay	Dense sand	Medium sand	Medium sand	Hard clay	Dense sand	Dense sand	Silt (F.C.)
23	Stiff clay	Stiff clay	Stiff clay	Stiff clay	W. bedrock	Firm clay	V. stiff clay	Stiff clay	Dense sand	Medium sand	Medium sand	Hard clay	Dense sand	Stiff clay	Silt (F.C.)
24	Stiff clay	Stiff clay	Stiff clay	Stiff clay	W. bedrock	Firm clay	V. stiff clay	Stiff clay	Dense sand	Medium sand	Medium sand	Hard clay	Dense sand	Stiff clay	Silt (F.C.)
25	[illegible]	[illegible]	Stiff clay	Stiff clay	W. bedrock	Firm clay	V. stiff clay	Stiff clay	Dense sand	Stiff clay	Medium sand	Hard clay	Dense sand	Stiff clay	Silt (F.C.)

Figure 5.40 Ground profile for 12 wind farm sites.

26	[illegible]	[illegible]	Stiff clay	Stiff clay	W. bedrock	Firm clay	V. stiff clay	Stiff clay	Dense sand	Stiff clay	Medium sand	Hard clay	Medium sand	Stiff clay	Silt (F.C.)
27	[illegible]	[illegible]	Stiff clay	Stiff clay	W. bedrock	Firm clay	V. stiff clay	Stiff clay	Dense sand	Dense sand	Medium sand	Hard clay	Medium sand	Stiff clay	Silt (F.C.)
28	[illegible]	[illegible]			W. bedrock	Firm clay	V. stiff clay	Stiff clay		Stiff clay	Medium sand	Hard clay	Medium sand	Stiff clay	Silt (F.C.)
29	[illegible]	[illegible]			W. bedrock	Firm clay	V. stiff clay	Stiff clay		Stiff clay	Medium sand	Hard clay	Medium sand	Stiff clay	Silt (F.C.)
30	[illegible]	[illegible]			W. bedrock	Firm clay	V. stiff clay	Stiff clay		V. stiff clay	Medium sand	Hard clay	Medium sand	Stiff clay	Silt (F.C.)
31							V. stiff clay	Stiff clay		V. stiff clay	Medium sand	Hard clay	Medium sand	Stiff clay	Silt (F.C.)
32							V. stiff clay	Stiff clay		V. stiff clay	Medium sand	Hard clay	Medium sand	Stiff clay	Silt (F.C.)
33							V. stiff clay	Stiff clay		V. stiff clay	Medium sand	Hard clay		Stiff clay	Silt (F.C.)
34							V. stiff clay	Stiff clay		Weak mudstone	Medium sand	Hard clay		Stiff clay	Silt (F.C.)
35							V. stiff clay	Stiff clay		Weak mudstone	Medium sand	Hard clay		Stiff clay	Silt (F.C.)
36							V. stiff clay			Weak mudstone	Medium sand			Stiff clay	Silt (F.C.)
37							V. stiff clay			Weak mudstone	Medium sand			Stiff clay	Silt (F.C.)
38							V. stiff clay			Weak mudstone	Medium sand			Stiff clay	Silt (F.C.)
39							V. stiff clay			Weak mudstone	Medium sand			Stiff clay	Silt (F.C.)
40							V. stiff clay			Weak mudstone	Medium sand			Stiff clay	Silt (F.C.)
41							V. stiff clay			Weak mudstone	Medium sand			Stiff clay	Silt (F.C.)
42							V. stiff clay			Weak mudstone	Medium sand			Stiff clay	Silt (F.C.)
43							V. stiff clay			Weak mudstone	Medium sand			Stiff clay	Silt (F.C.)
44							V. stiff clay			Weak mudstone	Medium sand			Stiff clay	Silt (F.C.)
45							V. stiff clay			Weak mudstone	Medium sand			Stiff clay	Silt (F.C.)

Figure 5.40 *(Continued)*

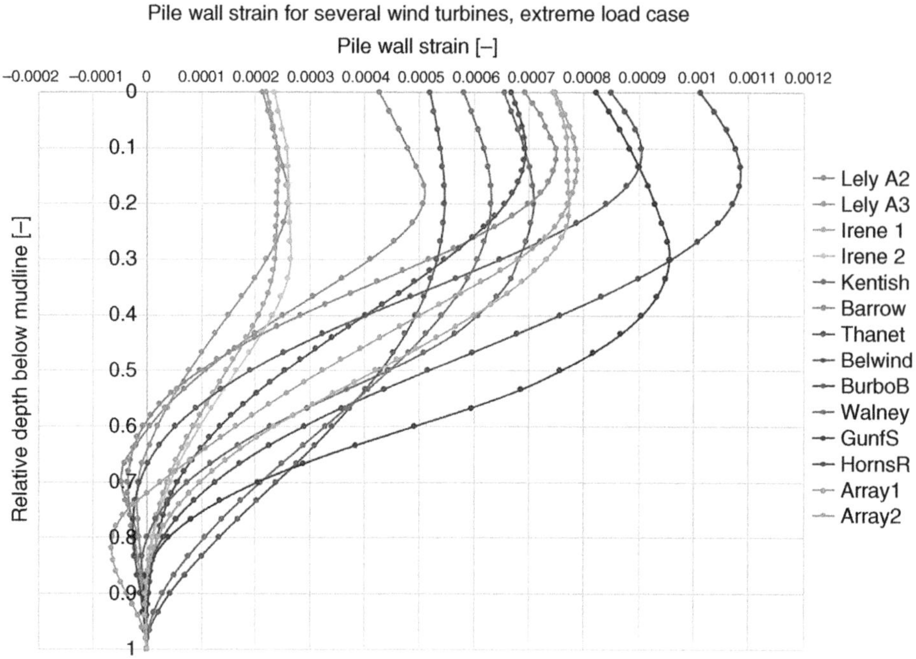

Pile wall strain for several wind turbines, extreme load case
Pile wall strain [–]
–0.0002 –0.0001 0 0.0001 0.0002 0.0003 0.0004 0.0005 0.0006 0.0007 0.0008 0.0009 0.001 0.0011 0.0012
0 0.1 0.2 0.3 0.4 0.5 0.6 0.7 0.8 0.9 1
Relative depth below mudline [–]
Lely A2
Lely A3
Irene 1
Irene 2
Kentish
Barrow
Thanet
Belwind
BurboB
Walney
GunfS
HornsR
Array1
Array2

Figure 5.41 Pile wall strain along the length of the pile (extreme load case).

Pile wall strain for several wind turbines, normal operation
Pile wall strain [–]
–0.0001 0 0.0001 0.0002 0.0003 0.0004 0.0005 0.0006 0.0007
0 0.1 0.2 0.3 0.4 0.5 0.6 0.7 0.8 0.9 1
Relative depth below mudline [–]
Lely A2
Lely A3
Irene 1
Irene 2
Kentish
Barrow
Thanet
Belwind
BurboB
Walney
GunfS
HornsR
Array1
Array2

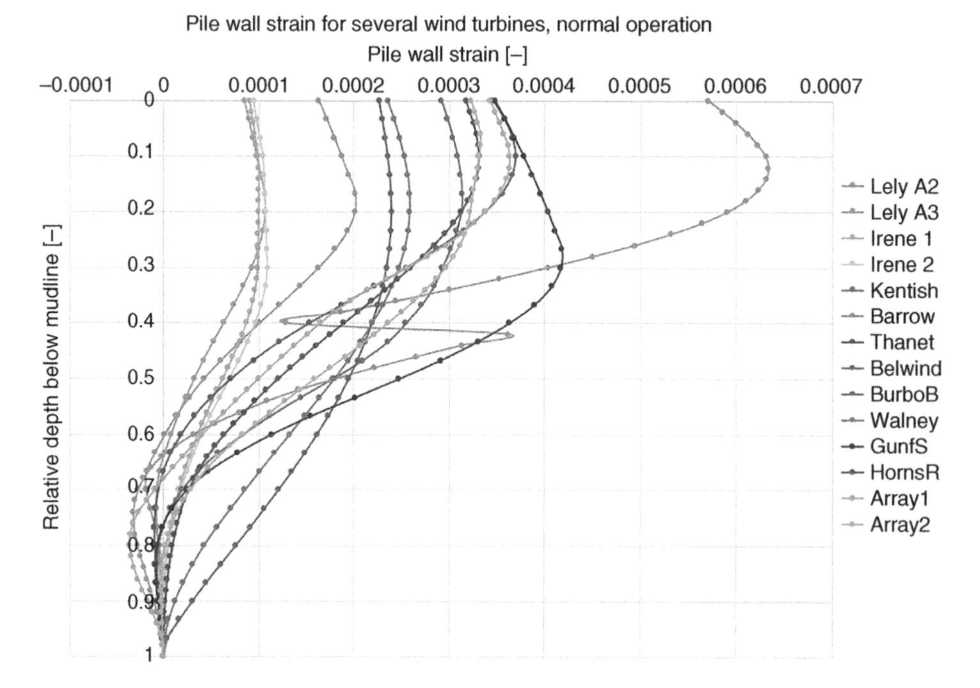

Figure 5.42 Pile wall strain (normal operating condition).

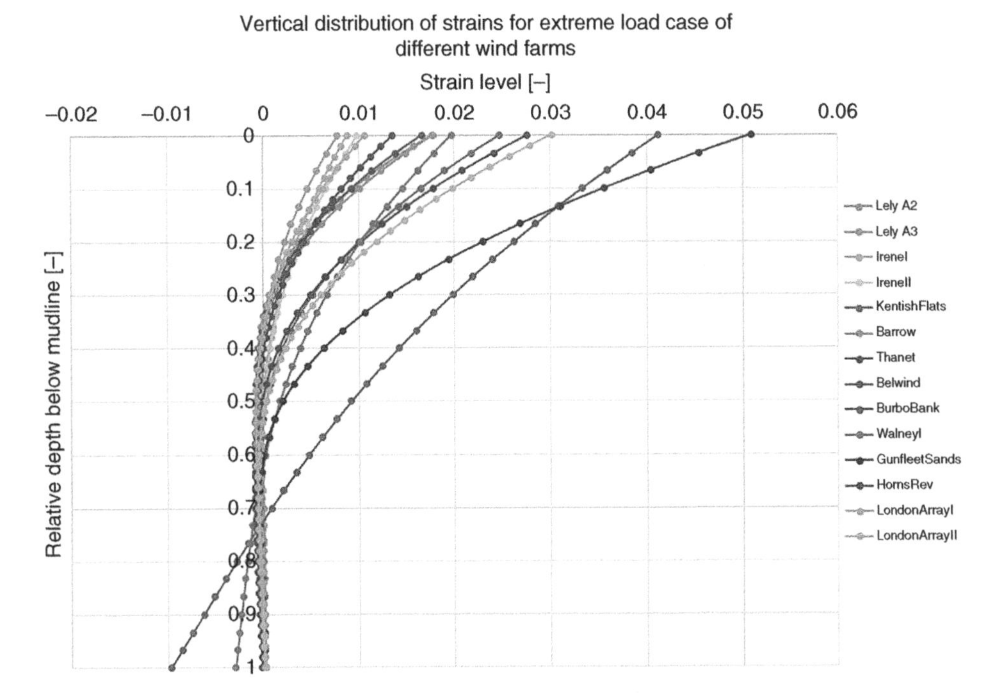

Figure 5.43 Soil strain (extreme load case).

enough, can cause degradation in soil stiffness. These shear strains may be estimated for normal working condition using the formulas given in Eq. (5.79). Figures 5.43 and 5.44 provide strain distribution in the soil along the length of the pile for two conditions.

There are therefore two opposing effects. The cyclic deflections may cause degradation in soil stiffness and may result in accumulated tilt as well as possible change in natural frequency. On the other hand, the small strains very close to the pile may cause compaction and may have a healing effect.

From the element test of a soil, it is known that strains below a certain limiting value do not cause degradation of the soil stiffness and do not change the behaviour significantly. Therefore, by estimating the strain magnitudes and comparing with results of element testing, one can determine whether degradation and thus accumulated tilt and changing natural frequency is to be expected.

5.6 Estimating the Number of Cycles of Loading over the Lifetime

One of the most challenging tasks in the analysis of the long-term behaviour of offshore wind turbines is the estimation of the number of cycles of loading of different magnitudes that will have an impact on the performance. This information is necessary to predict the fatigue life of the monopile, as well as to predict the accumulated mudline

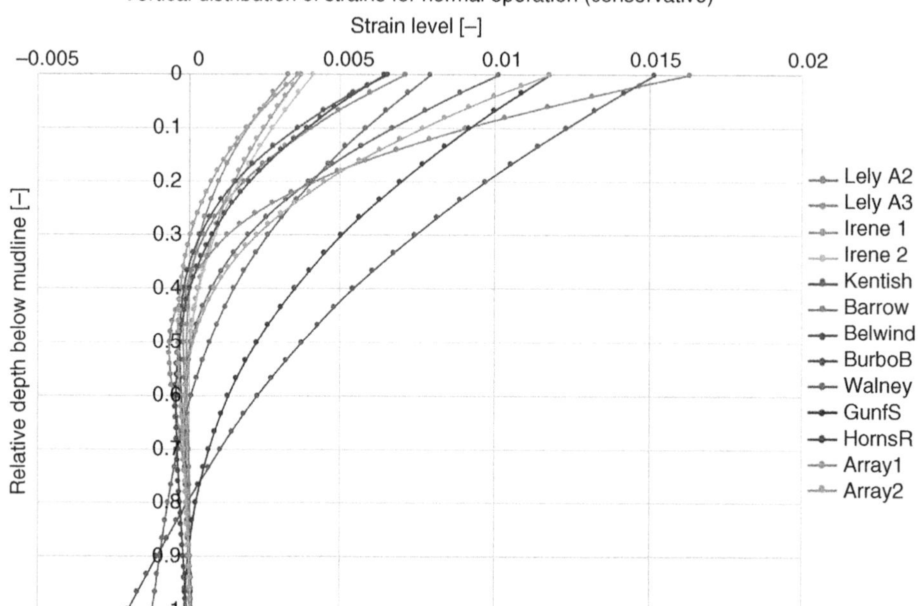

Figure 5.44 Soil strain along the length of the pile (normal operating condition).

Table 5.13 Soil parameters assumed for typical sandy soils in *p-y* analysis (Beam on Nonlinear Winkler Foundation approach).

Sand *p-y* parameters	Very loose	Loose	Medium	Dense	Very dense
Friction angle (ϕ')	28	30	33	38	40
Effective unit weight (γ' [kN m^{-3}])	6	7	8	9	10
Relative density (R_d)	15	20	40	60	80
Initial stiffness gradient (k) (kPa m^{-1})	3 000	8 000	16 000	30 000	40 000

Table 5.14 Soil parameters assumed for typical clayey soils in *p-y* analysis.

Clay *p-y* parameters	Soft	Firm	Stiff	Very stiff	Hard
Undrained shear strength (c_u)	25	50	100	200	400
J parameter (–)	0.25	0.25	0.25	0.25	0.25
Strain, 50% failure stress (ε_{50}[–])	0.007	0.006	0.005	0.004	0.003
Effective unit weight (γ' [kN m^{-3}])	7	8	9	10	12
Young's modulus (E_S [MPa])	1.25	2.5	5	10	20

Table 5.15 Parameters used for the calculations.

Parameter	Symbol (unit)	Lely A2	Lely A3	Irene Vorrink	Irene Vorrink	Blyth	Kentish I	Barrow II	Thanet III	Belwind IV	Burbo Bank	Walney I	Gunfleet Sands	Horns Rev	London Array 1	London Array 2
Reference turbulence intensity	I_{15} (%)	18.0	18.0	18.0	18.0	180	18.0	18.0	18.0	18.0	18.0	18.0	18.0	18.0	18.0	18.0
Rated wind speed	U_R (m s^{-1})	14.0	14.0	14.0	14.0	17.0	15.0	15.0	15.0	15.0	14.0	14.0	14.0	16.0	12.5	12.5
Rotor diameter	D (m)	43.0	43.0	43.0	43.0	66.0	90.0	90.0	90.0	90.0	107.0	107.0	107.0	80.0	120.0	120.0
Hub height above mean sea level	z_{hub} (m)	41.5	41.5	48.8	48.8	67.0	70.0	75.0	70.0	72.0	83.5	83.5	75.5	70.0	87.0	87.0
Mean water depth	S (m)	12.1	7.1	5.2	6.0	7.8	8.0	18.0	27.0	20.0	8.0	21.5	15	14.0	25.0	10.0
Maximum wave height before breaking	H_{break} (m)	9.4	5.5	4.1	4.7	8.6	6.2	14.0	21.1	15.6	6.2	16.8	11.7	10.9	19.5	7.8
Breaking wave time period	T_{break} (s)	10.9	8.3	7.1	7.7	10.4	8.9	13.3	16.3	14.0	8.9	14.5	12.1	11.7	15.6	9.9
50-yr maximum wave height	H_{E50} (m)	9.4	5.5	4.1	4.7	8.6	6.2	14.0	21.1	15.6	6.2	16.8	11.7	10.9	19.5	7.8
50-yr maximum wave period	T_{E50} (s)	10.9	8.3	7.1	7.7	10.4	8.9	13.3	16.3	14.0	8.9	14.5	12.1	11.7	15.6	9.9
50-yr significant wave height	H_{S50} (m)	5.1	3.0	2.2	2.5	4.6	3.4	7.5	11.3	8.4	3.4	9.0	6.3	5.9	10.5	4.2
50-yr peak wave period	T_{S50} (s)	8.0	6.1	5.2	5.6	7.6	6.5	9.7	11.9	10.3	6.5	10.6	8.9	8.6	11.5	7.3
1-yr maximum wave height	H_{E1} (m)	7.6	4.5	3.3	3.8	6.9	5.0	11.3	16.9	12.5	5.0	13.5	9.4	8.8	15.7	6.3
1-yr maximum wave period	T_{E1} (s)	9.8	7.5	6.4	6.9	9.3	7.9	11.9	14.6	12.6	7.9	13.0	10.9	10.5	14.0	8.9
1-yr significant wave height	H_{S1} (m)	4.1	2.4	1.7	2.0	3.7	2.7	6.0	9.1	6.7	2.7	7.2	5.0	4.7	8.4	3.4
1-yr significant wave period	T_{S1} (s)	7.1	5.5	4.7	5.0	6.8	5.8	8.7	10.7	9.2	5.8	9.5	8.0	7.7	10.3	6.5

deformations throughout the lifetime of the structure. To properly estimate the number of cycles at different load levels, a series of time domain simulations are necessary to statistically represent all operational states of the turbine in different environmental conditions. Rainflow counting (Matsuishi and Endo 1968) can be used to count the number of cycles from these time domain simulations. There are many other cycle counting methods available, such as peak counting, level-crossing counting, simple-range counting, range-pair counting, and reservoir counting. These methods, as well as rainflow counting, are defined in (ASTM 2005). Researchers have provided simple empirical formulae for estimating the accumulated rotation, and the simple linear damage accumulation rule (Miner 1945) can be used to assess the fatigue life. However, without detailed data about the wind turbine and sophisticated simulation tools, it is challenging to estimate the number of load cycles.

Soil behaviour is a function of strain level. Under large strains, soil behaves highly nonlinearly. Loose to medium dense sands may progressively build pore water pressure and liquefy. At certain threshold strains, some soils (clayey soil) may degrade while others may compact. With episodes of low strain level and under the action of tens of millions of load cycles, some soils may increase their stiffness. Therefore, to predict the long-term performance, one needs to know the corresponding wave height and wave period for maximum wave load calculation that will impose the largest moment in the foundation and the corresponding strain level. It is also necessary to estimate the number of cycles of loading that would influence the soil behaviour. Therefore, one needs to estimate the number of cycles of loading in a 3-hour sea state by calculating the worst-case scenario time period of wave loading. A method to predict the number of wave cycles is shown below in Section 5.6.1. The wind load cycles are typically at a much lower frequency than waves, but conservatively it is assumed that the wind and wave act at the same frequency, i.e. the frequency of the wave loading.

5.6.1 Calculation of the Number of Wave Cycles

In this step, a simplified estimation of the extreme wave height and the corresponding wave period for a given site is explained which involves the following sub-steps:

1. Obtain the relevant significant wave height H_S from a reliable source.
2. Calculate the corresponding range of wave periods, T_S.
3. Calculate the number of waves in a 3-hour period, N.
4. Calculate the maximum wave height, H_m.
5. Calculate the range of wave periods corresponding to the maximum wave height, T_m.

The sub-steps are now described in detail.

5.6.1.1 Sub-step 1. Obtain 50-Year Significant Wave Height

In absence of site-measured data, one can use data from offshore drilling stations or other sea state monitoring reports. For the United Kingdom, the document 'Wave mapping in UK waters', periodically prepared for the Health and Safety Executive, can be used (Williams 2008). In this document one can find nearby O&G stations or meteorological buoys and estimate the 50-year significant wave height at the wind farm site from that.

5.6.1.2 Sub-step 2. Calculate the Corresponding Range of Wave Periods

The range of wave periods for a given wave height can be estimated following DNV-OS-J101 (DNV 2014), and the following formula may be used:

$$11.1\sqrt{\frac{H_S}{g}} \leq T \leq 14.3\sqrt{\frac{H_S}{g}} \tag{5.81}$$

Typically, the most severe wave loads (following Morrison's equation or the McCamy-Fuchs diffraction solution) are produced by the lowest wave period, and the dynamic amplification is also highest since the frequency is closest to the natural frequency of the structure. Therefore, the peak wave period is taken as

$$T_S = 11.1\sqrt{\frac{H_S}{g}} \tag{5.82}$$

5.6.1.3 Sub-step 3. Calculate the Number of Waves in a Three-Hour Period

Typically, significant wave heights are given for a 3-hour period. In other words, this means that the significant wave height is calculated as the mean of the highest 1/3 of all waves. Therefore, many different wave heights occur within this 3-hour period, and the highest occurring wave height is called the maximum wave height H_m. To find this, one needs to know the number of waves in the 3-hour period, because the more waves there are, the higher the chance of higher waves occurring.

$$N = \frac{3hours}{T_S} = \frac{10800s}{T_S}(\approx 1000) \tag{5.83}$$

5.6.1.4 Sub-step 4. Calculate the Ratio of the Maximum Wave Height to the Significant Wave Height

The DNV code suggests taking the mode of the distribution of the highest wave heights, and thus:

$$H_m = H_S\sqrt{\frac{1}{2}ln(N)}(\approx 1.87H_S) \tag{5.84}$$

The maximum wave height may be taken conservatively as $H_m = 2H_S$. Please note that the water depth S may limit the maximum wave height. Typically, it is assumed that the breaking limit of waves (i.e. maximum possible wave height) in water depth S is $\overline{H}_m = 0.78S$. However, if the seabed has a slope, the wave may be higher than this limit, as was reported at the exposed site at Blyth (Camp et al. 2004). Therefore caution should be exercised when using this limit wave height.

5.6.1.5 Sub-step 5. Calculate the Range of Wave Periods Corresponding to the Maximum Wave Height

The same formulae can be used as in Sub-step 2.

$$11.1\sqrt{\frac{H_m}{g}} \leq T \leq 14.3\sqrt{\frac{H_m}{g}} \tag{5.85}$$

$$T_m = 11.1\sqrt{\frac{H_m}{g}} \tag{5.86}$$

The wave height and wave period combination of H_m, T_m can be used for maximum wave load calculation, incorporating dynamic amplification.

5.7 Methodologies for Long-Term Rotation Estimation

It may be noted that this is an area of active research. To fulfil the SLS requirements, the long-term behaviour of monopile foundations needs to be analysed according to DNV (2014). The main concern is the accumulated rotation $\Delta\theta$ and deflection $\Delta\rho$ (or equivalently the strain accumulation) at the mudline level. Even though the analysis is required by design standards, there is no consensus on an accepted methodology to carry out this analysis. Several approaches have been proposed based on extremely simple load scenarios, such as a cyclic excitation, which can be described by a mean load M_{mean}, a cyclic load magnitude $M_{\text{amp}} = M_{\max} - M_{\min}$, and number of cycles, N. The actual load acting on an OWT foundation, however, is extremely complicated and it is important to highlight the complexity:

1. The loading is not cyclic but in some case it may be dynamic. Loads applied are in a very wide frequency band ranging through the orders of magnitudes between 0.001–10 Hz, which includes the first few structural natural frequencies and blade natural frequencies. It is therefore difficult to estimate the numbers of load cycles. On the other hand, the frequency may be important in the determination of accumulated tilt. Even though it seems unlikely, it is not yet clear whether any significant excess pore pressures can occur and cause dynamic effects.
2. Some of the available methods were developed for very low number of cycles. Long and Vanneste (1994) points out that implicit numerical simulations typically allow for simulation of less than 50 cycles due to the accumulation of numerical errors. Furthermore, most reported tests they analysed were also carried out for 50 cycles or less and only one test had 500 cycles. The authors suggest caution when predicting the effects of very high numbers of load cycles. The numerical investigations and laboratory tests carried out by Achmus et al. (2009) and Kuo et al. (2012) go up to 10 000 cycles, and tests by Leblanc et al. (2010) have been carried out for up to 65 000 cycles. However, these are still orders of magnitudes below the expected number of load cycles of an OWT. Cuéllar (2011) has run four tests with different load scenarios for a remarkable five million cycles and identified qualitative behaviour of deformation accumulation for a high number of cycles.
3. The magnitude of dynamic loading also ranges from small to extreme loads, with load cycles ranging from a few to a few hundred cycles of extreme loads, and from millions to hundreds of millions of cycles of low-amplitude vibrations. In different states of the wind turbine different load magnitudes are expected.
4. The loading is not either one-way or two-way, but the whole range of load regimes are present at different times throughout the lifetime of the turbine. There is also disagreement in terms of whether one-way or two-way loading is more detrimental. Long and Vanneste (1994) suggest that one-way loading is the critical load scenario and two-way loading causes less accumulated strain. Achmus et al. (2009) and Kuo et al. (2012) also focus on one-way loading in their analysis. However, Leblanc et al.

(2010) found that the most critical scenario is two-way loading with $M_{min}/M_{max} = -0.5$.

5. The loading is not unidirectional, loads appear both in the along-wind (x) and cross-wind (y) directions during the operational life of the turbine, and cyclic vertical (z) loads are also present. Wind and waves are also not always collinear, causing multidirectional loading on the foundation.
6. The nacelle always turns into the wind, which means that the along-wind (x) and cross-wind (y) directions are not fixed in a global frame of reference but are turning as the yaw angle of the rotor changes. This means that the foundation is loaded both with along-wind and cross-wind loads throughout the lifetime of the turbine in all directions.

In pile design for offshore wind turbines, the *p-y* method is typically employed. Long and Vanneste (1994) give a good account of the research efforts into the analysis of piles under cyclic lateral loading by modified cyclic *p-y* curves. Improved *p-y* curves for cyclic lateral load were developed by Reese et al. (1974), O'Neill and Murchinson (1983), Little and Briaud (1988). More fundamental theoretical approaches have been attempted by Swane and Poulos (1982) as well as Matlock et al. (1978). As Long and Vanneste (1994) points out, these methods require parameters that are typically not available from site investigation.

The simplified approach of accumulated strain is often used in literature, which is equivalent to reduction of soil stiffness. The coefficient of subgrade reaction n_h can be reduced in order to account for the effects of cyclic loading. Such an approach was used by Prakash (1962), Davisson (1970), and Davisson and Salley (1970). Broms (1964) pointed out that the reduction of n_h depends on the density of the cohesionless soil. These studies suggest reducing n_h by a fixed percentage if a certain number of load cycles (~50) are expected (30% in Davisson (1970), 75% and 50% for dense and loose sand in Broms (1964)).

Logarithmic expressions for permanent strains of monopiles have also been proposed by Hettler (1981), Lin and Liao (1999), Verdure et al. (2003), Achmus et al. (2009), and Li et al. (2010). Power law expressions have been proposed for monopiles by Little and Briaud (1988), Long and Vanneste (1994), Leblanc et al. (2010), Klinkvort et al. (2010), and for caissons by Zhu et al. (2012) and Cox et al. (2014). Cuéllar (2011a,b) proposes a method by which three different curves are used to approximate the long-term accumulation.

Some important contributions are listed below.

5.7.1 Simple Power Law Expression Proposed by Little and Briaud (1988)

Little and Briaud (1988) proposed the simple power law expression for strain accumulation:

$$\varepsilon_N = \varepsilon_1 N^m \tag{5.87}$$

where ε_N is the strain after N cycles, ε_1 is the strain at the first load cycle and m is a constant that expresses dependence on soil and pile parameters, installation methods, and loading characteristics. Achmus et al. (2009) uses $m = 0.136$ for typical monopiles.

5.7.2 Degradation Calculation Method Proposed by Long and Vanneste (1994)

Long and Vanneste (1994) provide a simple approach for calculating the degradation of the coefficient of subgrade reaction determined from the analysis of 34 different load test scenarios. The degradation is expressed as

$$n_{hN} = n_{h1}N^{-t} \tag{5.88}$$

where n_{hN} is the coefficient after N cycles of loading, n_{h1} is the coefficient at the first cycle, and t is the degradation parameter, which can be calculated according to

$$t = 0.17F_rF_iF_d \tag{5.89}$$

F_r, F_i, F_d are parameters to take into account the cyclic load ratio, pile installation method and soil density, respectively. For a driven pile ($F_i = 1.0$) in medium sand ($F_r = 1.0$) under one-way loading ($F_r = 1.0$), the degradation parameter is $t = 0.17$.

Long and Vanneste (1994) emphasise that the load tests that serve as the basis of their analysis were mostly carried out for less than 50 load cycles, with only one test going up to 500. They suggest caution when applying these results for more than 50 load cycles, making the method hard to implement for OWT foundations where load cycles in the orders of magnitudes of $10^2 - 10^8$ are expected. Furthermore, Long and Vanneste (1994) point out that the most important factor is the cyclic load ratio M_{min}/M_{max}, however, in their formulation the reduction in the coefficient of subgrade reaction is not explicitly dependent of the cyclic load magnitude.

5.7.3 Logarithmic Method Proposed by Lin and Liao (1999)

Lin and Liao (1999) provide a logarithmic expression for strain accumulation:

$$\varepsilon_n = \varepsilon_1[1 + t\ \ln(N)] \tag{5.90}$$

$$t = 0.032L_P\sqrt[5]{\frac{n_h}{E_PI_P}}F_rF_iF_d \tag{5.91}$$

L_P is the embedded length of the pile, F_r, F_i, F_d are parameters to take into account the cyclic load ratio, pile installation method and soil density, respectively. In the basic case of a driven pile in dense sand in one-way loading, $t = 0.032L_P/T$ where $T = \sqrt[5]{(E_PI_P)/n_h}$ is the pile/soil relative stiffness ratio, with n_h being the coefficient of subgrade reaction.

Lin and Liao (1999) also provide a methodology to combine loads at different load levels into a single load case for the calculation of accumulated strain. This is achieved by converting all load cycles to a single load level by the method of equivalent accumulated strains. If there are two loads, say a and b, with t_a and t_b as degradation parameters and N_a and N_b as numbers of load cycles, respectively, then

$$\varepsilon_{N(a+b)} = \varepsilon_{1b}[1 + t_b\ ln(N_b^* + N_b)] \tag{5.92}$$

where N_b^* is the equivalent load cycle number of load a in terms of the degradation parameter t_b expressed as

$$N_b^* = e^{\frac{1}{t_b}\left[\frac{\varepsilon_{1a}}{\varepsilon_{1b}}(1+t_a ln(N_a))-1\right]} \tag{5.93}$$

5.7.4 Stiffness Degradation Method Proposed by Achmus et al. (2009)

Achmus et al. (2009) and Kuo et al. (2012) carried out laboratory tests and developed a numerical modelling procedure for analysing the long-term mudline deformations of rigid monopiles. According to their stiffness degradation method, the increase in plastic strain due to cyclic loading can be interpreted as a decrease in the soil's Young's modulus E_S:

$$\frac{\varepsilon_{p1}}{\varepsilon_{pN}} = \frac{E_{SN}}{E_{S1}} \tag{5.94}$$

where $\varepsilon_{pN=1}$ is the plastic strain at the first load cycle, ε_{pN} is the plastic strain at the Nth cycle, and similarly for the elastic modulus. The following semi-empirical approach for strain accumulation is used:

$$\varepsilon_n = \frac{\varepsilon_1}{(N)^{-b_1(x_c)^{b_2}}} \tag{5.95}$$

where:

$$b_1, b_2 = \text{model parameters,}$$

$$x_c = \text{the characteristic cyclic stress ratio ranging 0–1}$$

$$x_c = \frac{(\textit{cyclic stress ratio at loading}) - (\textit{cyclic stress ratio at unloading})}{1 - (\textit{cyclic stress ratio at unloading})} \tag{5.96}$$

Basic design charts for preliminary design are presented in Achmus et al. (2009) for the pile-head deflection; the rotation is only slightly touched on in the paper.

5.7.5 Accumulated Rotation Method Proposed by Leblanc et al. (2010)

Leblanc et al. (2010) carried out tests for rigid piles in sand to assess the long-term behaviour in terms of accumulated rotation at the mudline. Their tests were carried out using two main parameters:

$$\zeta_b = \frac{M_{max}}{M_R} \tag{5.97}$$

which describes the magnitude of loading with respect to the static moment resisting capacity of the pile M_R. The values are between 0 and 1, and

$$\zeta_c = \frac{M_{\min}}{M_{\max}} \tag{5.98}$$

which describes the nature of the cyclic loading. Values are between −1 for pure two-way loading and 1 for static load, with 0 being the pure one-way loading.

The load tests were carried out with two different values of relative density, $R_d = 4\%$ and $R_d = 38\%$. After up to 65 000 cycles of loading the accumulated tilt was found to be in the form

$$\theta_N = \theta_0 + \Delta\theta(N) \tag{5.99}$$

$$\Delta\theta(N) = \theta_S T_b(\zeta_b, R_d) T_c(\zeta_c) N^{0.31} \tag{5.100}$$

where R_d is the relative density of sand, θ_0 is the rotation at maximum load of the first load cycle, θ_S is the pile rotation under a static load equal to the maximum cyclic load. The functions $T_b(\zeta_b, R_d)$ and $T_c(\zeta_c)$ are given in graphs in Leblanc et al. (2010). A piecewise linear approximation for $T_c(\zeta_c)$ can be used for simplicity:

$$T_c = \begin{array}{lll} 13.71\zeta_c + 13.71 & for & -1 \leq \zeta_c < -0.65 \\ -5.54\zeta_c + 1.2 & for & -0.65 \leq \zeta_c < 0 \\ -1.2\zeta_c + 1.2 & for & 0 \leq \zeta_c < 1 \end{array} \tag{5.101}$$

Similarly, the equations for $T_b(\zeta_b, R_d)$ can be given for two values of R_d following Leblanc et al. (2010):

$$T_b = \begin{array}{lll} 0.4238\zeta_b - 0.0217 & for & R_d = 38\% \\ 0.3087\zeta_b - 0.0451 & for & R_d = 4\% \end{array} \tag{5.102}$$

5.7.6 Load Case Scenarios Conducted by Cuéllar (2011)

Cuéllar (2011) carried out lateral load tests of a rigid monopile, running four different load case scenarios for a remarkable 5 million load cycles. The goal of the analysis was to identify qualitative trends in the strain accumulation and to analyse densification of the soil due to cyclic lateral load on the pile. It was found in their study that the plot of the accumulation of permanent deformation against the number of cycles can be approximated by three different simplified curves.

1. In the first roughly 10^4 cycles, the accumulation follows a logarithmic curve in the form of $\rho_1 = C_1 + C_2 \log(N)$. A quick accumulation of permanent displacements likely due to the densification around the pile is followed by a region of stabilised cyclic amplitude.
2. Stabilised cyclic amplitude is characteristic of this intermediate region, where the accumulation is roughly linear with $\rho_2 = C_3 + C_4 N$.
3. The last section after the second inflection point at roughly 10^6 number of cycles can be approximated by a power law curve as $\rho_3 = C_5 N^{C_6}$.

The rate of accumulation never seems to fall to zero, but the accumulated rotation appears to increase indefinitely.

5.8 Theory for Estimating Natural Frequency of the Whole System

The structural model that can be used for simplified calculations is shown in Figure 5.18. The foundation is represented by three springs: lateral K_L, rotational K_R, and cross K_{LR} stiffness. The tower can be idealised by equivalent bending stiffness and mass per length. Two types of beam theory can be used: Euler-Bernoulli beam theory and improved Timoshenko beam theory, and partial differential equations need to be used. Timoshenko beam theory accounts for shear deformation and the effect of rotational inertia, see Figure 5.45. In both the models, the rotor-nacelle assembly (RNA) is modelled as a top head mass with mass moment of inertia.

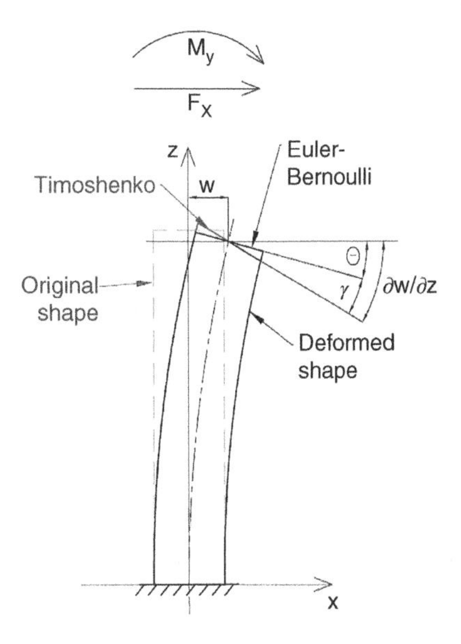

Figure 5.45 Timoshenko beam: the effect of rotary inertia and shear deformation are added to the Euler–Bernoulli beam model.

5.8.1 Model of the Rotor-Nacelle Assembly

The RNA is modelled as a top head mass M_2 with mass moment of inertia J, as shown in Figure 5.18. These parameters can be used to formulate the end boundary conditions of the PDEs of the motion of the tower. In addition, the mass M_2 exerts a downwards pointing force P due to gravity, which can be considered constant along the tower:

$$P = -M_2 g \tag{5.103}$$

The self-weight of the tower also produces an axial force on the sections below;

$$F_{M_3} = -mg(L - z) \tag{5.104}$$

where m is the average mass per length of the tower, L is the height of the tower, and z is the coordinate running along the tower. An approximate method is followed here as given in Adhikari and Bhattacharya (2011) and in Blevins (1979, p. 36):

$$P^* = -M_{corr}g = -(M_2 + C_M M_3)g \tag{5.105}$$

where M_{corr} is the corrected mass, M_2 is the mass of the nacelle, M_3 is the mass of the tower, and P^* is the modified constant axial force. The mass correction factor $C_M = 33/140 \approx 0.24$ for a cantilever beam is given in, e.g. Blevins (1979). The mass correction factor for flexible foundation was derived in Adhikari and Bhattacharya (2012). This expression uses the nondimensional variables given in Eq. (5.45).

$$C_M = \frac{3}{140} \cdot \frac{11\eta_R^2\eta_L^2 + 77\eta_L^2\eta_R + 105\eta_R^2\eta_L + 140\eta_L^2 + 420\eta_L\eta_R + 420\eta_R^2}{9\eta_R^2 + 6\eta_R^2\eta_L + 18\eta_R\eta_L + \eta_R^2\eta_L^2 + 6\eta_L^2\eta_R + 9\eta_L^2} \tag{5.106}$$

5.8.2 Modelling the Tower

Generally, if the beam is slender, the simple Euler-Bernoulli beam model can be used, which can be improved by taking into consideration the effect of rotational inertia

(Rayleigh beam). As a rule of thumb, the column can be considered as a slender beam if the length-to-diameter ratio is higher than about 20, and stocky when the ratio is smaller than 10, in which case the Timoshenko beam theory is necessary.

5.8.3 Euler-Bernoulli Beam – Equation of Motion and Boundary Conditions

The equation of motion can be written as

$$\frac{\partial^2}{\partial z^2}\left(EI(z)\frac{\partial^2 w(z,t)}{\partial z^2}\right) + m(z)\frac{\partial^2 w(z,t)}{\partial t^2} + \frac{\partial}{\partial z}\left(P^*\frac{\partial w(z,t)}{\partial z}\right) = p(z,t) \quad (5.107)$$

where $w(z,t)$ is the translational deflection of the beam at the cross-section z at time t, $EI(z)$ is the bending stiffness of the beam at cross section z, $m(z)$ is the mass per length of the beam at cross section z, P^* is the constant, corrected lumped mass at the top of the tower, and $p(z,t)$ is an arbitrary distributed force. In the following, an equivalent bending stiffness and a constant mass per length are assumed, the notations $w'(z,t)$ for derivative with respect to z and $\dot{w}(z,t)$ for the time derivative are introduced, the explicit notation of space and time dependence is omitted, and the variables are separating and a harmonic solution is assumed. Boundary conditions are determined considering the flexible foundation (K_L, K_R, and K_{LR}) and the lumped mass and inertia (M_2 and J). The readers are referred to Adhikari and Bhattacharya (2011), Arany et al. (2016), and Arany et al. (2015a,b) for details of the boundary conditions. Some details are provided in Appendices A to C.

5.8.4 Timoshenko Beam Formulation

The classical Euler-Bernoulli beam theory does not include the effects of shear deformation and rotary inertia, which should be included for stocky beams. The analysis was carried out for wind turbine towers for the sake of comparison to simpler beam models and to investigate whether the slender beam approximation is appropriate. The equation of motion was derived by Timoshenko and can be written with two equations as:

$$\begin{aligned} -GkA\left(\frac{\partial^2 w(z,t)}{\partial z^2} - \frac{\partial \Theta(z,t)}{\partial z}\right) + \rho A\frac{\partial^2 w(z,t)}{\partial t^2} &= p(z,t) \\ EI\frac{\partial^2 \Theta(z,t)}{\partial z^2} + GkA\left(\frac{\partial^2 w(z,t)}{\partial z^2} - \frac{\partial \Theta(z,t)}{\partial z}\right) - I\rho\frac{\partial^2 \Theta(z,t)}{\partial t^2} &= 0 \end{aligned} \quad (5.108)$$

where $w(z,t)$ is the transversal displacement in the x direction at the coordinate z at time t; ρ is the density of the material of the tower; A is the area of the cross section $A = (D_{outer}^2 - D_{inner}^2)\pi/4$; k is the Timoshenko shear coefficient; G is the shear modulus; $p(z,t)$ is the external force; EI is the bending stiffness of the tower; E is Young's modulus; $I = (D_{outer}^4 - D_{inner}^4)\pi/64$ is the area moment of inertia of the cross section; and $\Theta(z,t)$ is the angle due to pure bending. These two equations can be transformed into one. For the natural frequency, the excitation is taken to be zero:

$$EI\frac{\partial^4 w(z,t)}{\partial z^4} - \left(\frac{\rho EI}{Gk} + \rho I\right)\frac{\partial^4 w(z,t)}{\partial z^2 \partial t^2} + \frac{\rho^2 I}{Gk}\frac{\partial^4 w(z,t)}{\partial t^4} + \rho A\frac{\partial^2 w(z,t)}{\partial t^2} = 0 \quad (5.109)$$

Using separation of variables and assuming a harmonic solution:

$$\underset{I\cdot}{EI\frac{\partial^4 W(z)}{\partial z^4}} + \underset{II\cdot}{\omega^2\frac{\rho EI}{Gk}\frac{\partial^2 W(z)}{\partial z^2}} + \underset{III\cdot}{\omega^2\rho I\frac{\partial^2 W(z)}{\partial z^2}} + \underset{IV\cdot}{\omega^4\frac{\rho^2 I}{Gk}W(z)} + \underset{V\cdot}{\omega^2\rho AW(z)} = 0$$

I. Shear force: included in the Euler-Bernoulli beam theory.
II. Shear deformation: included in the Timoshenko beam theory. If shear deformation is to be excluded from the equation, it can be done by letting $G \to \infty$.
III. Rotary inertia: included in the Rayleigh beam theory. If rotary inertia is to be neglected, then terms containing ρI must be set equal to zero (but not ρIE).
IV. Coupling term between shear deformation and rotary inertia: it is only present if both effects are included.
V. Force by acceleration: included in the Euler-Bernoulli beam theory.

Using the nondimensional variable $\xi = z/L$ and incorporating the axial loading, the following equation can be written:

$$\underset{I\cdot}{\frac{EI}{L^4}\frac{\partial^4 W(\xi)}{\partial \xi^4}} + \underset{II\cdot}{\omega^2\frac{\rho IE}{L^2 Gk}\frac{\partial^2 W(\xi)}{\partial \xi^2}} + \underset{III\cdot}{\omega^2\frac{\rho I}{L^2}\frac{\partial^2 W(\xi)}{\partial \xi^2}} + \underset{IV\cdot}{\omega^4\frac{\rho^2 I}{Gk}W(\xi)} - \underset{V\cdot}{\omega^2\rho AW(\xi)}$$
$$= \frac{1}{L}\frac{\partial}{\partial x}\left(\frac{F_C(\xi)}{L}\frac{\partial W(\xi)}{\partial \xi}\right)$$
$$F_C(\xi) = -P - \rho gAL(1-\xi) \tag{5.110}$$

where $F_C(\xi)$ is the compressive force at each cross section.

$$W^{IV} + \left(\frac{P^*L^2}{EI} + \frac{\omega^2\rho IL^2}{EI} + \frac{\omega^2\rho L^2 EI}{EIGk}\right)W'' + \left(\frac{\omega^4\rho^2 IL^4}{EIGk} - \frac{\omega^2\rho AL^4}{EI}\right)W = 0 \tag{5.111}$$

Nondimensional parameters have been used by Arany et al. (2016) to study the problem and are shown in Table 5.16. The range of typical values (estimated based on four wind turbines) is also shown in Table 5.16 and 5.18.

The above-presented nondimensional numbers were determined for four offshore wind turbines, the measured and/or estimated natural frequencies of which are available in the literature. The information necessary for frequency estimation are given in Table 5.17, the nondimensional variables are calculated in Table 5.18 and the results of the proposed model are compared to available values in Table 5.18. The approximations tend to underestimate the natural frequency except for the Lely A2 wind farm, even if the foundation is infinitely stiff (fixed-base natural frequency). It is to be noted here that the model uses a constant average thickness for the towers, determined based on the range of thickness available in the literature and the masses of the towers. Real towers tend to have thicker walls in the bottom section and thinner ones in the upper sections, which increases the tower's natural frequency as compared to a constant thickness. As the natural frequency was found to be highly sensitive to the chosen wall thickness, this parameter may be the reason behind the underestimation. There are many other sources of uncertainty, including but not limited to damping (soil, structural, aerodynamic, hydrodynamic), lengths, soil parameters and foundation stiffness, flexibility of support structure connections, and mass moment of inertia (Table 5.19).

Table 5.16 Nondimensional groups: definitions and practical range.

Dimensionless group	Formula	Typical values
Nondimensional lateral stiffness	$\eta_L = \dfrac{K_L L^3}{EI}$	2 500–12 000
Nondimensional rotational stiffness	$\eta_R = \dfrac{K_R L}{EI}$	25–80
Nondimensional cross stiffness	$\eta_{LR} = \dfrac{K_{LR} L^2}{EI}$	(−515) to (−60)
Nondimensional axial force	$v = \dfrac{P^* L^2}{EI}$	0.005–0.1
Mass ratio	$\alpha = \dfrac{M_2}{M_3}$	0.75–1.2
Nondimensional rotary inertia	$\beta = \dfrac{J}{mL^2}$	a)
Nondimensional shear parameter	$\gamma = \dfrac{E}{Gk}$	~4.5 (for steel tubular towers)
Nondimensional radius of gyration	$\mu = \dfrac{r}{L}$	0.008–0.025
Frequency scaling parameter	$c_0 = \sqrt{\dfrac{EI}{M_3 L^3}}$	~1–5
Nondimensional rotational frequency	$\Omega_k = \dfrac{\omega_k}{c_0} = \omega_k \sqrt{\dfrac{M_3 L^3}{EI}}$	—

K_L, K_R, K_{LR} are the lateral, rotational, and cross stiffness of the foundation, respectively; EI is the equivalent bending stiffness of the tapered tower; L is the height of the tower; P^* is the modified axial force (see Eq. 4); M_2 is the total top mass (blades, hub, nacelle, etc.); M_3 is the mass of the tower; J is the rotary inertia of the top mass; m is the equivalent mass per length of the tower; $r = \sqrt{I/A}$ is the radius of gyration of the tower; ω_k is the kth natural frequency of the tower.

a) The rotary inertia is taken to be zero for all wind turbines considered, as information is not available in the referenced literature.

5.8.5 Natural Frequency versus Foundation Stiffness Curves

Figure 5.46 plots the relationship between two non-dimensional support stiffness (η_L and η_R) and the relative frequency for various values of cross-coupling stiffness (K_{LR}). It can be concluded from these graphs that with the increase of the cross-spring stiffness, the OWT is closer to the high-slope zone of the curves. In such circumstances, the changes in the stiffness parameters introduce a higher change in natural frequency. In Figures 5.47–5.49, the same four turbines are placed in the relative frequency figures with respect to the three stiffness parameters. Each line style shows the relative frequency curve of a given OWT, and the estimated values on the curves are also shown for each turbine. In general, the sensitivity of the natural frequency to each parameter may be characterised by the slope of the curve at the estimated stiffness values.

One of the main purposes of SSI analysis is to study the effect of change of soil parameters during the operation of the turbine, i.e. how much the frequency will change if soil softens or stiffens due to repeated cyclic loading. This change in foundation stiffness has

Table 5.17 Information required for frequency estimation.

Wind turbine	Lely A2: NM41 2-bladed, 500 kW turbine for study purposes	North Hoyle Vestas V80 2 MW industrial wind turbine	Irene Vorrink 600 kW study purpose wind turbine	Walney 1 Siemens SWT-3.6-107 3.6 MW industrial wind turbine
Given and calculated geometric and material data				
Equivalent bending stiffness – EI (GNm)	22	133	21.5	274
Young's modulus of the tower material – E (GPa)]	210	210	210	210
Shear modulus of the tower material – G (GPa)	79.3	79.3	79.3	79.3
Tower height – L (m)	41.5	70	51	83.5
Bottom diameter – D_b (m)	3.2	4.0	3.5	5.0
Top diameter – D_t (m)	1.9	2.3	1.7	3.0
Tower wall thickness range – t (mm)	~12	~35	...14	20.80
Lateral foundation stiff-ness – K_L (GN m^{-1})	0.83	3.1...3.5	0.76	3.65
Rotational foundation stiffness – K_R (GNm rad^{-1})	20.6	33.8...62.1	15.5	254.3
Cross-coupling foundation stiffness – K_{LR} (GN)	−2.22	−1.71	−2.35	−20.1
Top mass (rotor-nacelle assembly) – M_2(kg)	32 000	100 000	35 700	234 500
Tower mass – M_3 (kg)	31 440	130 000	31 200	260 000
Average wall thickness – t_h (mm)	12	35	11	40
Shear coefficient – k (–)	0.532 8	0.532 8	0.532 6	0.532 7

been reported in small-scale tests and observed in the case of the Hornsea Met Mast Twisted Jacket foundation. A few design pointers may be deduced from the study:

- From a design point of view, it is important to note that the nondimensional stiffness values have a certain region where the frequency function becomes flat, i.e. any change in foundation stiffness will have very little impact on the natural frequency. The designer may wish to choose the tower stiffness (EI) and foundation stiffness (K_L,

Table 5.18 Nondimensional parameters.

Wind turbine	Lely A2	North Hoyle	Irene Vorrink	Walney 1e
Nondimensional parameters				
Nondimensional lateral stiffness – η_L	2698	11775	5880	7763
Nondimensional rotational stiffness – η_R	38.88	28	39.64	77.49
Nondimensional cross-coupling stiffness – η_{LR}	−174	−63	−284	−511.7
Nondimensional axial force – v	0.033	0.011	0.030	0.043
Mass ratio – α	1.018	0.760	1.144	0.9
Nondimensional rotary inertia – β	0	0	0	0
Frequency scaling parameter – c_0	3.130	1.323	2.035	1.3454
Nondimensional radius of gyration – μ	0.0214	0.0161	0.01795	0.0168
Nondimensional shear parameter – γ	4.970	4.970	4.970	4.97

Table 5.19 Frequency results.

Wind turbine	Lely A2	North Hoyle	Irene Vorrink	Walney 1
Natural frequency results				
Measured value – f_m (Hz)	0.634	N/A	0.546	0.35
Result produced by [28] – f_S (Hz)	0.740	0.345	0.457	—
Present study (Euler-Bernoulli) – f_{E-B} (Hz) (error [%])	0.735 (15.9%)	0.345 (N/A)	0.456 (16.5%)	0.331(5.9%)
Present study (Timoshenko) – f_T (Hz) (error[%])	0.734 (15.8%)	0.345 (N/A)	0.456 (16.5%)	0.331(5.9%)
Fixed-base frequency (infinitely stiff foundation)	0.765	0.364	0.475	0.345

K_R, and K_{LR}) such as to remain in the safe region to ensure that even if the stiffness parameters change during the lifetime of the turbine, the natural frequency will not be greatly affected.

- It can be observed that the structures are relatively insensitive to the lateral stiffness (K_L) and the most important factor is the rotational stiffness (K_R). In general, one can conclude that the change in rotational stiffness of the foundation causes the greatest change in natural frequency.
- The cross stiffness (K_{LR}) also has important effects and may not be neglected.

5.8.6 Understanding Micromechanics of SSI

The readers are referred to the work of Cui and Bhattacharya (2017) and Lopez Querol et al. (2017) where discrete element modelling of the monopile-soil interaction has been studied.

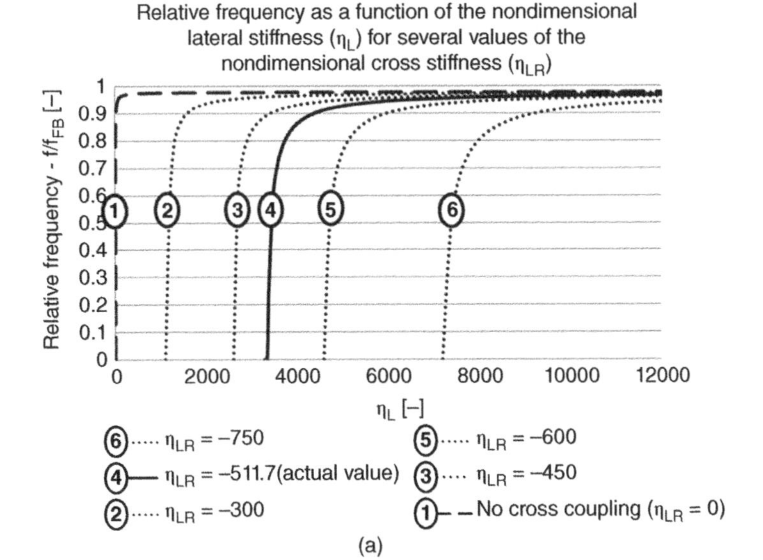

(a)

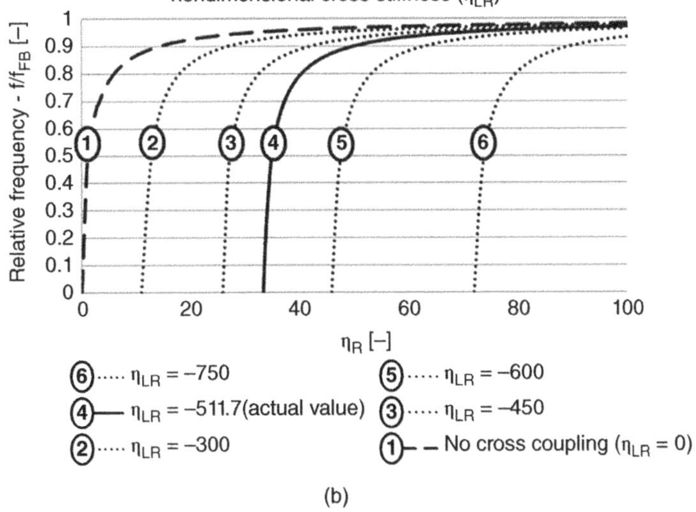

(b)

Figure 5.46 .Relative frequency as a function of (a) the nondimensional lateral stiffness, (b) the nondimensional rotational stiffness for several values of the nondimensional cross stiffness.

Chapter Summary and Learning Points

1. SSI governs the long-term performance of the wind turbine system. The issues affecting a monopile foundation are the extreme wind and wave loading, cyclic and dynamic loading, and soil erosion.
2. The change in natural frequencies of the wind turbine system may be affected by the choice of foundation system, i.e. deep foundation or multiple pods on shallow

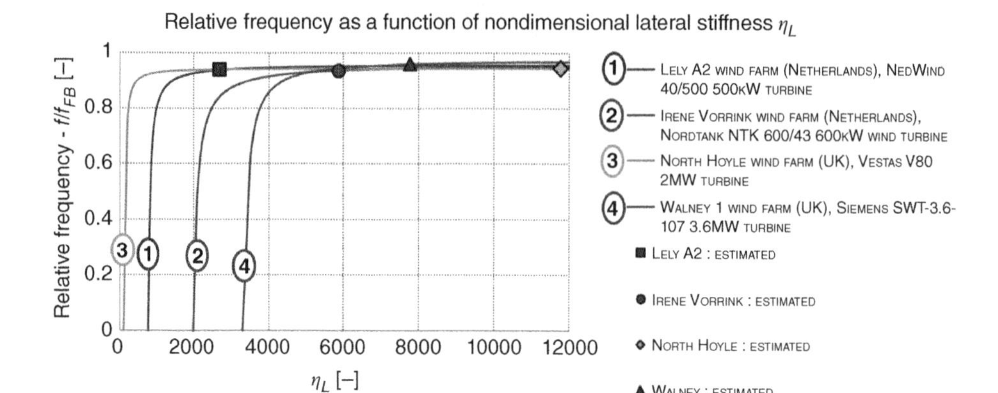

Figure 5.47 Relative frequency $f_r = \frac{f}{f_{FixedBase}}$ as a function of the nondimensional lateral stiffness $\eta_L = \frac{K_L L^3}{EI}$.

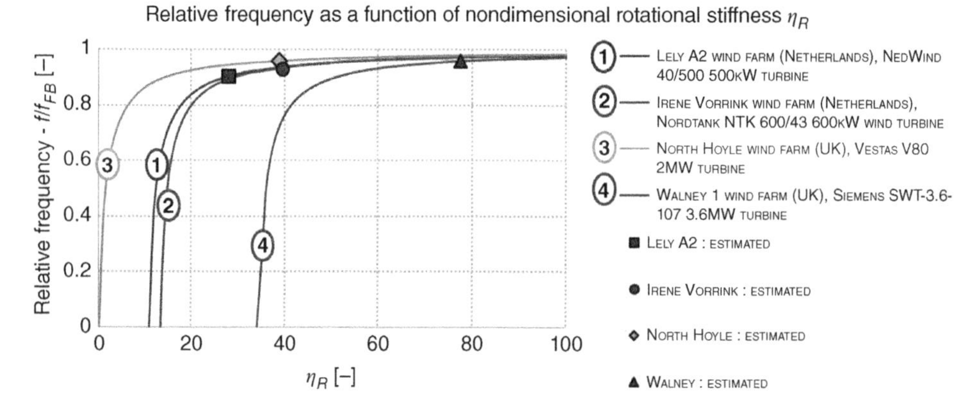

Figure 5.48 Relative frequency $f_r = \frac{f}{f_{FixedBase}}$ as a function of the nondimensional rotational stiffness $\eta_R = \frac{K_R L}{EI}$.

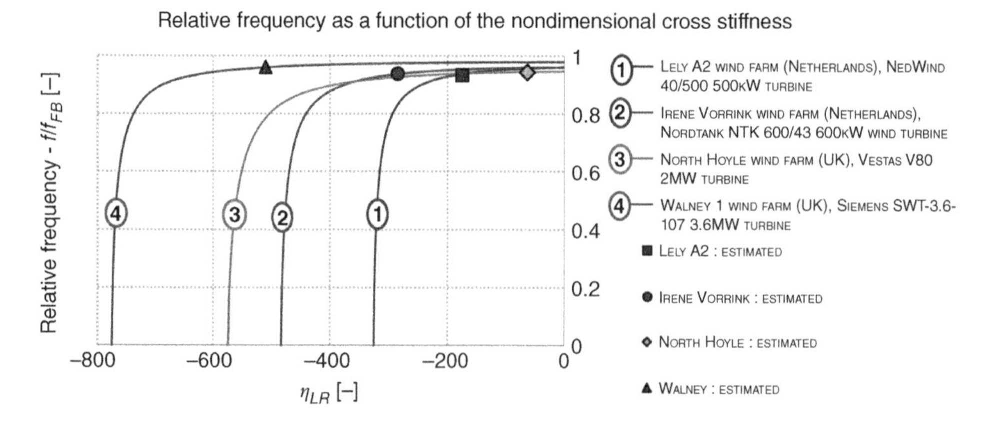

Figure 5.49 Relative frequency $f_r = f/f_{FB}$ as a function of the nondimensional cross stiffness $\eta_{LR} = K_R L^2/EI$.

foundations. Deep foundations such as monopiles will exhibit sway-bending mode, i.e. the first two modes are widely spaced. However, multiple pod foundations supported on shallow foundations (such as tetrapod or tripod on suction caisson) will exhibit rocking modes in two principal planes.

3. The natural frequencies of wind turbine systems change with repeated cyclic/dynamic loading. In the case of a strain-hardening site (such as loose to medium-dense sandy site) the natural frequency is expected to increase, and for strain-softening site (such as normally consolidated clay) the natural frequency will decrease.
4. Due to the rocking modes of vibration, there will be a 'two-peak' response, i.e. two closely spaced frequencies for multipod foundations (tripod/tetrapod).

6

Simplified Hand Calculations

Learning Objectives

1. The aim of this chapter is to carry out hand calculations relevant to different aspects of foundation design based on methods presented in Chapters 2 and 5.
2. It is often necessary to cross-check high-fidelity finite element calculations. The methods presented in the chapter can be used for such purposes.
3. For financial viability of a project, a quick estimation of the sizes of the foundation are needed without the recourse to expensive numerical modelling. The calculations presented in this chapter will be helpful for such purpose.

6.1 Flow Chart of a Typical Design Process

It is often necessary to take a bird's-eye view of a design process highlighting the different ingredients. Figure 6.1 shows a typical flowchart of a foundation design.

As may be observed, the main ingredients for foundation design are:

(a) Loads on the foundation under different scenarios
(b) Site investigation, i.e. ground conditions
(c) Criteria for design

This chapter provides the following examples:

1. Target frequency of a turbine
2. Stiffness of a monopile foundation using three types of method: simplified, standard, and advanced methods
3. Stiffness of a mono-caisson
4. Estimation of loads on a monopile foundation through the use of spreadsheet type program
5. Natural frequency of a monopile-supported wind turbine
6. Design of a monopile-supported wind turbine
7. Design of jacket-type wind turbine system
8. Design of a spar buoy type wind turbine

The aim of this chapter is to present some example problems on the different concepts.

Design of Foundations for Offshore Wind Turbines, First Edition. Subhamoy Bhattacharya.

Companion website: www.wiley.com/go/bhattacharya/offshorewindturbines

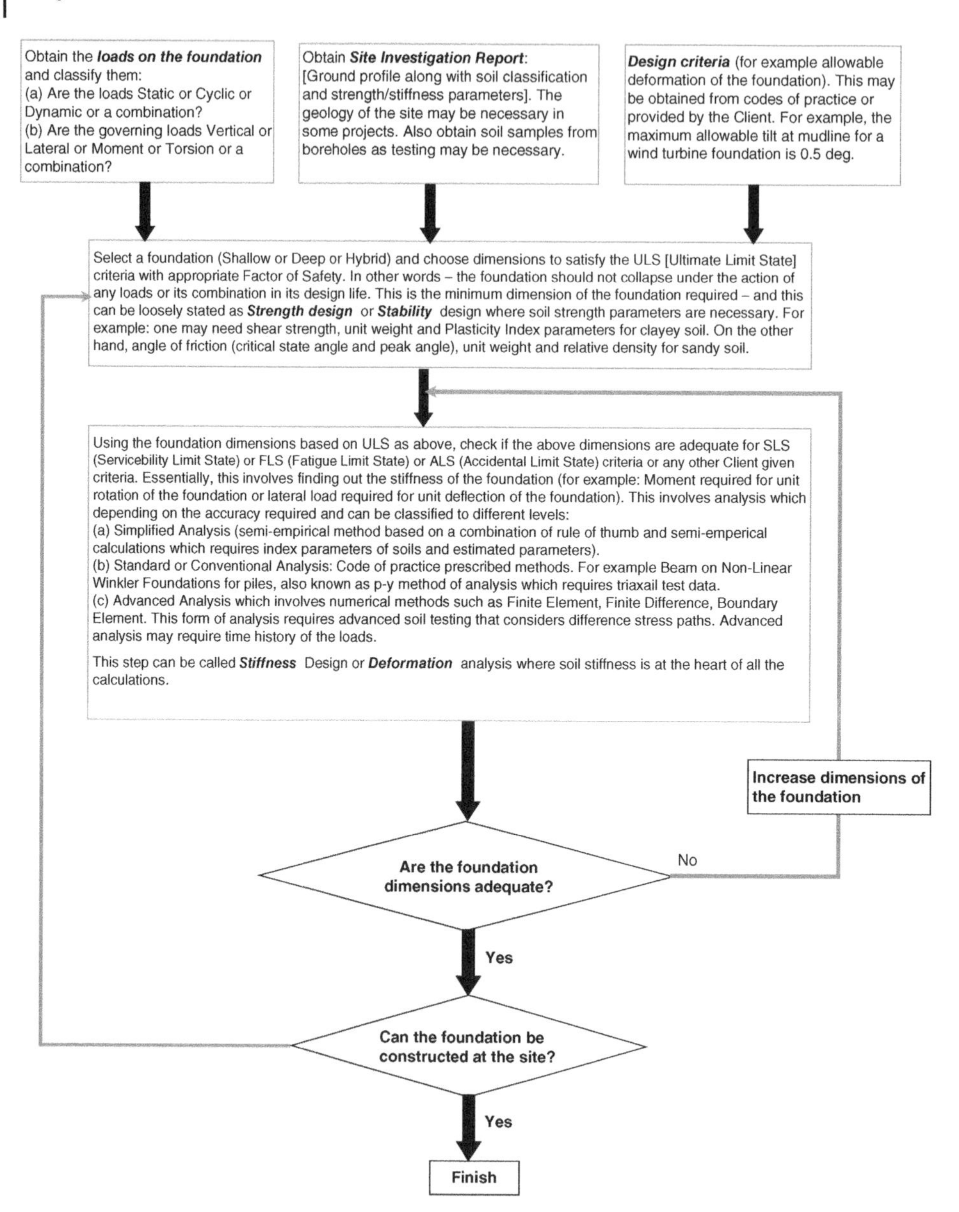

Figure 6.1 Flowchart of a design process.

6.2 Target Frequency Estimation

Example 6.1 ***Target Frequency for SWT-4-130 Turbines in the China Sea*** The readers are referred to Chapter 2 for theoretical aspects. The calculation involves construction of wind and wave spectrum and using the information on the rotor frequency (1P) range. An example is taken from proposed wind farm in Fujian Pingtan Dalian Island; see Figure 6.2 for the location. This 300 MW offshore wind turbine farm

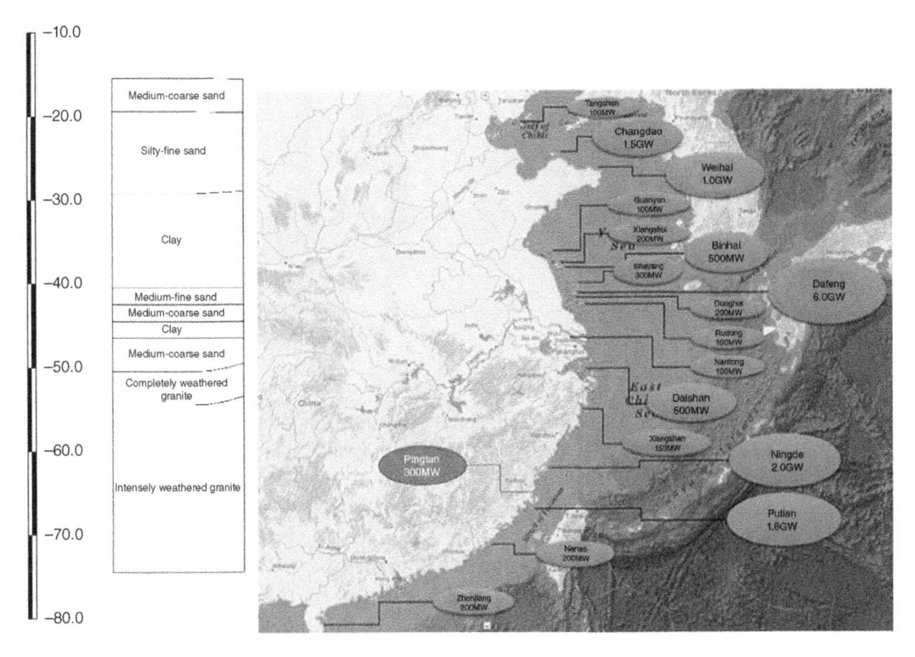

Figure 6.2 Typical ground profile and location of Pingtan Wind Farm. [Photo courtesy: Yupei Huang ZEPDI].

will have 4 MW capacity turbines. Table 6.1 summarises the data used to estimate the target frequency.

Figure 6.3 plots the wind spectra using Kaimal spectrum and wave using JONSWAP spectra and the formulas are provided in Chapter 2. Detailed calculation on the spectral estimates are discussed in Chapter 2 considering the example of Walney Wind Farm. Also in the figure, 1P (rotational frequency range) and 3P (blade passing frequency) are plotted. JONSWAP spectra may not be applicable for this offshore China location, and region-specific spectra would be necessary. However, in the absence of such information, JONSWAP has been used to show the application of the methodology.

Following Figure 6.3, the gap between upper bound of 1P and lower bound of 3P is narrow and therefore in soft-stiff design, choosing of target frequency is challenging.

Table 6.1 Properties of the turbine and site characterises.

Turbine characteristics	3-bladed horizontal axis pitch regulated 4 MW turbine (Siemens SWT-4-130); RNA mass of 240 tonnes total comprising of 100 tonnes rotor; 130 m rotor, 1P range (5-14RPM)
Site characteristics (wind)	Mean wind speed at hub level is 9.3 m s^{-1}, extreme wind speed (50-year return period) is 48 m s^{-1}, turbine intensity (10%)
Site characteristics (wave)	Design wave height (10 m) and peak wave period is 10.5 s; water depth is 28 m
Cut in and cut out wind speed	Cut-in wind speed 3–5 m s^{-1} and cut-out wind speed 32 m s^{-1}, nominal; power at 11–12 m s^{-1}

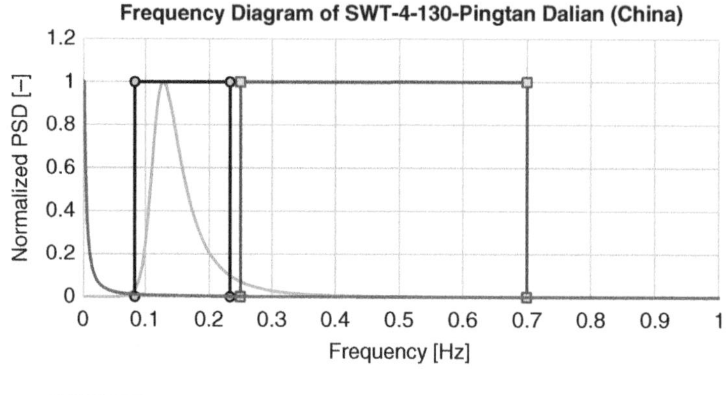

Figure 6.3 Frequency diagram.

Figure 6.4 Structural model of the support structure. (Source: Photo Courtesy: Dr Yu).

Any value in this narrow gap will be within 10% of 1P and 3P and therefore does not conform to the DNV guidelines. In such cases, engineering judgement is necessary and the guiding principle is: *which target frequency will cause the least amount of fatigue damage over the design lifetime*. Therefore, it is necessary to carry out fatigue load mitigation measures either through control system design (leapfrog certain 1P frequency) or designing structural sections considering fatigue damage. As wave loads have significant amount of energy and circa few orders of magnitudes more than 1P and 3P, it is necessary to avoid waves frequencies. In this case, the target range between 0.25 and 0.35 may be acceptable.

The ground condition is shown in Figure 6.2. A suitable design in this scenario may be a jacket on long slender piles. One of the option designs is presented here: 40 m height four-legged jacket supported on piles. The jacket is 14 m × 14 m at the top and flares symmetrically to 26 m × 26 m at the bottom. The piles are 3 m diameter steel tubular shape having length of 50 m. A structural model is shown in Figure 6.4.

6.3 Stiffness of a Monopile and Its Application

To estimate foundation stiffness, the following methods are used:

Simplified method. Closed-form solutions for foundation stiffness can be obtained for simple ground profiles. Any spreadsheet programs or even a simple calculator can be used to compute them and typically take only few minutes. This method can be useful in the preliminary design and during the optimization stage or even financial feasibility study.

Standard method. In this method, pile-soil interaction is represented by a set of discrete Winkler springs where the spring stiffness is obtained through *p-y* curves available in different design standards. Standard software is available at reasonable costs to carry the analysis and will only take few hours to carry out an analysis. Complex ground soil profiles can be analysed. Few soil parameters are required, and the effects of cyclic load are taken empirically.

Advanced methods. These are continuum models and advanced 3D finite element (FE) software packages or programs are necessary. Such packages are expensive, computationally demanding, and require experienced engineer to carry out the simulations. However, such models are versatile and can model complex ground profile and any type of soils with different constitutive relations. Cyclic loading can also be applied and pore-water pressure accumulation can also be accounted for.

Example 6.2 ***Estimation of Stiffness of Monopile Using a Closed-Form Solution***
Figure 6.5 shows the stiffness variation of the ground with depth and the possible idealizations for using closed-form solutions. The example closely follows the profile of Gunfleet sand site where the monopile is 4.7 m diameter, 38 m long, having average wall thickness of 94 mm. From the figure, it appears that a parabolic profile is suitable as an idealisation for the presented ground profile. Moreover, due to low soil stiffness in the upper layers, it is apparent that this pile will behave as a rigid pile. Therefore, foundation stiffness is estimated using closed-form solutions of rigid piles in a parabolic ground profile.

Based on the formulas presented in Chapter 5, we can obtain the three foundation stiffness terms: K_L, K_R, and K_{LR} using formulation presented by Shadlou and Bhattacharya (2016).

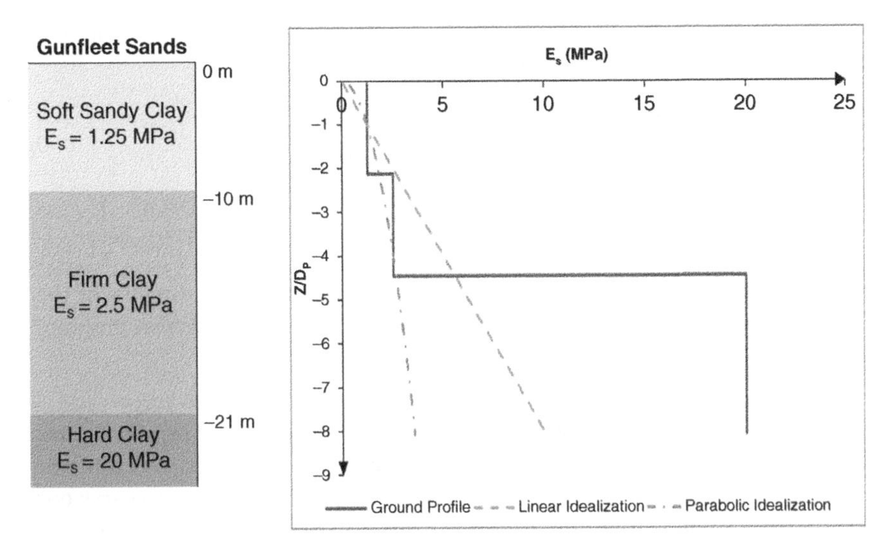

Figure 6.5 Ground profile.

Assuming Poisson's ratio is 0.4 and stiffness at $1D_P$ diameter is 1.25 MPa:

$$^{*}f_{(vs)} = 1 + 0.6|\upsilon_s - 0.25| = 1 + 0.6|0.4 - 0.25| = 1.09$$

$$K_L = 2.65\left(\frac{L}{D_P}\right)^{1.07} f_{(vs)}E_{SO}D_P = 2.65\left(\frac{38}{4.7}\right)^{1.07} \times 1.09 \times 1.25 \times 4.7$$
$$= 158.81\ \text{MN/m}$$

$$K_{LR} = -1.8\left(\frac{L}{D_P}\right)^{2} f_{(vs)}E_{SO}D_P{}^{2} = -1.8\left(\frac{38}{4.7}\right)^{2} \times 1.09 \times 1.25 \times 4.7^2$$
$$= -3541.41\ \text{MN}$$

$$K_R = 1.63\left(\frac{L}{D_P}\right)^{3} f_{(vs)}E_{SO}D_P{}^{3} = 1.63\left(\frac{38}{4.7}\right)^{3} \times 1.09 \times 1.25 \times 4.7^3$$
$$= 121863\text{MNm/rad} = 121.86\text{GNm/rad} \quad (6.1)$$

Application of Foundation Stiffness Terms for Deformation Prediction

(1) These values of K_L, K_{LR}, and K_R can be used to obtain the natural frequency of the wind turbine system. Example 6.5 shows step-by-step application of the methodology.
(2) For serviceability limit state (SLS) requirements, it is necessary to predict the deflections. It has been estimated that for a particular load case, lateral load on the pile head is 3.55 MN (i.e. $H = 3.55$ MN) and the maximum overturning moment is 165 MNm (i.e. $M = 165$ MNm). Using the stiffness estimates, and assuming linearity in load-deflection and moment-rotation relations, pile-head deflection and rotation can be estimated as follows (see Figure 6.6):

$$\begin{bmatrix} H \\ M \end{bmatrix} = \begin{bmatrix} K_L & K_{LR} \\ K_{LR} & K_R \end{bmatrix} \begin{bmatrix} \rho \\ \theta \end{bmatrix} \Rightarrow \begin{bmatrix} \rho \\ \theta \end{bmatrix} = \begin{bmatrix} K_L & K_{LR} \\ K_{LR} & K_R \end{bmatrix}^{-1} \begin{bmatrix} H \\ M \end{bmatrix} \quad (6.2)$$

$$\begin{bmatrix} K_L & K_{LR} \\ K_{LR} & K_R \end{bmatrix}^{-1} \begin{bmatrix} H \\ M \end{bmatrix} = \begin{bmatrix} 158.81 & -3541.41 \\ -3541.41 & 121863 \end{bmatrix}^{-1} \begin{bmatrix} 3.55 \\ 165 \end{bmatrix} = \begin{bmatrix} 0.154\ \text{m} \\ 0.00575\ \text{rads} \end{bmatrix}$$

0.00575 rad is 0.329°, which can be compared with the intended performance of the structure. The above calculations can easily be carried out using a spreadsheet program or even a pocket calculator.

Example 6.3 ***Monopile Stiffness (K_L, K_R, and K_{LR}) Based on Standard and Advanced Methods*** The problem statement is described schematically in Figure 6.6. Geotechnical engineers often use beam on nonlinear Winkler foundation to carry out load-deflection and moment-rotation analysis. This is also known as *p-y*, *t-z*, and *q-z* analysis and is a standard and efficient method. This can also be carried out using the finite element method. The aim of this example is to show how the stiffness can be extracted from such analysis.

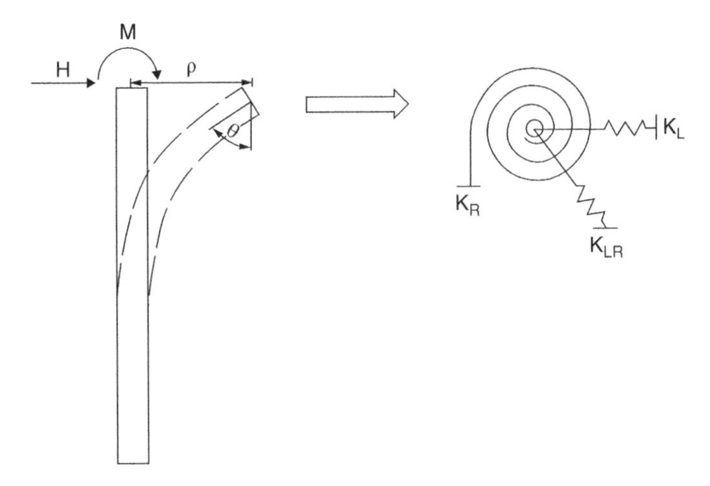

Figure 6.6 Converting FE analysis or *p-y* analysis into a pile-head stiffness.

Table 6.2 Information on 3 WTG structure for two sites.

Properties	Gunfleet sand site	Horns Rev
m_{RNA} (tons) – Mass of RNA	234.5	100
m_T(tons) – Mass of the tower	193	130
L_T (m) – Length of the tower	60	70
Length of the substructure L_S (m)	28	0
Diameter at the top of the tower D_t (m)	3	2.3
Diameter at the bottom of the Tower D_b (m)	5	4
Average thickness of the tower t_T (m)	0.033	0.035
Young's modulus of the tower material E_T (GPA)	210	210
Pile length L_p (m)	38	21.9
Pile diameter D_p (m)	4.7	4
Pile wall thickness t_p (m)	0.094	0.05
Young's modulus of the pile material E_p (GPa)	210	210

The case study of Horns Rev 1 is used to show the application of the methodology and various information are given in Table 6.2. The ground information is obtained from Augustensen et al. (2009) and is shown in Figure 6.7. The monopile foundation supports a 60 m long tower carrying a Vestas V80 2 MW wind turbine. The pile has a diameter of 4.0 m and varying wall thickness (WT in Figure 6.7). The reported ultimate loads on the pile head were $H = 4.6$ MN and $M = 95$ MNm. Table 6.3 provides description of the soil layers, which was obtained through an extensive test program including geotechnical borings, cone penetration tests (CPTs), and triaxial tests given in Augustensen et al. (2009).

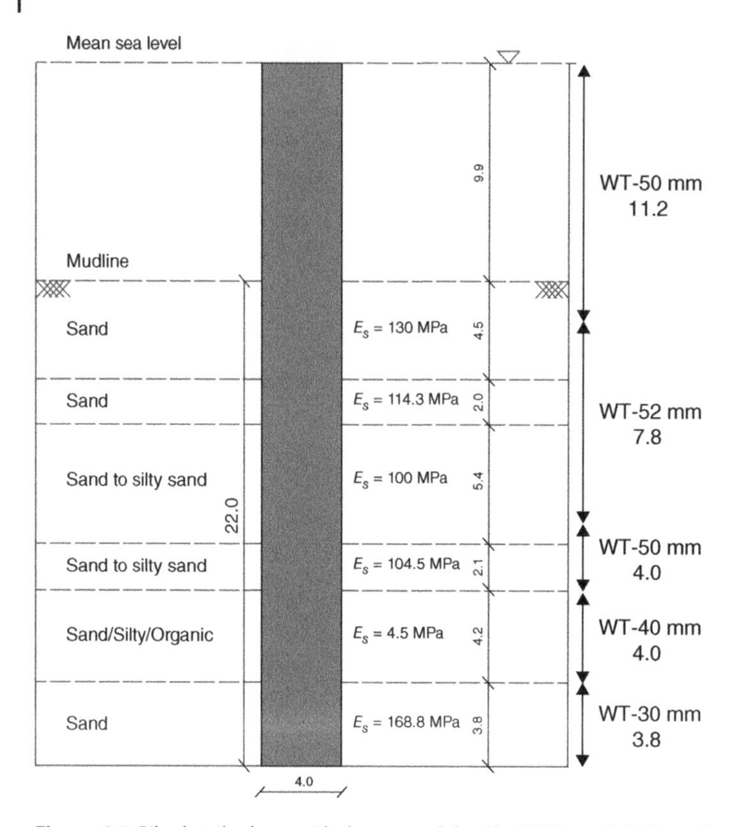

Figure 6.7 Pile details along with the ground details. (WT is wall thickness).

Table 6.3 Soil properties.

Soil layer	Soil type	Depth/m	E_S/MPa	ϕ/degrees	γ/(kN m^{-3})
1	Sand	0–4.5	130	45.4	10
2	Sand	4.5–6.5	114.3	40.7	10
3	Sand to silty sand	6.5–11.9	100	38	10
4	Sand to silty sand	11.9–14.0	104.5	36.6	10
5	Sand/silt/organic	14.0–18.2	4.5	27	7
6	Sand	18.2–>	168.8	38.7	10

ALP (Oasys) has been used for the *p-y* analysis of the monopile, which can compute deflections, rotations, and bending moments along the pile. Several other software packages (e.g. PYGM, SAP2000, LPILE) can also be used for this type of analysis. In this study, *p-y* curves along different depths are modelled in two ways:

1) Elastic-plastic springs by taking the soil properties given in Table 6.3
2) API recommended *p-y* springs

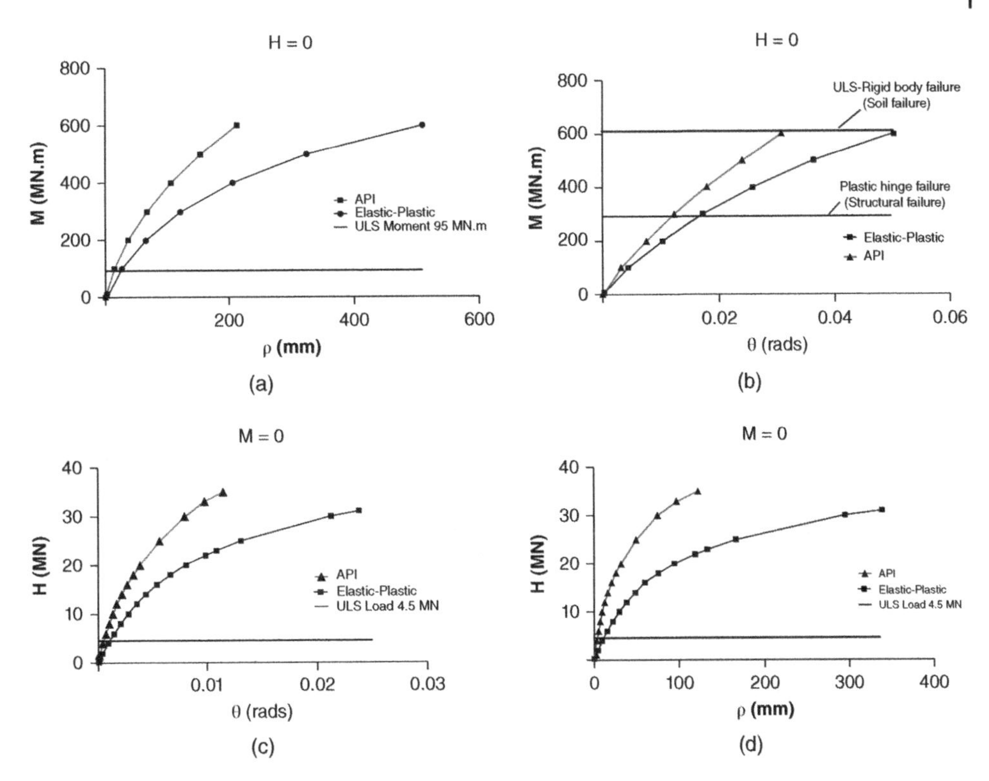

Figure 6.8 Results of the analysis. (a) moment-deflection curve; (b) moment-rotation; (c) lateral load-rotation curve; (d) lateral load-deflection curve.

The API code provides the following formulations for sandy soils:

$$p = Ap_u \tanh\left(\frac{kx}{Ap_u}y\right) \tag{6.3}$$

where A is a factor that depends on the type of loading and p_u depends on the depth and angle of internal friction ϕ'.

However, for the elastic-plastic model used in the study, the following equations were used to construct the p-y curves.

$$p = ky,\ k = (E_s h) \tag{6.4}$$

where h is the distance between the midpoint of the elements immediately above and below the spring under consideration i.e. influence area.

The plastic phase of the curve is given by:

$$F_p = (K_q\sigma_v' + cK_c)hD_P \tag{6.5}$$

where K_q and K_c are factors that depend on depth and ϕ'. Figures 6.8a–d shows the results obtained from the analysis. The curves in the figure plots the following:

(a) Mudline moment versus mudline deflection (δ) for zero horizontal load ($H = 0$) at the pile head. For comparison purpose, a line showing the ULS moment is also shown.

Table 6.4 Results from the analysis and estimation of foundation stiffness.

Model	Loads	ρ (mm)	θ (rad)	I_L	I_{LR}	I_{RL}	I_R
Elastic-plastic	H = 0.2 MN m M = 0	0.43	4.1E−05	2.15E−03	2.1E−04	–	–
API code	H = 0.2 MN m M = 0 MN m	0.15	2.3E−05	7.60E−03	1.1E−04	–	–
Elastic-plastic	M = 2 MN m H = 0 MN	0.41	8.3E−05	–	–	2.1E−04	4.2E−05
API code	M = 2 MN m H = 0 MN m	0.223	6.0E−05	–	–	1.1E−04	3.0E−5

(b) Mudline moment versus mudline rotation (θ) for zero horizontal load ($H = 0$) at the pile head. In this graph, the moment of resistance of the pile is shown for two cases: (a) Structural failure of the pile i.e. a plastic hinge is formed. Readers are referred to Example 6–10 where the methods to calculate are shown; (b) Soil fails i.e. the pile fails as a rigid body whereby the surrounding soils yields.
(c) Horizontal load versus mudline rotation (θ) for zero moment ($M = 0$) at the pile head.
(d) Horizontal load versus mudline deflection (δ) for zero moment ($M = 0$) at the pile head.

It is evident from Figures 6.8a–d, that the ultimate limit state (ULS) loads ($H = 4.5$ MN, $M = 95$ MNm) lie within the linear range. This means the deflections and rotations arising from such forces can be estimated using K_L, K_R, and K_{LR}. Table 6.4 summarises the results for the analysis where two load cases have been applied: (a) $H = 0.2$ MN and $M = 0$; (b) $H = 0$ and $M = 0.2$ MNm where the major steps are also shown.

Equation (6.2) can be rewritten as Eq. (6.6) through matrix operation where I (impedance matrix) is a 2×2 matrix given by Eq. (6.7). A summary of the equations is given by Eq. (6.8):

$$\begin{bmatrix}\rho\\ \theta\end{bmatrix} = [I] \times \begin{bmatrix}H\\ M\end{bmatrix} \tag{6.6}$$

$$I = \begin{bmatrix}I_L & I_{LR}\\ I_{RL} & I_R\end{bmatrix} \tag{6.7}$$

$$K = \begin{bmatrix}K_L & K_{LR}\\ K_{RL} & K_R\end{bmatrix} = I^{-1} = \begin{bmatrix}I_L & I_{LR}\\ I_{RL} & I_R\end{bmatrix}^{-1} \tag{6.8}$$

It may also be noted that Figure 6.8 also plots the ultimate capacity of the pile based on two considerations: (i) soil fails and the pile fails as rigid body failure; (ii) pile fails by forming plastic hinge.

Using the equations, the stiffness terms are predicted as follows:

Elastic-Plastic formulation

$$I = \begin{bmatrix}0.00215 & 0.00021\\ 0.00021 & 0.000042\end{bmatrix} => I^{-1} = K = \begin{bmatrix}K_L & K_{LR}\\ K_{RL} & K_R\end{bmatrix} = \begin{bmatrix}894.1 & -4451.3\\ -4451.3 & 46252.1\end{bmatrix}$$

$$K_L = 894.1\frac{MN}{m}\ K_{LR} = -4451.3MN\ K_R = 46252.1\frac{MN.m}{Rad} \tag{6.9}$$

API formulation

$$I = \begin{bmatrix} 0.0076 & 0.00011 \\ 0.00011 & 0.00003 \end{bmatrix} => I^{-1} = K = \begin{bmatrix} K_L & K_{LR} \\ K_{RL} & K_R \end{bmatrix} = \begin{bmatrix} 3102 & -11823.7 \\ -11823.7 & 78275.5 \end{bmatrix}$$

$$K_L = 3102\frac{MN}{m}\ K_{LR} = -11823.7MN\ K_R = 78275.5\frac{MN.m}{Rad} \tag{6.10}$$

The readers are also referred to Jalbi et al. (2017) for detailed discussion on the above method, including the physical meaning of the cross-coupling term. It may be noted that the stiffness terms are quite different in the two formulations given by Eqs. (6.9) and (6.10), as also reflected in Figure 6.8. One of the many reasons is that API formulation is empirical and is calibrated against small-diameter piles and the extrapolation to large-diameter piles is not validated or verified.

API *p-y* formulation doesn't take into account the resistance offered by the base of the pile as well as the resistance due to the skin friction along the two sides. This has been discussed and mathematically shown in Shadlou and Bhattacharya (2016) and is further explored in PISA project.

Advanced Method

FE package PLAXIS 3D was also used to evaluate the deflection and rotation due to the pile-head loads. To save computational efforts space, half the pile was modelled, see Figure 6.9. A classical Mohr-Coulomb material model was set for the soil with the same stiffness and strength properties provided in Table 6.3. For the pile material, elastic perfectly plastic model was used. Ten node tetrahedral elements were assigned to the soil volume and six-node triangular plate elements for were used for the pile. The interface between the soil and the pile was modelled with double-noded elements. The soil extents were set as $20D_p$, and a medium-dense mesh was used. The pile was extended by 21 m above the ground to simulate the effect of the applied moment. Similar plots such as that shown in Figure 6.8 were obtained. Figure 6.10 plots the deflection along the length of the pile obtained for $H = 4.6$ MN and $M = 95$ MN m.

Similar method as presented by Eqs. (6.9) and (6.10) can be used to obtain foundation stiffness values for natural frequency estimations.

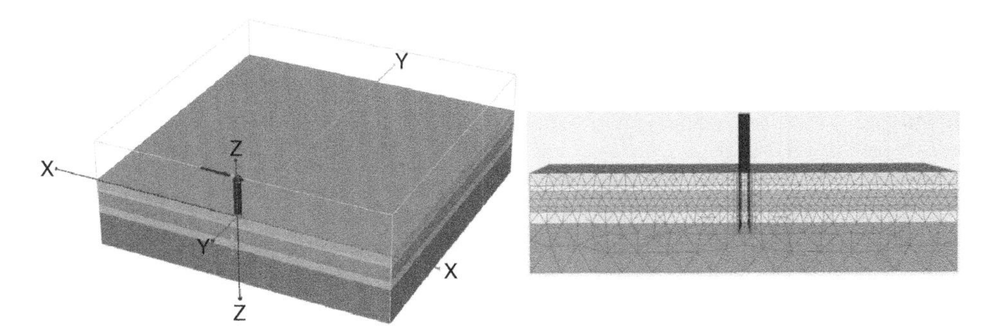

Figure 6.9 Geometry used in Plaxis model.

Example 6.4 ***Pile-Head Deflections and Rotations*** Figure 6.6 schematically shows the essence of the matrix operations presented in the earlier section. Essentially, the foundation is replaced by a set of springs that are obtained from lateral load/moment analysis of piles. This is quite similar to a substructure approach.

Equation (6.9) can be used to predict pile-head deflections and rotations, as shown in Eq. (6.11). It may be noted from Figure 6.8a–d that the ultimate loads were within the linear range. All units are in MN and m.

Elastic plastic

$$\begin{bmatrix} \rho \\ \theta \end{bmatrix} = \begin{bmatrix} 894.1 & -4451.3 \\ -4451.3 & 46252.1 \end{bmatrix}^{-1} \begin{bmatrix} 4.6 \\ 9.5 \end{bmatrix} = \begin{bmatrix} 0.03 \\ 4.8E-03 \end{bmatrix}$$

API formulation

$$\begin{bmatrix} \rho \\ \theta \end{bmatrix} = \begin{bmatrix} 3102 & -11823.7 \\ -11823.7 & 78275.5 \end{bmatrix}^{-1} \begin{bmatrix} 4.6 \\ 9.5 \end{bmatrix} = \begin{bmatrix} 0.014 \\ 3.5E-03 \end{bmatrix} \quad (6.11)$$

The results predict a deflection of 30 mm and a rotation of 0.280° (degree) using the elastic plastic model. On the other hand, the results predict 14 mm of deflection and 0.20° using API model. The advanced model based on PLAXIS 3D shown in Figure 6.10 predict a deflection of 42.5 mm and rotation of 0.3°.

Example 6.5 ***Prediction of the Natural Frequency of the Horns Rev Wind Turbine Structure*** The foundation stiffness estimated in examples 6.3 and 6.4 are used to predict natural frequency of a wind turbine system using methods presented in Chapter 5. The example of Horns Rev is taken to show the calculation procedure. To supplement unavailable data, engineering judgement used is based on a similar 2 MW wind turbine and is summarised in Table 6.5.

The steps are also shown below.

Step 1: (Fixed-Based Natural Frequency of the Tower)

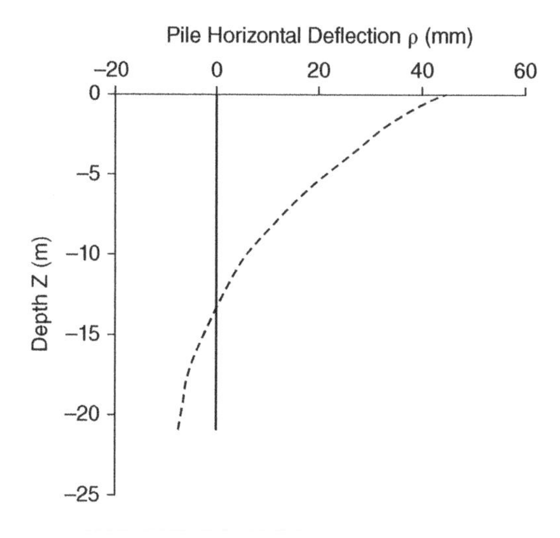

Figure 6.10 Plots the deflection of the pile obtained from the PLAXIS analysis for pile head loads of H = 4.6 MN and M = 95 MN m.

Table 6.5 Tower properties.

Top diameter of tower(m)	2.3
Bottom diameter of tower (m)	4.0
Average wall thickness (mm)	35
Tower height (m)	70
Mass of RNA (Tonnes)	100
Mass of tower (Tonnes)	130

Calculate the bending stiffness ratio of tower to the pile:

$$\chi = \frac{E_T I_T}{E_P I_P} \tag{6.12}$$

Calculate the substructure/tower length ratio:

$$\psi = \frac{L_S}{L_T} \tag{6.13}$$

Calculate the substructure flexibility coefficient to account for the enhanced stiffness of the transition piece:

$$C_{MP} = \sqrt{\frac{1}{1 + (1 + \psi^3)\chi - \chi}} \tag{6.14}$$

Obtain the fixed-base natural frequency of the tower:

$$f_{FB} = \frac{1}{2\pi} C_{MP} \sqrt{\frac{3E_T I_T}{\left(m_{RNA} + \frac{33m_T}{140}\right) {L_T}^3}} \tag{6.15}$$

The cross-sectional properties of the tower can be calculated as in Eq. (6.16):

$$D_T = \frac{D_b + D_t}{2} \; I_T = \frac{1}{8}(D_T)^3 t_T \pi \tag{6.16}$$

Step 2: Calculate the nondimensional foundation stiffness parameters.

The equivalent bending stiffness of tower needed for this step:

$$q = \frac{D_b}{D_T} f(q) = \frac{1}{3} \times \frac{2q^2(q-1)^3}{2q^2 \ln q - 3q^2 + 4q - 1} \; EI_\eta = EI_{top} \times f(q) \tag{6.17}$$

where I_{top} is the second moment of area of the top section of the tower: Derivation of Eq. (6.17) is provided in Appendix C.

$$\eta_L = \frac{K_L L^3}{EI_\eta} \tag{6.18}$$

$$\eta_{LR} = \frac{K_{LR} L^2}{EI_\eta} \tag{6.19}$$

$$\eta_R = \frac{K_R L}{EI_\eta} \tag{6.20}$$

Step 3: Calculate the foundation flexibility factors.

$$C_R(\eta_L, \eta_{LR}, \eta_R) = 1 - \frac{1}{1 + 0.6\left(\eta_R - \frac{\eta^2_{LR}}{\eta_L}\right)} \tag{6.21}$$

$$C_L(\eta_L, \eta_{LR}, \eta_R) = 1 - \frac{1}{1 + 0.5\left(\eta_L - \frac{\eta^2_{LR}}{\eta_R}\right)} \tag{6.22}$$

Step 4: Calculate the flexible natural frequency of the OWT system.

$$f_0 = C_L C_R f_{FB} \tag{6.23}$$

The readers are referred to Arany et al. (2015b) and Arany et al. (2016) for detailed derivation of the above method.

For simplicity, the substructure flexibility coefficient (C_{MP}) is considered to be 1. Therefore, following Eqs. (6.16) and (6.17), we obtain:

$$D_T = \frac{D_b + D_t}{2} => \frac{4 + 2.3}{2} = 3.15 \; I_T = \frac{1}{8}(3.15)^3 \times 0.035 \times \pi = 0.415 \text{ m}^4$$

$$f_{FB} = \frac{1}{2\pi}\sqrt{\frac{3 \times 210E9 \times 0.415}{\left(100000 + \frac{33\times130000}{140}\right)70^3}} = 0.385 \text{ Hz} \tag{6.24}$$

The fixed-base frequency is therefore 0.385 Hz.

$$q = \frac{4}{2.3} = 1.74 \quad f(q) = \frac{1}{3} \times \frac{2 \times 1.74^2(1.74 - 1)^3}{2 \times 1.74^2 \ln 1.74 - 3 \times 1.74^2 + 4 \times 1.74 - 1} = 3.56$$

$$EI_\eta = 210 \times \frac{1}{8}(2.3)^3 \times 0.035 \times \pi \times 3.56 = 119.4 \text{ GNm}^2 \tag{6.25}$$

The nondimensional groups are:

$$\eta_L = \frac{0.8941 \times 70^3}{119.4} = 2568 \; \eta_{LR} = \frac{-4.45 \times 70^2}{119.4} = -182.6 \; \eta_R = \frac{46.25 \times 70}{119.4} = 27.1$$

The foundation flexibility coefficients are given as follows:

$$C_R(\eta_L, \eta_{LR}, \eta_R) = 1 - \frac{1}{1 + 0.6\left(27.1 - \frac{-182.6^2}{2568}\right)} = 0.894$$

$$C_L(\eta_L, \eta_{LR}, \eta_R) = 1 - \frac{1}{1 + 0.5\left(2568 - \frac{-182.6^2}{27.1}\right)} = 0.999$$

The natural frequency is therefore given by:

$$f_0 = 0.894 \times 0.999 \times 0.385 = 0.344 \; Hz \tag{6.26}$$

Figure 6.11 First mode of vibration.

6.3.1 Comparison with SAP 2000 Analysis

The system was modelled using SAP 2000 and a modal analysis is carried out to obtain the natural frequency. Nonprismatic beam elements of varying diameter were assigned to the tower while the soil structure interaction was represented by discrete linear Winkler springs. The spring stiffness values were taken from the soil properties (Table 6.2) and are a function of modulus of elasticity at the location of the spring and the spacing between two adjacent springs. A lumped mass was assigned at the tower head to model the dead mass of the RNA (rotor-nacelle assembly). The first natural frequency recorded was 0.351 Hz. Figure 6.11 shows the first mode of vibration.

6.4 Stiffness of a Mono-Suction Caisson

Example 6.6 ***Stiffness of Mono-Caisson and Application*** The wind turbine used for this example is the 5 MW reference wind turbine provided by National Renewable Energy Laboratory (NREL), see (Jonkman et al. 2009), and the details of the turbine support structure are summarised in Table 6.6. It is necessary to check the adequacy of a 12 m diameter caisson with 6 m skirt length given the subsurface ground is homogeneous, having $E_S = E_{SO} = 40$ MPa, see Figure 6.12. The ULS loads are estimated as for the site as $H = 4$ MN and overturning moment (M) of 200 MNm. The example shows the estimation of caisson stiffness: K_L, K_R, and K_{LR}. It is assumed that the transition piece has same cross-sectional properties as the bottom diameter of the tower.

Table 6.6 Details of the OWT support structure.

Top diameter of the tower (m)	3.87
Bottom diameter of the tower (m)	6.0
Wall thickness of the tower (mm)	27
Height of the tower (m)	87.6
Platform height (transition piece) (m)	30
Mass of RNA (tons)	350
Mass of tower (tons)	347.5
Rated rotor speed (RPM)	6.9–12.1

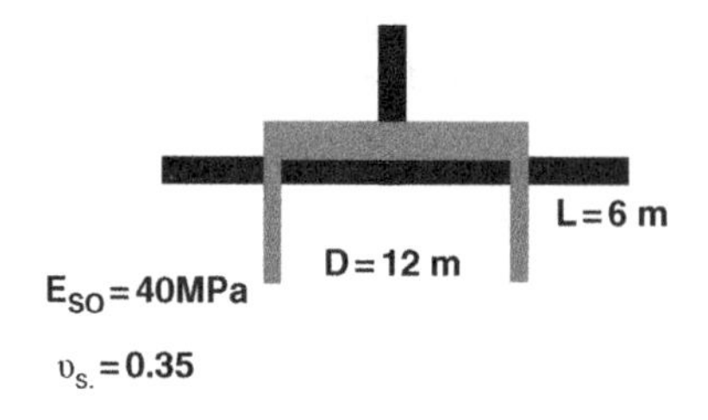

Figure 6.12 Example problem.

The closed-form solution provided in Chapter 5 (Table 5.6) developed in Jalbi et al. (2018) is used and reproduced in Table 6.7.

$$f(v_s) = 1.1 \times \left(0.096\left(\frac{L}{D}\right) + 0.6\right) v_s^2 - 0.7v_s + 1.06$$
$$= 1.1 \times [0.096(0.5) + 0.6] \times 0.35^2 - (0.7 \times 0.35) + 1.06 = 0.9$$

$$K_L = 2.91\left(\frac{L}{D}\right)^{0.56} DE_{SO}f(v_s) = 2.94 \times (0.5)^{0.56} \times 12 \times 40 \times 10^{-3} \times 0.9 = 0.86\ \frac{\text{GN}}{\text{m}}$$

$$K_{LR} = -1.87\left(\frac{L}{D}\right)^{1.47} D^2E_{SO}f(v_s) = -1.87 \times (0.5)^{1.47} \times 12^2 \times 40 \times 10^{-3} \times 0.9$$
$$= -3.5\ \text{GN}$$

$$K_R = 2.7\left(\frac{L}{D}\right)^{1.92} D^3E_{SO}f(v_s) = 2.7 \times (0.5)^{1.92} \times 12^3 \times 40 \times 10^{-3} \times 0.9 = 44\ \frac{\text{GNm}}{\text{rad}} \tag{6.27}$$

Assuming that the foundation behaviour (load-deflection or moment-rotation) is in the linear range, the deflections and rotations can be estimated as follows:

$$\begin{bmatrix} H \\ M \end{bmatrix} = \begin{bmatrix} K_L & K_{LR} \\ K_{LR} & K_R \end{bmatrix} \begin{bmatrix} \rho \\ \theta \end{bmatrix} \Rightarrow \begin{bmatrix} \rho \\ \theta \end{bmatrix} = \begin{bmatrix} 0.86 & -3.5 \\ -3.5 & 44 \end{bmatrix}^{-1} \begin{bmatrix} 0.004\ \text{GN} \\ 0.2\ \text{GNm} \end{bmatrix} = \begin{bmatrix} 0.034\ \text{m} \\ 0.007\ \text{rads} \end{bmatrix} \tag{6.28}$$

The predicted rotation is 0.4°.

The natural frequency of the system is calculated in the next section.

Step 1: Calculate the fixed-based natural frequency of the tower.

Calculate the bending stiffness ratio of tower to the pile.

$$\chi = \frac{E_T I_T}{E_P I_P}$$

Calculate the substructure (platform)/tower length ratio.

$$\psi = \frac{L_S}{L_T}$$

Calculate the substructure flexibility coefficient to account for the enhanced stiffness of the transition piece:

$$C_{MP} = \sqrt{\frac{1}{1 + (1 + \psi^3)\chi - \chi}}$$

Table 6.7 Stiffness table.

Ground profile	$\frac{K_L}{DE_{SO}f(v_s)}$	$\frac{K_{LR}}{D^2E_{SO}f(v_s)}$	$\frac{K_R}{D^3E_{SO}f(v_s)}$
Homogeneous	$2.91\left(\frac{L}{D}\right)^{0.56}$	$-1.87\left(\frac{L}{D}\right)^{1.47}$	$2.7\left(\frac{L}{D}\right)^{1.92}$

Obtain the fixed-base natural frequency of the tower:

$$f_{FB} = \frac{1}{2\pi} C_{MP} \sqrt{\frac{3E_T I_T}{\left(m_{RNA} + \frac{33m_T}{140}\right) L_T{}^3}} \tag{6.29}$$

Where the cross-sectional properties of the tower can be calculated as

$$D_T = \frac{D_b + D_t}{2} \quad I_T = \frac{1}{8}(D_T)^3 t_T \pi$$

Step 2: Calculate the nondimensional foundation stiffness parameters.

The equivalent bending stiffness of tower needed for this step:

$$q = \frac{D_b}{D_T} \quad f(q) = \frac{1}{3} \times \frac{2q^2(q-1)^3}{2q^2 \ln q - 3q^2 + 4q - 1} \quad EI_\eta = EI_{top} \times f(q)$$

where I_{top} is the second moment of area of the top section of the tower:

$$\eta_L = \frac{K_L L^3}{EI_\eta}$$

$$\eta_{LR} = \frac{K_{LR} L^2}{EI_\eta}$$

$$\eta_R = \frac{K_R L}{EI_\eta}$$

Step 3: Calculate the foundation flexibility factors.

$$C_R(\eta_L, \eta_{LR}, \eta_R) = 1 - \frac{1}{1 + 0.6\left(\eta_R - \frac{\eta^2_{LR}}{\eta_L}\right)}$$

$$C_L(\eta_L, \eta_{LR}, \eta_R) = 1 - \frac{1}{1 + 0.5\left(\eta_L - \frac{\eta^2_{LR}}{\eta_R}\right)}$$

Step 4: Calculate the flexible natural frequency of the OWT system.

$$f_0 = C_L C_R f_{FB}$$

Numerical calculation:

$$D_T = \frac{3.87 + 6}{2} = 4.935 \text{ m}$$

$$I_T = \frac{1}{8}(4.935)^3 \times 0.027 \times \pi = 1.25 \text{ m}^4$$

$$\chi = \frac{210 \times \pi \times (4.935)^3 \times 0.027 \times \frac{1}{8}}{210 \times \pi \times (6)^3 \times 0.027 \times \frac{1}{8}} = 0.55$$

$$\psi = \frac{30}{87.6} = 0.34$$

$$C_{MP} = \sqrt{\frac{1}{1 + (1 + 0.34)^3 \times 0.55 - 0.55}} = 0.75$$

$$f_{FB} = \frac{1}{2\pi} \times 0.75 \times \sqrt{\frac{3 \times 210 \times 10^9 \times 1.25}{\left(350000 + \frac{33 \times 347460}{140}\right) \times 87.6^3}} = 0.26 \text{ Hz}$$

The fixed-base frequency is therefore 0.26 Hz.

$$q = \frac{6}{3.87} = 1.55$$

$$f(q) = \frac{1}{3} \times \frac{2 \times 1.55^2 (1.55 - 1)^3}{2 \times 1.55^2 \ln 1.55 - 3 \times 1.55^2 + 4 \times 1.55 - 1} = 2.71$$

$$EI_\eta = 210 \times \frac{1}{8}(3.87 - 0.027)^3 \times 0.027 \times \pi \times 2.71 = 342 \text{ GNm}^2$$

The nondimensional groups are:

$$\eta_L = \frac{0.86 \times 87.6^3}{342} = 1690$$

$$\eta_{LR} = \frac{-3.5 \times 87.6^2}{342} = -78$$

$$\eta_R = \frac{44 \times 87.6}{342} = 11$$

The foundation flexibility coefficients are given as follows:

$$C_R(\eta_L, \eta_{LR}, \eta_R) = 1 - \frac{1}{1 + 0.6\left(11 - \frac{-78^2}{1690}\right)} = 0.82$$

$$C_L(\eta_L, \eta_{LR}, \eta_R) = 1 - \frac{1}{1 + 0.5\left(1690 - \frac{-78^2}{11}\right)} = 0.999 \tag{6.30}$$

The natural frequency is therefore given by:

$$f_0 = 0.82 \times 0.999 \times 0.26 = 0.17 \text{ Hz}$$

6.5 Mudline Moment Spectra for Monopile Supported Wind Turbine

In the frequency diagram (Figure 2.13) shown in Chapter 2 and Example 6.1, the PSD magnitudes are normalised to unity as they have different units and thus the magnitudes are not directly comparable. In Chapter 2, a generalised method is presented to evaluate the relative magnitudes of four loadings (wind, wave, 1P, and 3P) by transforming them to bending moment spectra using site and turbine specific data. This formulation can be used to construct bending moment spectra at the mudline, i.e. at the location where the highest fatigue damage is expected. Equally, this formulation can also be tailored to find the bending moment at any other critical cross section, e.g. the transition piece (TP) level. This example case study is considered to demonstrate the application of the proposed methodology. The constructed spectra may serve as a basis for frequency based fatigue estimation methods available in the literature.

Example 6.7 ***Application of the Mudline Moment Spectra for Walney 1 Wind Turbine*** The method is applied to an actual Siemens industrial wind turbine of 3.6 MW rated power at the Walney 1 wind farm site. The necessary turbine and site information are available in the Siemens brochure, website of DONG Energy

Table 6.8 Relevant data of the Siemens SWT-3.6-107 wind turbine and the Walney 1 site.

Turbine data	
Turbine type	Siemens SWT-3.6-107
Turbine power	3.6 MW
Turbine rotational speed	5–13 rpm
Operational wind speed range	4–25 m s^{-1}
Number of blades	3
Tower and support structure data	
Hub height from mean sea level	$H = 83.5$ m
Tower top diameter	$D_t = 3$ m
Tower bottom diameter	$D_b = 5$ m
Monopile/substructure diameter	$D_P = 6$ m
Rotor and blade data	
Turbine rotor diameter	$D = 107$ m
Rotor overhang	$b = 4$ m
Blade root diameter	$B_{root} = 4$ m
Blade tip chord length	$B_{tip} = 1$ m
Blade length	$L = 52$ m
Site data	
Mean sea depth	21.5 m
Average distance from closest shore	19 km
Yearly mean wind speed	9 m s^{-1}
Dominant wind direction	West/South-West
Estimated fetch	60 km

(the developer of the wind farm), website of the Lindoe Offshore Renewables Center, lorc.dk and Arany et al. (2015a). All necessary information is presented in Table 6.8.

In this section, the mudline moment spectra are derived for each loading and some load values are estimated for several typical operational conditions of the particular turbine. Dynamic amplification is taken into account.

Walney 1 Wind Farm Site

The Walney site is located 14 km off the coast of Walney Island in the Irish Sea (UK). The first phase (Walney 1) of the wind farm contains 51 Siemens SWT-3.6-107 type wind turbines of 3.6 MW rated power. The average wind speed at the site is 9 m s^{-1}, the dominant wind direction is west/southwest. This location in the Irish Sea is relatively sheltered as the shores are relatively close in most directions. The average fetch is estimated at 60 km and this value is used to calculate the JONSWAP spectrum. The significant wave heights are limited at the site and the highest waves are in the range of a few metres. The water depth ranges between 19 and 23 m at the site, and in the calculations an average value of 21.5 m is used. A conservative upper-bound estimate of 16% is assumed for the site turbulence intensity (IEC high turbulence site).

The offshore wind turbine's (OWT's) cut-in wind speed is 4 m s^{-1} and its cut-out speed is 25 m s^{-1}. The rated wind speed is 13–14 m s^{-1}. The turbines are pitch-regulated variable-speed turbines. The rotational speed ranges between 5 and 13 rpm. The OWTs have a rotor diameter of 107 m, and the hub height is 83.5 m above mean sea level. A tapered tubular tower is assumed with linearly varying diameter between the bottom and top diameters of 5 m and 3 m, respectively. The OWTs are installed on 6 m diameter monopile foundations (this value is used for calculating the wave load on the substructure). The natural frequency is also necessary for the calculation of dynamic amplification factors (DAFs) and it is estimated as 0.335 Hz following Arany et al. (2015a).

Wind Load Spectrum

The wind moment spectrum is given by Eq. (6.31). To construct the moment, one needs the diameter of the rotor ($D = 107$ [m]), density of air $\rho_a = 1.225$ [kg/m^3], hub height $H = 83.5$ [m], mean sea level ($MSL = 21.5$ [m]), the turbulence intensity ($I = 16\%$). The mean wind speed is taken as the yearly mean wind speed at the site $\overline{U} = 9$ [m/s], and this way the thrust coefficient $C_T = 7/\overline{U} = 7/9$ [−] and the standard deviation of wind speed $\sigma_U = I\overline{U}$ can be calculated. The Kaimal spectrum is constructed using the standard value of the integral length scale as given in the DNV code $L_k = 340.2$ [m]:

$$\widetilde{S}_{uu}(f) = \frac{\frac{4L_k}{U}}{\left(1 + \frac{6fL_k}{U}\right)^{\frac{5}{3}}} = \frac{\frac{4\cdot 340.2}{9}}{\left(1 + \frac{6\cdot 340.2f}{9}\right)^{\frac{5}{3}}} = \frac{151.2}{(1 + 226.8f)^{\frac{5}{3}}} \tag{6.31}$$

Following the discussion in Section 2.6.2 (Chapter 2), the mudline moment spectrum (spectral density of mudline bending moment) can be written as:

$$S_{MM,wind}(f) = 1.337 \cdot 10^{12} \cdot \widetilde{S}_{uu}(f) = \frac{2.022 \cdot 10^{14}}{(1 + 226.8f)^{\frac{5}{3}}} \tag{6.32}$$

The mudline moment spectrum is plotted in Figure 6.13 for $\overline{U} = 9$ [m/s]. The fore–aft bending moment at mudline due to turbulence is approximated by equations given in Chapter 2. The static component is calculated based on the formulation in Chapter 2 using the mean wind speed. In the Kaimal spectrum, most of the power is concentrated in the very low frequencies; therefore, the dynamic part of the loading is approximated by placing the total standard deviation of the bending moment, which is the integral under the spectrum curve, at the peak frequency $f_p = 0.0017$ [Hz].

Table 6.9 shows the calculated forces and moments for several different mean wind speeds. The DAF is about 1 for such low-frequency excitations.

Wave Load Spectrum

The wave load spectrum is given by Eq. (6.22). To construct the spectrum, one needs the inertia coefficient of the substructure $C_m = 2$ (based on DNV), the density of sea water $\rho_w = 1030$ [kg/m^3], the diameter of the substructure $D_P = 6$ [m], the mean sea depth $S = 21.5$ [m], and the wave number k. The latter can be determined from the dispersion relation (Eq. (6.12)), using the angular frequency of the waves. The deep-water assumption $\lambda < 2S$ does not hold for all important frequencies at the

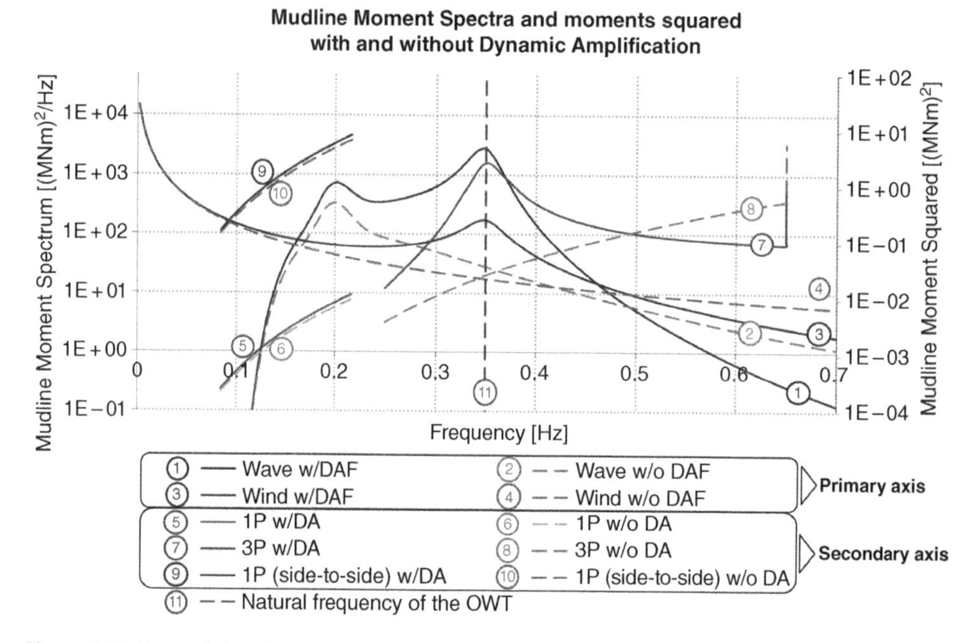

Figure 6.13 Fore–aft bending moment spectrum at mudline – Siemens SWT-3.6-107 at the Walney 1 wind farm. The amplitudes of the 1P and 3P moment squares are to be compared to the integral under the wind and wave spectrum curves, not directly to the spectra.

Table 6.9 Wind loading (static and fluctuating components) for several wind speed values.

Wind speed $\overline{U}[m/s]$	Thrust force on the hub, static component $Th_{stat}[MN]$	Thrust force on the hub, dynamic component $Th_{dyn}[MN]$	Fore–aft bending moment at mudline level, static component $M_{wind,stat}[MNm]$	Fore–aft bending moment at mudline level, dynamic component $M_{wind,dyn}[MNm]$
5	0.193	0.066	20.24	7.00
9	0.347	0.120	36.43	12.59
15	0.578	0.200	60.72	20.99
20	0.771	0.266	80.96	27.98

Walney site. The JONSWAP spectrum can be determined as shown in Eq. (6.33). The fetch is estimated at $F = 60$ [km], the mean wind speed is $\overline{U} = 9$ [m/s], the intensity of spectrum is $\alpha = 0.011$, the peak enhancement factor is $\gamma = 3.3$ and the peak frequency is $f_p = 0.197$ [Hz], the spectrum is given as:

$$S_{ww}(f) = \frac{6.8 \cdot 10^{-4}}{f^5} e^{-\frac{5}{4}\left(\frac{0.197}{f}\right)^4} \gamma^r$$

$$r = e^{-\frac{(f-0.197)^2}{0.0776\sigma^2}} \quad \sigma = \begin{cases} 0.07 & f \leq f_p \\ 0.09 & f > f_p \end{cases} \tag{6.33}$$

The readers are referred to Section 2.4 for the relevant theory. The background behind the method is presented in Arany et al. (2015a). The mudline moment spectrum can be numerically calculated, and is presented in Figure 6.13. The magnitude of the mudline bending moment due to wave loading is estimated for several wind speeds using a single linear wave approximation of the spectrum. The significant wave height $H_{1/3}$ and peak wave period T_p are determined from the JONSWAP spectrum. The moment load is calculated based on the equation presented in Chapter 2. The horizontal wave force can be calculated as shown in Eqs. (6.34) and (6.35):

$$F_{T,max} = M_{I,max} = \int_{-S}^{\eta} dF_{I,max} dz \tag{6.34}$$

with $\eta = 0$ (upper limit of the integral is the surface elevation) one can write

$$F_{T,max} = C_m \rho_w \frac{D_P^2 \pi^3}{2} \frac{H_{1/3}}{T^2 \sinh(kS)} \int_{-S}^{0} \cosh(k(S+z))dx = C_m \rho_w \frac{D_P^2 \pi^3}{2} \frac{H_{1/3}}{T^2 k} \tag{6.35}$$

The results for the significant wave height, peak wave period and force and bending moment loads for some wind speed values are shown in Table 6.10. The DAFs along with the increased dynamic loads are also given in Table 6.10.

1P Loading

To determine the 1P moment spectrum, one needs a typical value of the mass imbalance of the rotor. Discussion is presented in Chapter 2 (Section 2.71) and the Walney 1 example is considered together with the calculations of moment for different mean wind speed at the hub. Therefore, they are not presented here and the readers are referred to Table 2.2 (Chapter 2).

The maximum of the fore–aft bending moment caused by the imbalance can be written as:

$$M_{1P} = 4 \cdot 4\pi^2 \cdot 2000 \cdot f^2 = 3.1583 \cdot 10^5 \cdot f^2 \tag{6.36}$$

The maximum bending moment occurs at the highest rotational speed of $\Omega = 13$ [rpm], that is $f = 0.2167$ [Hz], its value is $M_{1P} = 0.015$ [MNm]. The spectrum of the 1P loading is a Dirac-delta function. Figure 6.13 shows the 1P loading as a function of the Hertz frequency of the rotation f (on the secondary axis).

Table 6.10 Wave parameters and wave loading for several wind speed values.

Wind speed $\bar{U}[m/s]$	Significant wave height $H_{1/3}[m]$	Peak wave period $T_p[s]$	Peak frequency $f_p[Hz]$	DAF [–]	Horizontal wave force, dynamic component $F_T[MN]$/(with DAF)	Mudline moment, dynamic component $M_T[MNm]$/(with DAF)
5	0.64	4.17	0.240	2.03	0.146/(0.296)	3.15/(6.39)
9	1.15	5.08	0.197	1.52	0.328/(0.499)	5.1/(7.76)
15	2.05	6.02	0.166	1.32	0.540/(0.7128)	7.6/(10.06)
20	2.56	6.62	0.151	1.25	0.706/(0.8825)	9.45/(11.84)

Even though the side-to-side direction is not considered here, it is important to mention that the 1P side-to-side mudline bending moments are significantly higher than the fore–aft components. This is because of the large lever arm of the force i.e. the distance between the hub height and the mudline ($H + MSL$). Another factor that may become important is the proximity to the natural frequency of the structure in the side-to-side direction coupled with negligible aerodynamic damping. It may be noted that for side-to-side vibrations, the damping is mainly due to structural damping, which is approximately 0.5%. In the example considered, the difference in DAF for fore–aft and side-to-side excitation is not very significant, but close to resonance the maximum DAF in fore–aft response is 10, while that of the side-to-side response is 100. Table 2.2 in Chapter 2 shows the values of 1P fore–aft and side-to-side moments from the imbalance. It can be seen that after reaching the rated rotational speed of 13 rpm (at about 14 m s^{-1}), the 1P load does not increase.

3P Loading

The 3P moment can be determined by estimating the total drag moment on the top part of the tower which is covered by the downward pointing blade and then reducing this moment by the ratio of the face area of the blade and the face area of the top part of the tower. Discussion and methodology to calculate are presented in Chapter 2, taking the example of Walney site. The 3P forces and moments are estimated in Table 2.3 for several values of the mean wind speed. Note that in the vicinity of 6.125 m s^{-1} (~6.7 rpm rotational speed), the DAF gets very high and the 3P moment is an order of magnitude higher than without DAF. The effects of DAF on 1P and 3P loading are shown in Figure 6.13. It can be seen that after reaching the maximum rotational speed (13 rpm) at the mean wind speed of 14 m s^{-1}, the load keeps increasing as the tower drag increases, but the frequency of excitation remains constant at the 3P value corresponding to 13 rpm $f_{3P\,@\,13\,\text{rpm}} = 0.65$ Hz.

Summary of Loads and Comparison

The static and dynamic loads are summarised in Table 6.11 for two different wind speeds with dynamic loads in bold font. For the sake of completeness, static current load is included in the table. The current load can be calculated as a drag force on the substructure using Morison's equation. Data about the current and tidal stream conditions at the Walney site are not available, however, using the tidal atlas of the Irish Sea available (on the internet at www.visitmyharbour.com) one can see that the tidal stream velocity rarely exceeds 1 knot (approximately 0.514 m s^{-1}). Using this value, Morison's equation estimates a static force of 0.0176 MN and a static mudline bending moment of 0.189 MNm due to currents acting on the substructure. The effect of dynamic amplification on the spectra is shown in Figure 6.13 and the effect on the 1P and 3P loads is shown separately in Figure 6.14.

Limitations of the Spectral Method

The methodology presented in this study provides mudline moment spectra as a basis for frequency-based fatigue damage estimation. As a tool for the early design phase, the methodology has limitations that are important to note and are listed below:

Table 6.11 Estimated static and dynamic loads (values with dynamic amplification in brackets).

Load	$\bar{U} = 9\,[m/s]$ Fore–aft bending moment [MNm]/(w/DAF)	$\bar{U} = 9\,[m/s]$ Force (parallel to axis x) [MN] (w/DAF)	$\bar{U} = 20\,[m/s]$ Fore–aft bending moment [MNm] (w/DAF)	$\bar{U} = 20\,[m/s]$ Force (parallel axis x) [MN] (w/DAF)
Wind load (static)	36.4	0.35	80.96	0.771
Wind load (dynamic)	**12.6 (12.6)**	**0.12 (0.12)**	**27.98**	**0.27 (0.27)**
Current/tidal stream	0.19	0.018	0.19	0.018
Wave loading	**5.1 (7.76)**	**0.33 (0.50)**	**9.45 (11.84)**	**0.71 (0.88)**
1Ploading	**0.01 (0.01)**	–	**0.025 (0.043)**	–
(3P loading)	*0.225 (0.275)*	*0.003 (0.004)*	*1.11 (0.40)*	*0.014 (0.005)*
Tower drag (static)	0.57	0.006	2.82	0.031
Total static load	37.2	0.37	84.0	0.82
Total dynamic load	**17.7 (20.4)**	**0.45 (0.62)**	**37.5 (39.9)**	**0.97 (1.15)**
Total maximum load	54.9 (57.6)	0.82 (0.99)	121.5 (123.9)	1.79 (1.97)

The wind load was calculated using the peak frequency of 0.0017 [Hz]; therefore, the dynamic amplification factor (DAF) is one for the wind loading.
The total static load is calculated as the sum of the static wind load, the current load, and the tower drag load.
The total dynamic load is calculated as the sum of the dynamic wind load, the wave loading, and the 1P loading. The 3P cyclic loading is basically a load loss; therefore, it is not to be added.
However, the wave loads are significantly lower, since the wind farm location in the Irish Sea is somewhat sheltered and the fetch is limited; therefore, very high waves do not commonly develop.

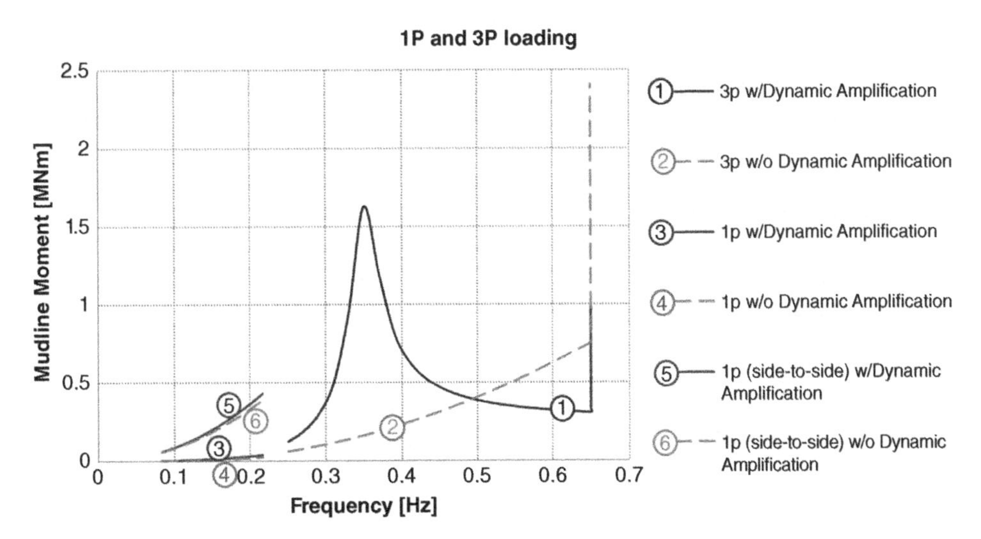

Figure 6.14 1P fore–aft bending moment, 1P side-to-side bending moment, and 3P bending moment at mudline with and without dynamic amplification.

1. The analysis takes into account only the power production stage of the wind turbine. However, fatigue damage also occurs during start-up, shutdown, parked state, and other scenarios.
2. Due to long-term wind speed variations, the static component of the wind produces stress cycles with high amplitude but low frequency. Even though the number of cycles is much lower than for the loads addressed in this study, the high amplitudes may contribute significantly to the fatigue damage.
3. The quasi-steady approximation of wind speed fluctuations gives a rough estimation of the fluctuating wind load. Present formulation is a simplified tool to precede detailed time domain analysis. For detailed fatigue analysis a nonlinear time domain approach (like the blade element momentum (BEM) theory is suitable. However, the BEM method requires a significant amount of information about the blade design as well as a detailed description of the aerofoil characteristics of each radial section.
4. The proposed formulation assumes collinear wind and wave directions, i.e. that there is no misalignment between the wind speed and the propagation direction of waves. This is often true for winds blowing onshore, but not common for winds blowing offshore. This means that the estimation given by summing the individual contributions is a conservative upper bound estimate of the fatigue.
5. Theoretical spectra should be used with care. Standards usually suggest using site-specific spectra whenever possible. However, for the method shown here, theoretical spectra are sufficient as a first estimate.
6. The simplistic formula for the thrust coefficient used is a rough upper-bound estimate that was tested for several wind turbines. A better way of calculating the thrust coefficient of the wind turbine is applying the BEM theory (or other more refined methods).
7. The estimation of fetch is an uncertain process, and the resulting wave heights and wave periods should be compared to typical values at a given site. Alternatively, the formulation based on $H_{1/3}$ and T_p may be used. The first-order Airy wave approximation of the wave particle kinematics is only valid for deeper waters, and it may be necessary to use higher-order methods. High waves and severe wave impacts of low cycle number but high load magnitude may produce significant fatigue damage as well.
8. The derivations were carried out for three-bladed wind turbines because they are dominating the offshore wind industry. However, the calculation methods are similar for two-bladed turbines as well. For the same rotational speed range, the 1P–2P gap is smaller than the 1P–3P gap, although two-bladed turbines tend to spin somewhat faster.

Concluding Remarks and Applicability to Fatigue Calculation

An attempt has been made to provide a quick and simple methodology to estimate the fore–aft mudline bending moment spectra of offshore wind turbines for the four main types of dynamic loads: wind, wave, rotor mass imbalance (1P), and blade passage (3P). An example calculation is presented for an operational industrial Siemens SWT-107-3.6 wind turbine at the Walney 1 site and the mudline moment spectra are plotted (taking into account dynamic amplifications).

The motivation behind this simplified methodology is to provide a basis for a quick frequency domain fatigue damage estimation in the preliminary design phase of these structures, which is otherwise a very lengthy process usually done in time domain. This would also encourage integrated design of OWTs incorporating the dynamics and fatigue analysis in the early stages of structural design. Therefore, it was also an objective to use as little site-specific information of the wind turbine as possible. Information such as the control parameters, the blade design, aerofoil characteristics, generator characteristics, etc. are not necessary in this formulation. Frequency domain fatigue damage estimation methods such as the Dirlik, Tovo-Benasciutti, and Zhao-Baker methods are available, and the formulation presented can be used.

6.6 Example for Monopile Design

Monopiles are the most commonly used foundations. This section of the chapter provides a example based on a method proposed by Arany et al. (2017) titled 'Design of monopiles for offshore wind turbines in 10 steps'. The method can be presented in the form of a flowchart (see Figure 6.15).

Example 6.8 ***Design Steps Using a Typical UK Site*** The outer Thames Estuary (Eastern Coast of the UK) is a typical offshore wind farm. The chosen turbine is Siemens SWT-3.6-120. The site is a shallow-water site with water depth ranging from intertidal (occasionally no water) to 25 m mean water depth. Depending on the location of the WTG within the wind farm, several different pile designs are required. In this example, the deepest water (25 m) is considered. The turbine is at the edge of the wind farm, and fatigue loads due to turbulence generated by other turbines is neglected in this analysis. The soil at the site is predominantly London clay with sands and gravels in the uppermost layers. This section of considers the application of the simplified design procedure:

Establishing Design Criteria

The design criteria are typically established based on four factors:

1. *Design codes.* The most important ones are design requirements by IEC defined in IEC-61400-1(IEC 2005), IEC-61400-3(IEC 2009a), DNV-OS-J101 'Design of Offshore Wind Turbine Structures' (DNV 2014) and the Germanischer Lloyd Windenergie's 'Guideline for the Certification of Offshore Wind Turbines' (Germanischer Lloyd 2005). For fatigue analyses, DNV-RP-C203 'Fatigue Design of Offshore Steel Structures' (DNV 2005) is relevant. For the assessment of environmental conditions DNV-RP-C205 'Environmental Conditions and Environmental Loads' (DNV 2010a,b) may need to be consulted. The API code of practice 'Recommended Practice for Planning, Designing and Constructing Fixed Offshore Platforms – Working Stress Design' (API 2005) may be relevant.
2. *Certification body.* Typically, a certification body allows for departure from the design guidelines if the design is supported by sound engineering and sufficient evidence/test results.

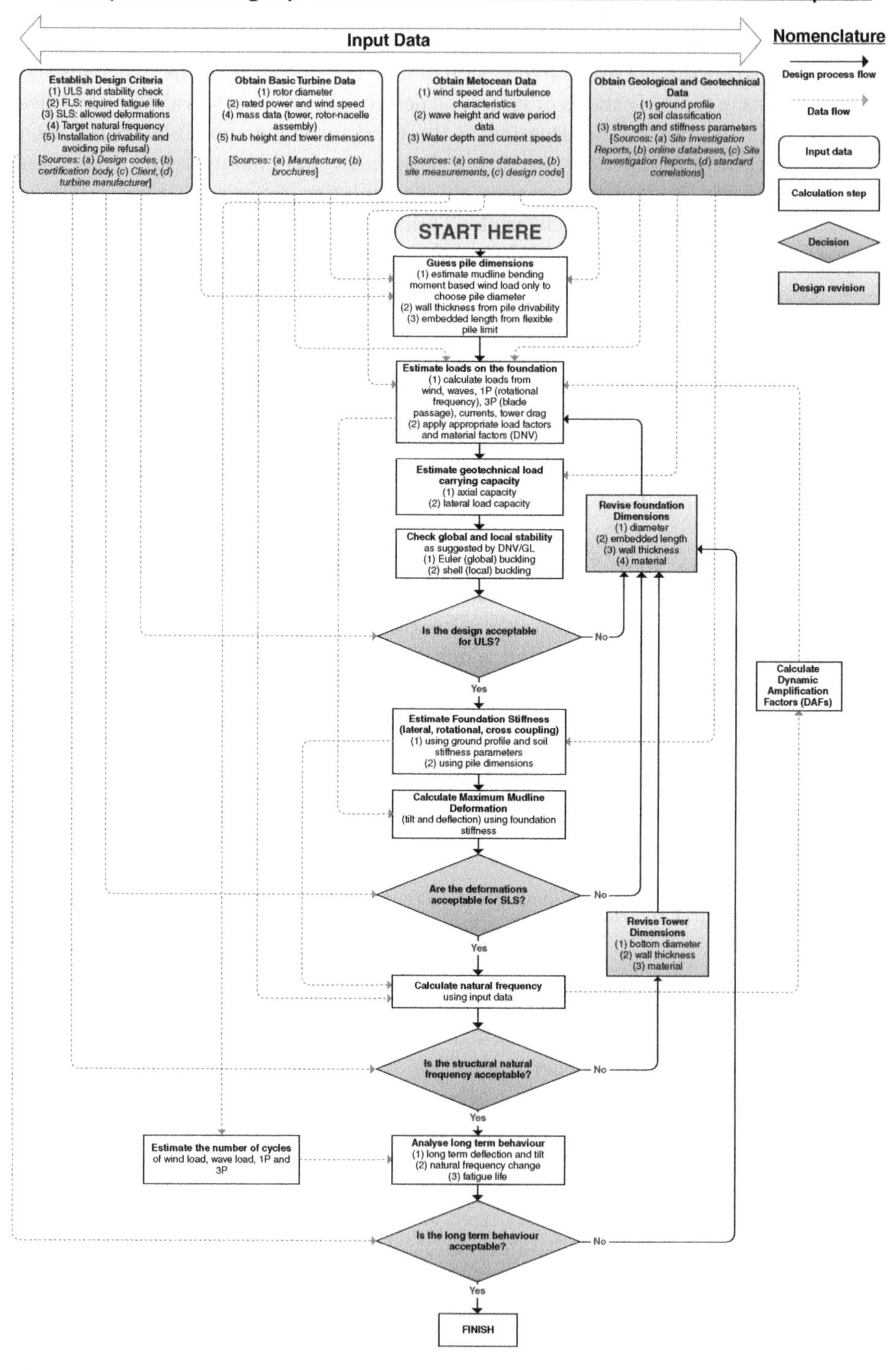

Figure 6.15 Flowchart.

3. *Client*. Occasionally, the client may pose additional requirements based on their appointed consultant.
4. *Turbine manufacturer*. The manufacturer of the wind turbine typically imposes strict SLS requirements. In addition, the expected hub height is a requirement for the turbine type and the site. The tower dimensions are also often inputs to foundation design.

The requirements are summarised in Table 6.12.

Table 6.12 Design basis or criteria for design.

#		Category	Description	Limit
R1	R1.A	ULS	Foundation's load-carrying capacity has to exceed the maximum load (for horizontal and vertical load, and overturning moment).	$M_{ULS} < M_f$ $F_{ULS} < F_f$ $V_{ULS} < V_f$
	R1.B	ULS	The pile's yield strength should exceed the maximum stress.	$\sigma_m < f_{yk}$
	R1.C	ULS	Global (Euler type or column) buckling has to be avoided.	
	R1.D	ULS	Local (shell) buckling has to be avoided.	
R2		FLS	The lifetime of the foundation should be at least 50 years.	$T_L > 50$ yrs
R3	R3.A	SLS	Initial deflection must be less than 0.2 m.	$\rho_0 < 0.2$ m
	R3.B	SLS	Initial tilt must be less than 0.5°.	$\theta_0 < 0.5^{\circ}$
	R3.C	SLS	Accumulated deflection must be less than 0.2 m.	$\rho_{acc} < 0.2$ m
	R3.D	SLS	Accumulated tilt must be less than 0.25°.	$\theta_{acc} < 0.25$
R4		SLS (natural frequency)	The structural natural frequency of the wind turbine-tower-substructure-foundation system has to avoid the frequency of rotation of the rotor (1P) by at least 10%.	$f_0 > 1.1 f_{1P,max}$ $= 0.24$ Hz
R5		Installation	Pile wall thickness (initial guess)	$t_p \geq 6.35 + \frac{D_P}{100}$ [mm]

Obtain Input Data

This simplified analysis aims to use a minimal amount of information about the turbine and the site, in order to enable the designer of monopiles to find the necessary pile dimensions quickly and easily for feasibility studies, tender design and early design phases. The necessary data are given below by data groups.

Basic Turbine Data

The basic turbine data required for these analyses are listed in Table 6.13. They are typically obtained from the manufacturer of the turbine; however, a large portion of the data can be found in brochures and online databases, such as 4COffshore.com (4C Offshore Limited 2016) or LORC.dk (Lindoe Offshore Renewables Center 2011).

Metocean Data

The most important Metocean data for this simplified analysis are summarised in Table 6.14. These are wind speed and turbulence characteristics, wave characteristics, water depth at the site, and maximum current speed at the site. These data are typically obtained from measurements, either at the site or close to the site location, taken over many months or even several years. The wind speed data are of key importance for the estimation of energy production potential (and thus, the profitability) of the offshore wind farm, and is typically readily available by the time the design of the wind farm starts. Wave data can be obtained from measurement data by government agencies, as well as from oil and gas production stations (see e.g. (Williams 2008)). The relevant data for the example site for the current simplified analysis are given in Table 6.14.

Geological and Geotechnical Data

The geological and geotechnical data are the most challenging as well as most expensive to obtain and require site investigation. A good source of information is the British

Table 6.13 Turbine data and chosen pile material parameters.

Parameter	Symbol	Value	Unit
Hub height	z_{hub}	87	m
Rotor diameter	D	120	m
Tower height	L_T	68	m
Tower top diameter	D_t	3	m
Tower bottom diameter	D_b	5	m
Tower wall thickness	t_T	0.027	m
Density of the tower material	ρ_T	7860	kg/m^3
Tower mass	m_T	250	tons
Rated wind speed	U_R	12	m/s
Mass of the rotor-nacelle assembly (RNA)	m_{RNA}	243	tons
Operational rotational speed range of the turbine	Ω	5–13	rpm

Table 6.14 Metocean data.

Parameter	Symbol	Value	Unit
Wind speed Weibull distribution shape parameter	s	1.8	[−]
Wind speed Weibull distribution scale parameter	K	8	m/s
Reference turbulence intensity	I	18	%
Turbulence integral length scale	L_k	340.2	m
Density of air	ρ_a	1.225	kg/m^3
Significant wave height with 50-year return period	H_S	6.6	m
Peak wave period	T_S	9.1	s
Maximum wave height (50-year)	H_m	12.4	m
Maximum wave peak period	T_m	12.5	s
Maximum water depth (50-year high water level)	S	25	m
Density of sea water	ρ_w	1030	kg/m^3

Geological Survey, which contains data from around the United Kingdom. In the worst-case scenario, a first estimation can be carried out by just knowing the basic site classification such as stiff clay or dense sand.

The geotechnical data necessary for the analysis include:

- *Ground profile.* For the example site, it is assumed that the uppermost layers (roughly the upper 20 m) are loose to medium-dense sand and silt overlying layers of London clay.
- *Strength and stiffness parameters.* Loose to medium sand/silt in the upper layers have a submerged unit weight of $\gamma' = 9$ [kN/m^3] and the friction angle in the range of $\phi' = 28 - 36^{\circ}$.

The modulus of subgrade reaction is chosen following Terzaghi (1955). The soil's modulus of subgrade reaction is approximated as linearly increasing, with coefficient of subgrade reaction

$$n_h = \frac{A \cdot \gamma'}{1.35} \approx \frac{600 \cdot 9000}{1.35} = 4 \left[\frac{\text{MN}}{\text{m}^3}\right] \tag{6.37}$$

where $A = 300 - 1000$ for medium dense sand and $A = 100 - 300$ for loose sand, $\gamma' = 9$ [kN/m^3] as given above. The geotechnical data are summarised in Table 6.15.

Pile and Transition Piece

The pile's material is chosen as the industry standard S355 structural steel. The important properties of this material are: Young's modulus E_P, the density ρ_P and the yield strength f_{yk}, which are given in Table 6.15. The use of higher-strength steel may be considered for foundation design; however, cost constraints typically result in S355 being used. The total width of the grout and the transition piece together is taken as $t_{TP} + t_G = 0.15$[m], which results in the substructure diameter.

Table 6.15 Geotechnical, pile material, and transition piece data.

Parameter	Symbol	Value	Unit
Soil's submerged unit weight	γ'	9	kN/m^3
Soil's angle of internal friction	ϕ'	28–36	°
Soil's coefficient of subgrade reaction	n_h	4000	kN/m^3
Pile wall material – S355 steel – Young's modulus	E_P	200	GPa
Pile wall material – S355 steel – density	ρ_P	7860	kg/m^3
Pile wall material – S355 steel – Yield stress	f_{yk}	355	MPa
Grout and transition piece combined thickness	$t_G + t_{TP}$	0.15	[m]

Guess Initial Pile Dimensions

The initial pile dimensions are guessed based on the ULS design load. The load calculations are carried out based on the equations presented in Chapter 2. A spreadsheet can be used to carry out these calculations. The wind load on the rotor can already be calculated in the first step; however, the wave loading depends on the monopile diameter, and therefore it can only be calculated after the initial pile dimensions are available.

Calculate Highest Wind Load

The wind load for ULS is determined from the 50-year extreme operating gust (EOG), which is assumed to produce the highest single occurrence wind load; this is wind scenario (U-3), see Section 2.6.1 for further details. The procedure outlined in Section 2.6.2 (Chapter 2) is used to estimate the wind load for this scenario. First the EOG wind speed is calculated using data from Tables 6.13 and 6.14.

$$U_{10,50-year} = 35.7\,[\text{m/s}]$$

$$U_{10,1-year} = 28.6\,[\text{m/s}]$$

$$\sigma_{U,c} = 3.15\,[\text{m/s}]$$

$$u_{EOG} = 8.1[\text{m/s}]$$

Using this, the total wind load is estimated as

$$Th_{wind,EOG} \approx 1.63\,[\text{MN}] \tag{6.38}$$

and using the water depth $S = 25$ [m] and the hub height above mean sea level $z_{hub} = 87$ [m]

$$M_{wind,EOG} = Th_{wind,EOG}(S + z_{hub}) \approx 182\,[\text{MNm}] \tag{6.39}$$

Applying a load factor of $\gamma_L = 1.35$ the total wind moment is ~246 [MNm]. Calculation of the wind load for the other load cases is omitted here for brevity; however, it is found that the EOG at U_R (U-3) gives the highest load.

Calculate Initial Pile Dimensions

The initial value of wall thickness may be chosen according to API (2005) as

$$t_P \geq 6.35 + \frac{D_P}{100}[\text{mm}] \tag{6.40}$$

This wall thickness value may not necessarily provide sufficient stability to avoid local or global buckling of the pile, or to ensure that the pile can be driven into the seabed with the simplest installation method avoiding pile tip damage leading to early refusal. Therefore, these issues need to be addressed separately, as well as fatigue design of the pile, which may require additional wall thickness. Figure 6.16 shows the wall thickness for installed offshore wind turbines of different monopile diameters. As can be seen, some piles have wall thicknesses significantly higher than the API-required thickness. For details on buckling-related issues (global buckling, avoiding local pile buckling or propagating pile tip damage due to installation), see Bhattacharya et al. (2005), Aldridge et al. (2005). For practical reasons, the wall thickness is typically chosen based on standard plate thickness values to optimise manufacturing.

Using the pile thickness formula of API (2005) given by Eq. (6.40), the following can be written for the area moment of inertia of the pile cross section:

$$I_P = \frac{1}{8}(D_P - t_P)^3 t_P \pi = \frac{1}{8}\left(D_P - 6.35 - \frac{D_P}{100}\right)^3 \left(6.35 + \frac{D_P}{100}\right)\pi \tag{6.41}$$

The following has to be satisfied to avoid pile yield with material factor $\gamma_M = 1.1$:

$$\sigma_m = \frac{M_{wind}, EOG}{I_P}\frac{D_P}{2} < \frac{f_{yk}}{\gamma_M} \approx 322\,[\text{MPa}] \tag{6.42}$$

from which the required diameter is determined as

$$\frac{D_P}{I_P} < \frac{2f_{yk}}{\gamma_M M_{wind,EOG}} \tag{6.43}$$

This results in an initial pile diameter of $D_P = 4.5$ [m] with a wall thickness of $t_P \approx 51$ [mm].

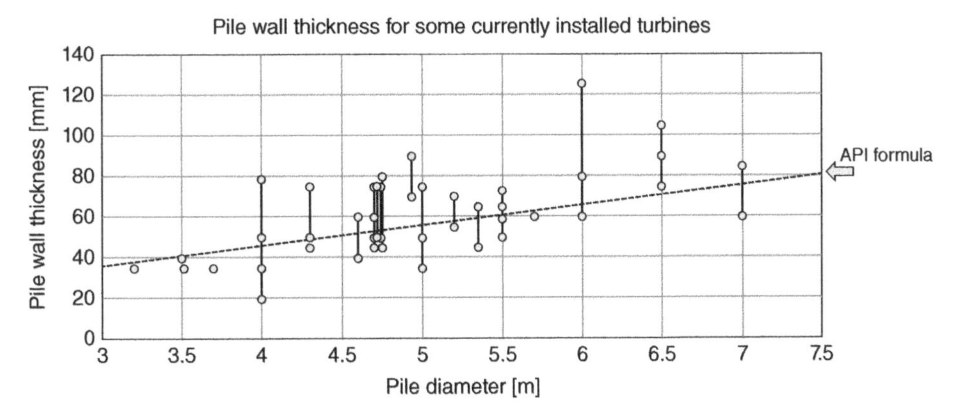

Figure 6.16 Wall thickness of several currently installed wind turbines. The case studies used for the plot are collated in Table 6.16.

The embedded length is determined next.

The formula of Poulos and Davis (1980) given below can be used to estimate the required embedded length

$$L_P = 4.0\left(\frac{E_P I_P}{n_h}\right)^{\frac{1}{5}} \approx 39\,[\mathrm{m}] \tag{6.44}$$

The initial pile dimensions are then

$$D_P = 4.5\,[\mathrm{m}] \;\; t_P = 0.051\,[\mathrm{m}] L_P = 39\,[\mathrm{m}] \tag{6.45}$$

Estimate Loads on the Foundation

Now that an initial guess for the pile dimensions is available, the wave load can be calculated. For the combination of wind and wave loading, many load cases are presented in design standards. Five conservative load cases are considered in Table 6.17. Among the potential severe load cases not covered by these scenarios are shutdown events of the wind turbine, as these situations require detailed data about the wind turbine (rotor, blades, control system parameters, generator, etc.), but are likely to provide a lower foundation load than the scenarios in Table 6.17.

Calculate Wind Loads (Other Wind Scenarios)

The wind loads on the structure are independent of the substructure diameter, and therefore the wind loads can be evaluated before the pile and substructure design are available. Section 2.6.2 is used to determine the turbulent wind speed component and through that the thrust force and overturning moment. Table 6.17 summarises the important parameters and presents the wind loads for the different wind scenarios. Note that the mean of the maximum and minimum loads are not equal to the mean force without turbulent wind component. This is because the thrust force is proportional to the square of the wind speed.

Calculate Critical Wave Loads

The wave load is first calculated only for the most severe wave scenarios used for Load Cases E-2 and E-3, i.e. wave scenario (W-2) and (W-4), the 1-year and 50-year extreme wave heights (EWHs). The methodology described in Section 2.6.3 is used to calculate the wave loading, and the substructure diameter, which in this case is $D_S = 4.8$ [m]. The relevant 50-year wave height and wave period are taken from Table 6.16. The 1-year equivalents are calculated following (DNV 2014) from the 50-year significant wave height according to

$$H_{S,1} = 0.8 H_{S,50} = 5.3\,[\mathrm{m}] \quad T_{S,1} = 11.1\sqrt{H_{S,1}/g} = 8.1\,[\mathrm{s}] \tag{6.46}$$

and then the procedure in Section 5.6 is used to determine the 1-year maximum wave height and period:

$$H_{m,1} = 10.1\,[\mathrm{m}]\text{ and } T_{m,1} = 11.2\,[\mathrm{s}] \tag{6.47}$$

The wave heights and wave periods are summarised for all wave scenarios (W-1) to (W-4) in Table 6.18.

Table 6.16 Pile diameters and wall thicknesses of monopiles shown in Figure 6.16.

#	Wind farm and turbine	Pile diameter [m]	Wall thickness range [mm]		
1	Lely, Netherlands, A2 turbine	3.2	35	–	35
2	Lely, Netherlands, A3 turbine	3.7	35	–	35
3	Irene Vorrink, Netherlands	3.515	35	–	35
4	Blyth, England, UK	3.5	40	–	40
5	Kentish Flats, England, UK	4.3	45	–	45
6	Barrow, England, UK	4.75	45	–	80
7	Thanet, England, UK	4.7	60	–	60
8	Belwind, Belgium	5	50	–	75
9	Burbo Bank, England, UK	4.7	45	–	75
10	Walney, England, UK	6	60	–	80
11	Gunfleet Sands, England, UK	5	35	–	50
12	London Array, England, UK	4.7	50	–	75
13	Gwynt y Mór, Wales, UK	5	55	–	95
14	Anholt, Denmark	5.35	45	–	65
15	Walney 2, England, UK	6.5	75	–	105
16	Sheringham Shoal, England, UK	5.7	60	–	60
17	Butendiek, Germany	6.5	75	–	90
18	DanTysk, Germany	6	60	–	126
19	Meerwind Ost/Sud, Germany	5.5	50	–	65
20	Northwind, Belgium	5.2	55	–	70
21	Horns Rev, Denmark	4	20	–	50
22	Egmond aan Zee, Netherlands	4.6	40	–	60
23	Gemini, Netherlands	5.5	59	–	73
24	Gemini, Netherlands	7	60	–	85
25	Princess Amalia, Netherlands	4	35	–	79
26	Inner Dowsing, England, UK	4.74	50	–	75
27	Rhyl Flats, Wales, UK	4.72	50	–	75
28	Robin Rigg, Scotland, UK	4.3	50	–	75
29	Teesside, England, UK	4.933	70	–	90

The maximum of the inertia load occurs at the time instant $t = 0$ when the surface elevation $\eta = 0$ and the maximum of the drag load occurs when $t = T_m/4$ and $\eta = H_m/2$.

The maximum drag and inertia loads for wave scenario (W-2) with the 1-year EWH are then

$$
\begin{aligned}
F_{D,max} &= 0.65\ [\mathrm{MN}] \\
M_{D,max} &= 17.1\ [\mathrm{MNm}] \\
F_{I,max} &= 1.48\ [\mathrm{MN}] \\
M_{I,max} &= 36.7\ [\mathrm{MNm}]
\end{aligned} \tag{6.48}
$$

Table 6.17 Load and overturning moment for wind scenarios (U-1) – (U-4).

Parameters	Symbol [unit]	Wind scenario (U-1)	Wind scenario (U-2)	Wind scenario (U-3)	Wind scenario (U-4)
Standard deviation of wind speed	σ_U [m/s]	2.63	3.96	–	–
Standard deviation in f > f_{1P}	$\sigma_{U,f>f_{1P}}$ [m/s]	0.73	1.22	–	–
Turbulent wind speed component	u [m/s]	0.94	2.44	8.1	4.86
Maximum force in load cycle	F_{max} [MN]	0.68	0.84	1.63	0.40
Minimum force in load cycle	F_{min} [MN]	0.49	0.37	0.39	0.25
Mean force without turbulence	F_{mean} [MN]	0.58	0.58	0.58	0.28
Maximum moment in load cycle	M_{max} [MN]	75.8	94.4	182.8	44.6
Minimum moment in load cycle	M_{min} [MN]	55.4	41.4	43.7	28.1
Mean moment without turbulence	M_{mean} [MN]	65.2	65.2	65.2	31.3

Table 6.18 Wave heights and wave periods for different wave scenarios.

Parameters	Symbol [unit]	Wave scenario (W-1)	Wave scenario (W-2)	Wave scenario (W-3)	Wave scenario (W-4)
Wave height	H [m]	5.3	10	6.6	12.4
Wave period	T [s]	8.1	11.2	9.1	12.5

The maxima of the wave loads and moments may be conservatively taken as

$$F_{wave,W-2} = 2.13\,[\text{MN}] \text{ and } M_{wave,W-2} = 53.8\,[\text{MNm}] \tag{6.49}$$

Similarly, for the 50-year EWH in wave scenario (W-4) the drag and inertia loads are given as

$$\begin{aligned} F_{D,max} &= 1.07\,[\text{MN}] \\ M_{D,max} &= 23.8\,[\text{MNm}] \\ F_{I,max} &= 1.7\,[\text{MN}] \\ M_{I,max} &= 43.9\,[\text{MNm}] \end{aligned} \tag{6.50}$$

and the maxima of the wave load

$$F_{wave,W-4} = 2.77\,[\text{MN}]\ M_{wave,W-4} = 67.7\,[\text{MNm}] \tag{6.51}$$

Load Combinations for ULS

The most severe load cases in Table 6.19 for ULS design are E-2 and E-3, the extreme wave scenario (50-year EWH) combined with extreme turbulence model (ETM) and the extreme operational gust (EOG) combined with the yearly maximum wave height (1-year EWH). A partial load factor of $\gamma_L = 1.35$ has to be applied for ULS environmental

Table 6.19 ULS load combinations.

Load	Extreme Wave Scenario (E-2)ETM (U-2) and 50-year EWH (W-4)	Extreme Wind Scenario (E-3)EOG at U_R (U-3) and 1-year EWH (W-2)
Maximum wind load [MN]	0.84	1.63
Maximum wind moment [MNm]	94.4	182.6
Maximum wave load [MN]	2.77	2.13
Maximum wave moment [MNm]	67.7	53.8
Total load [MN]	3.61	3.79
Total overturning moment [MNm]	162.1	236.4

loads according to DNV (2014) and IEC (2009a). Table 6.19 shows the ULS loads for the two load combinations, and it is clear from the table that for this particular example the driving scenario is (E-3), since the overturning moment is dominated by the wind load.

The new total loads in Table 6.19 are used to recalculate the required foundation dimensions. This will result in an iterative process of finding the necessary monopile size for the ULS load, which can be easily solved in a spreadsheet. The analysis results in the following dimensions:

$$D_P = 4.9\,[\mathrm{m}]\; t_P = 56\,[\mathrm{mm}]\; L_P = 42\,[\mathrm{m}] \tag{6.52}$$

The stability analysis has to be carried out following the Germanischer Lloyd (2005) Chapter 6 on the design of steel support structures.

Estimate Geotechnical Load Carrying Capacity

In typical scenarios, the limiting case for maximum lateral load results from the yield strength of the pile. However, a check has to be performed to make sure that the foundation can take the load, that is, that the soil does not fail at the ULS load. Based on the standard methods outlines in Section 5.3.3.4, the ultimate horizontal load bearing capacity and the ultimate moment capacity of the pile are established as $F_R = 38MN$ and $M_R = 2275MNm$, respectively. These are well above the limit.

In terms of vertical load, it is expected that failure due to lateral load occurs first and that stability under lateral load ensures the pile's ability to take the vertical load imposed mainly by the deadweight of the structure. The analysis of vertical load-carrying capacity is therefore omitted here, but has to be performed in actual design.

Estimate Deformations and Foundation Stiffness

Many methods as explained in Examples can be used to evaluate foundation. In this example, the method by Poulos and Davis (1980) is used. The method requires the modulus of subgrade reaction for cohesive (clayey) and the coefficient of subgrade reaction for cohesionless (sandy) soils. The upper layers are dominant for the calculation of deflections and stiffness. The sand and silt layers were approximated here with the coefficient of subgrade reaction $n_h = 4$ $[MN/m^3]$ following Terzaghi (1955). The foundation stiffness calculations use the Poulos and Davis (1980) formulae for flexible piles in medium sand.

$$K_L = 1.074 n_h^{\frac{3}{5}} (E_P I_P)^{\frac{2}{5}} \quad K_{LR} = -0.99 n_h^{\frac{2}{5}} (E_P I_P)^{\frac{3}{5}} \quad K_R = 1.48 n_h^{\frac{1}{5}} (E_P I_P)^{\frac{4}{5}} \tag{6.53}$$

and their values are

$$K_L = 0.57\,[\mathrm{GN/m}] \quad K_{LR} = -5.9\,[\mathrm{GN}] \quad K_R = 99.3\,[\mathrm{GN/rad}] \tag{6.54}$$

The deflections and rotations are calculated following equations in Chapter 5 and calculations presented in

$$\rho = 10.4[\mathrm{cm}]\ \theta = 0.569[^\circ] \tag{6.55}$$

The pile tip deflection is acceptable but the rotation exceeds 0.5°.

Again, an iterative process is necessary by which the necessary pile dimensions are obtained. This can be done by the following iterative steps:

1. The foundation dimensions are increased.
2. Recalculate the foundation loads.
3. The foundation stiffness parameters are recalculated based on appropriate equations.
4. The mudline deformations are recalculated.
5. The process is repeated until the deflection and rotation are both below the allowed limit.

A spreadsheet can be used to easily obtain the necessary dimensions:

$$D_P = 5.2[\mathrm{m}] \quad t_P = 59[\mathrm{mm}] \quad L_P = 43[\mathrm{m}] \tag{6.56}$$

and the deformations are now

$$\rho_0 = 0.095[\mathrm{m}] \quad \theta_0 = 0.495[^\circ] \tag{6.57}$$

Calculate Natural Frequency and Dynamic Amplification Factors

The natural frequency is calculated following the method given in Chapter 5 and developed in Arany et al. (2015a) and Arany et al. (2016). The first natural frequency and the damping of the first mode in the along-wind and cross-wind directions are used to obtain the DAFs that affect the structural response.

Calculate Natural Frequency

The structural natural frequency of the turbine-tower-substructure-foundation system is given by $f_0 = C_L C_R C_S f_{FB}$, where C_S, C_R, and C_L are the substructure flexibility coefficient and the rotational and lateral foundation flexibility coefficients, respectively. The fixed-base natural frequency is

$$f_{FB} = \frac{1}{2\pi}\sqrt{\frac{3E_T I_T}{L_T^3\left(m_{RNA\,RNA} + \frac{33}{140}m_T\right)}} = 0.379[Hz] \tag{6.58}$$

The substructure flexibility coefficient C_S is calculated by assuming that the monopile goes up to the bottom of the tower. The tower is 68 m tall, the hub height is 87 m, the nacelle is ~5 m tall, so the distance between the mudline and the bottom of the tower is about $L_S = 41.5[m]$ (this is the platform height). The foundation flexibility is expressed in terms of two dimensionless parameters, the bending stiffness ratio $\chi = E_T I_T/(E_P I_P) = 0.214$ and the length ratio $\psi = L_S/L_T = 0.6104$.

$$C_S = \sqrt{\frac{1}{1+(1+\psi)^3\chi - \chi}} = 0.773 \tag{6.59}$$

The nondimensional foundation stiffnesses are calculated:

$$\eta_L = \frac{K_L L_T^3}{EI_\eta} = 978 \quad \eta_{LR} = \frac{K_{LR} L_T^2}{EI_\eta} = -149 \quad \eta_R = \frac{K_R L_T}{EI_\eta} = 36.9 \tag{6.60}$$

and the foundation flexibility coefficients are calculated:

$$C_R = 1 - \frac{1}{1+0.6\left(\eta_R - \frac{\eta_{LR}^2}{\eta_L}\right)} = 0.895 \quad C_L = 1 - \frac{1}{1+0.5\left(\eta_L - \frac{\eta_{LR}^2}{\eta_R}\right)} = 0.995 \tag{6.61}$$

The natural frequency is then

$$f_0 = 0.995 \times 0.895 \times 0.773 \times 0.379 = 0.6884 \times f_{FB} = 0.261 \text{ Hz} \tag{6.62}$$

This is acceptable, as the condition was that $f_0 > 0.24[\text{Hz}]$.

Calculate Dynamic Amplification Factors

The dynamic amplification of the wave loading is calculated using the peak wave frequency and an assumed damping ratio. The total damping ratios for the along-wind (x) and cross-wind (y) directions are chosen conservatively as 3% and 1%, respectively. The along-wind damping is larger due to the significant contribution of aerodynamic damping. In real cases, the aerodynamic damping depends on the wind speed, and the along-wind value may be between 2% and 10%. The chosen value is conservatively small for the relevant wind speed ranges, see e.g. Camp et al. (2004), or the discussion on damping in Arany et al. (2016). The DAFs are calculated as

$$DAF = \frac{1}{\sqrt{\left(1-\left(\frac{f}{f_0}\right)^2\right)^2 + \left(2\xi\frac{f}{f_0}\right)^2}} \tag{6.63}$$

where f is the excitation frequency, f_0 is the Eigen frequency and ξ is the damping ratio. The DAFs for all wave scenarios are presented in Table 6.20. The difference in DAFs in the along-wind (x) and cross-wind (y) directions is apparently negligible for this example, and in Table 6.21 the higher value is used when loads with DAF are calculated.

Recalculate Wave Loads and Foundation Dimensions

The ultimate load case and the deformations have to be checked again to include dynamic amplification of loads. The wave loads recalculated for the increased pile diameter and are presented in Table 6.20. The updated values of foundation dimensions

Table 6.20 Dynamic amplification factors and wave loads.

Parameters	Symbol [unit]	Wave scenario (W-1)	Wave scenario (W-2)	Wave scenario (W-3)	Wave scenario (W-4)
Wave period	T *[s]*	8.1	11.2	9.1	12.5
Wave frequency	f *[Hz]*	*0.123*	*0.089*	*0.110*	*0.080*
Dynamic amplification – along-wind	DAF_x [–]	*1.285*	*1.131*	*1.215*	*1.103*
Dynamic amplification – cross-wind	DAF_y [–]	*1.288*	*1.133*	*1.215*	*1.104*
Total wave load	F_w [MN]	1.41	2.77	1.77	3.56
Total wave moment	M_w [MNm]	32.1	69.8	40.0	97.2
Total wave load with DAF	$F_{w,DAF}$ [MN]	*1.82*	*3.14*	*2.15*	*3.93*
Total wave moment with DAF	$M_{w,DAF}$[MNm]	*41.4*	*79.1*	*48.6*	*107.3*

Table 6.21 Calculated loads with dynamic amplification factors.

Parameter	Normal operation E-1	Extreme wave scenario E-2	Extreme wind scenario E-3	Cut-out wind + extreme wave E-4	Wind-wave misalignment E-5
Mean wind load [MNm]	65.2	65.2	65.2	31.2	65.2
Maximum wind load [MNm]	75.8	94.4	182.6	44.6	94.4
Minimum wind load [MNm]	55.4	30.7	43.7	28.1	30.7
Maximum wave load [MNm]	*41.4*	*107.3*	*79.1*	*107.3*	*107.3*
Minimum wave load [MNm]	*−41.4*	*−107.3*	*−79.1*	*−107.3*	*−107.3*
Combined maximum load [MNm]	117.2	201.7	**261.7**	151.9	142.9
Combined minimum load [MNm]	14	−76.6	−35.4	−79.2	30.7
Cycle time period [s]	8.1	12.5	11.2	12.5	12.5
Cycle frequency [Hz]	0.123	0.080	0.089	0.080	0.080
Maximum stress level [MPa]	96.8	166.6	216.1	125.5	166.6
Maximum cyclic stress amplitude [MPa]	131.0	255.2	281.5	214.1	255.2

are obtained through an iterative process as before, easily calculable in a spreadsheet. The final dimensions are

$$D_P = 5.2[m] \quad t_P = 59[mm] \quad L_P = 43[m] \quad f_0 = 0.261[Hz] \tag{6.64}$$

The final loads for each load scenario (E-1)–(E-4) are given in Table 6.21. The table also contains the maximum stresses and cyclic stress amplitudes for each load case.

Long-Term Natural Frequency Change

The dynamic stability of the structure can be threatened by changing structural natural frequency over the lifetime of the turbine. Resonance may occur with environmental and mechanical loads resulting in catastrophic collapse or reduced fatigue life and

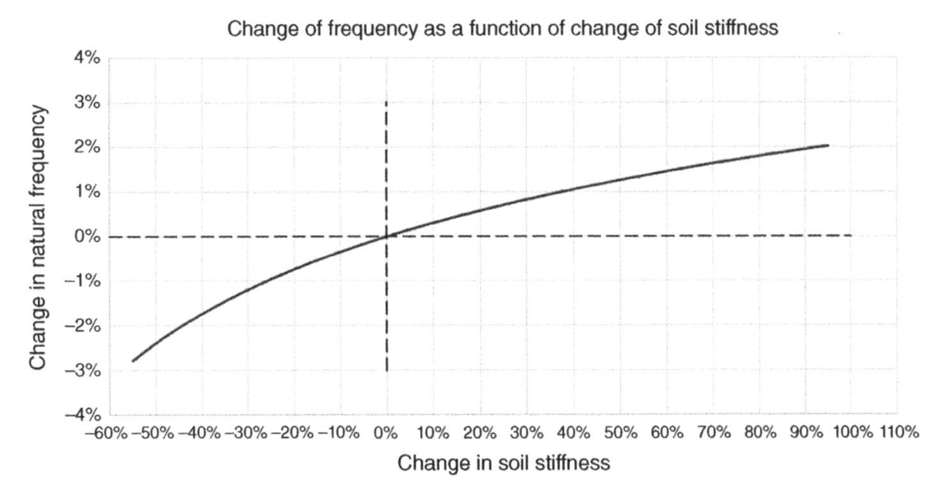

Figure 6.17 Frequency change due to change in soil stiffness during the lifetime of the turbine.

serviceability. Therefore, it is an important aspect to see the effects of changing soil stiffness on the natural frequency of the structure. Figure 6.17 shows the percentage change in natural frequency against the percentage change in the soil stiffness (coefficient of subgrade reaction n_h). It can be seen from the figure that a change of 30% in soil stiffness produces less that 1.5% change in the natural frequency. It is also apparent that degradation is more critical than stiffening from the point of view of frequency change.

Long-Term Deflection and Rotation

The rotation prediction is typically the critical aspect in monopile design, as opposed to the prediction of deflection. An attempt has been made to use the method of Leblanc et al. (2010) for the prediction of the long term tilt. In addition to problems listed in Chapter 5 another practical problem occurs when using this approach. Leblanc et al. (2010) investigated the ultimate moment capacity of the pile by experiments and noted that a clear point of failure could not be established from the tests, and thus they defined the ultimate moment capacity – somewhat arbitrarily – as the bending moment that causes 4° of mudline rotation. However, this value was found to be relatively close to the value calculated by the method given in Poulos and Davis (1980).

The approach of Poulos and Davis (1980) gave the ultimate moment capacity as $M_{R,P\text{–}D} = 2275$ [MNm] while the 4° rotation approach gave $M_{R,4^\circ} = 2029$ [MNm] based on linearity of K_R. Note that both these values are significantly higher than the maximum moment that is expected ($M_{ULS} = 261.7$ [MNm], and also than the pile yield bending moment $M_f \approx 390$ [MNm].

The tests carried out by Leblanc et al. (2010) for rigid piles utilise relatively high levels of loading to establish the long-term rotation as a function of the number of cycles. This is likely due to the high levels of ultimate moment capacity $M_{R,4^\circ}$ estimated by their approach as compared to typical pile yield failure limits M_f. This resulted in test scenarios with significantly higher levels of loading than those expected for an actual offshore wind turbine. The test scenarios have been carried out with maximum load magnitude to load capacity ratios $\zeta_b = M_{max}/M_R$ between 0.2 and 0.53 for a relative density of $R_d = 4\%$, and between 0.27 and 0.52 for the relative density of $R_d = 38\%$. A linear curve has been fitted to the test results by Leblanc et al. (2010) in a graph, and approximate equations have been given in this Chapter 5. Using these linear expressions, the test results can be extrapolated beyond the range of measured results, which is undesirable. However, the linear equations cross the abscissa at ~0.15 and ~0.06 for $R_d = 4\%$ and $R_d = 38\%$, respectively, and below these values the equation takes negative values, which is unrealistic.

Using the maximum bending moments calculated conservatively in Table 6.22 for the design load cases, the ratio is only $\zeta_b = 0.13$ even for the most severe 50-year maximum ULS load $M_{ULS} = 261.7$ [MNm], which is expected to occur only once in the lifetime of the turbine. The angle of internal friction for the soil in the chosen site is about 28°–36°, with the average in the upper regions being around 30°; therefore, the $R_d = 4\%$ curve is assumed to be more representative. It is clear from Figure 6.18 that the actual load magnitudes expected throughout the lifetime of the turbine are in the range where the linear extrapolation of the test results of Leblanc et al. (2010) would give unrealistic negative values for the rotation accumulation. Most of the likely lifetime load cycles for typical turbines would have magnitudes in the region below the range of available scale test results (i.e. $\zeta_b < 0.15$).

It is clear that it is hard to arrive at a conclusion about the accumulation of pile-head rotation following this method when the expected load cycle magnitudes in practical problems are below the lower limit of the scale tests. Guidance is not given regarding these load scenarios in Leblanc et al. (2010), and it is not known whether it is safe to assume no rotation accumulation below the point where the linear approximation curve reaches zero (i.e. below ~0.15 for $R_d = 4\%$ and below ~0.06 for $R_d = 38\,\%$). The methodology cannot be used for such scenarios due to a lack of data for relevant load levels.

If the load levels predicted are out of range, it is suggested that this analysis be complemented with the calculation of relevant strain levels in the soil due to the pile deformation. The long-term behaviour can then be based on the maximum strain levels expected for the type of soils at the site. Resonant column test or cyclic simple shear test or cyclic triaxial test of soil samples can be carried out to predict the long-term behaviour using

Table 6.22 Number of cycles survived at different extreme load scenarios.

Parameter	Normal operation E-1	Extreme wave scenario E-2	Extreme wind scenario E-3	Cut-out wind extreme wave E-4	Wind-wave misalignment E-5
Maximum stress level σ_m [MPa] as defined in Eq. (5.49)	162	87	214	123	101
Maximum cyclic stress amplitude σ_c [MPa] as defined in Eq. (5.50)	108	86.5	112	92	47.5
Number of cycles the monopile may survive	1.55×10^6	3.02×10^6	1.38×10^6	2.45×10^6	2.63×10^7

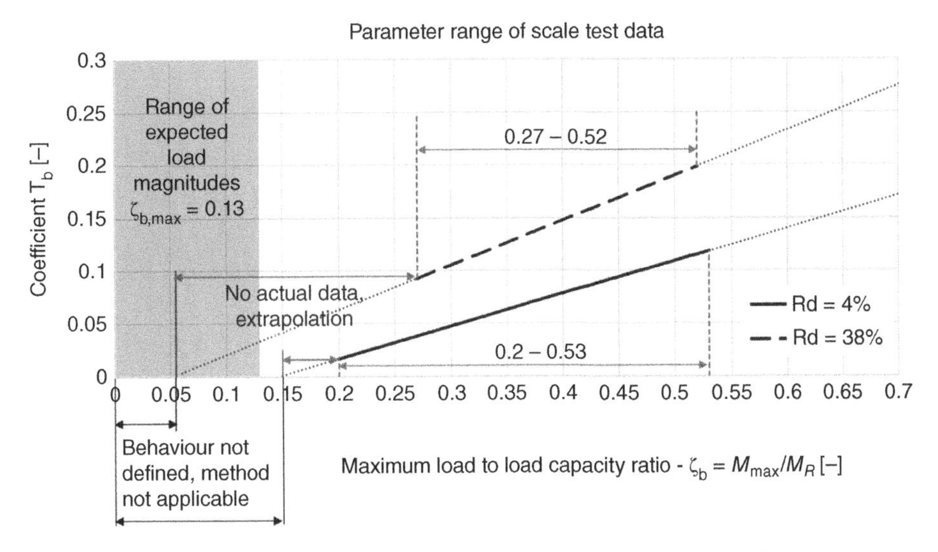

Figure 6.18 Range of ζ_b values in the scale tests of Leblanc et al. (2010) and values expected for the example case.

the concept of threshold strain, see Lombardi et al. (2013) for monopiles in cohesive soils. The readers are referred to Chapter 5 for the fundamentals of this approach.

Fatigue Life

The fatigue analysis of the structural steel and the weld of the flush ground monopile is carried out using the methodology described in Chapter 5. Material factor of $\gamma_M = 1.1$ is used, and the yield strength of the S355 structural steel used for the monopile is thus reduced to $\sigma_y = 322$ [MPa]. A load factor of $\gamma_L = 1.0$ is applied. With these, the maximum stress levels caused by the load cases can be calculated and the results are given in Table 6.22. It was found that the highest stress amplitude observed is $\sigma_{m,\,max} = 216.1$ [MPa] and the maximum cyclic stress amplitude is $\sigma_{c,\,max} = 281.5$ [MPa].

In a study by Kucharczyk et al. (2012) it was identified that the fatigue endurance limit of the S355 steel is $\sigma_{end} = 260[MPa]$. Fatigue endurance limit of the material means that under stress cycles with a magnitude lower than this value, the material can theoretically withstand any number of cycles. The highest load case of $\sigma_{c,\,max} = 281.5$ [MPa] is expected to occur only once in 50 years (extremely low number of cycles), and it is a safe assumption that the fatigue life of the structural steel is satisfactory.

The fatigue analysis of welds of the flush ground monopile is carried out following DNV (2005), using the C1 category of S-N curves, as suggested in, e.g. Brennan and Tavares (2014). In Table 6.22, Load Case E-3 can be described as the 50-year ultimate load scenario, while Load Case V represents an estimate of the 1-year highest. Using the thickness correction factor, the representative S-N curve is shown in Figure 6.19. Table 6.22 shows how many cycles the monopile can survive under different stress cycle amplitudes for different load cases.

The simplified procedure arrived at the pile dimensions as follows: 5.2 m diameter and 44.5 m long and wall thickness of 59 mm. It is of interest to compare this to the

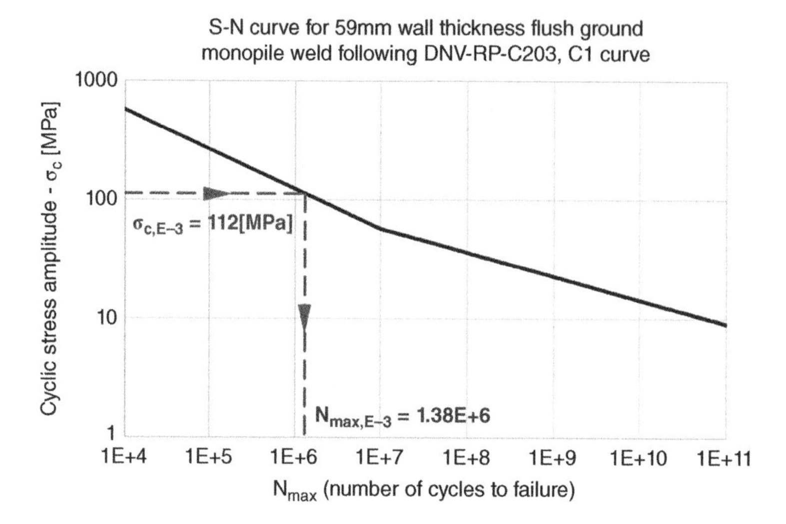

Figure 6.19 S-N curve for 59 mm wall thickness flush ground monopile weld following (DNV 2005).

actual foundation dimensions for the London Array wind farm, which is installed in a site with similar conditions to those used in the example. The wind farm comprises of 175 turbines (Siemens SWT-3.6-120) in water depths ranging from 0 to 25 m. The monopile diameters are between 4.7 and 5.7 m with wall thickness ranging from 44 to 87 mm. The piles were hammered up to 40 m into the seabed.

Example 6.9 ***Jacket-Type and Seabed-Frame Structure*** The jacket structure design is in many ways similar to offshore oil and gas jackets. The readers are referred to Section 2.5 where the load transfer mechanism of the superstructure loads to the foundation are shown. Essentially, for multiple supported structure the loads are transferred as push-pull action. One of the first steps is finding the support reaction at the support foundation and subsequently choosing the piles. Figure 6.20 shows an example where the loads on a jacket are shown. The loads and moments at the bottom of the tower can be applied together with the wave loads at the jacket nodes.

For jacket-supported structures, small-diameter pile will suffice and API *p-y* formulation can be used for such purpose. Certain software allows the application of wave loads directly and a typical model is shown in Figure 6.21. Once the foundation type and size are chosen, it is required to obtain the foundation stiffness, and as expected, the vertical stiffness of the foundation is critical. The next step is the estimation of the natural frequency of the system, and Figure 6.22 shows examples from a jacket problem.

The readers are referred to Jalbi and Bhattacharya (2018) for closed-form solutions for obtaining natural frequency of jacket. The method presented in the paper can also be used to optimise the jacket dimensions.

Choosing Vertical Stiffness of Foundations for No-Rocking Condition

Rocking modes of vibration are low frequency and may interact with the wave loads, causing a resonance-type condition. While typical piles are stiff axially, other types of foundations such as suction caissons may not be sufficiently stiff and may promote

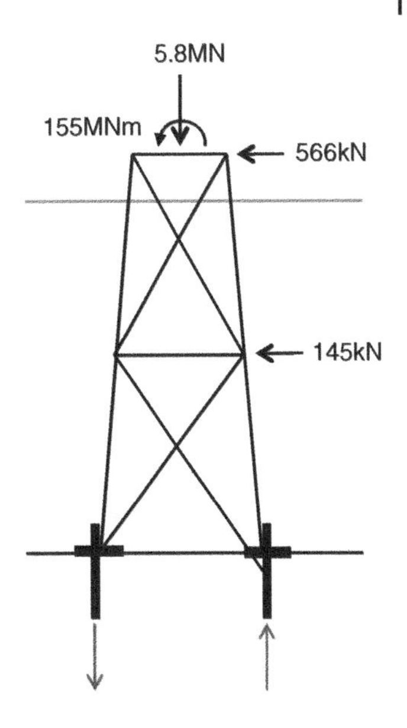

Figure 6.20 Forces acting on the jacket inputted to GSA model. Red arrows indicate the direction of axial forces in piles.

rocking-type vibration. In this section, this important design consideration is presented by considering a 5 MW jacket. The properties and parameters are given in Table 6.23. Essentially, this is a four-legged jacket structure based on EU-funded UPWIND project supporting a 5 MW wind turbine in deeper waters and the details can be found in UPWIND report. The 5 MW reference wind turbine is NREL type (see Figure 6.23).

The jacket-supported system can be analysed for different values of support stiffness k_v and parametric study is conducted to understand the variation of first fundamental frequency (f_0) with increasing vertical stiffness of the springs k_v. For easy understanding of the concepts, the analysis is presented in a nondimensional way:

(1) The results are plotted using the term k_V/k_t where k_t is the lateral stiffness of the tower in the equivalent mechanical model. This can be estimated by applying a unit load at the tower tip.
(2) The natural frequency is plotted using nondimensional group f_0/f_{fb}.

Two types of behaviour can be observed as shown in Figure 6.24: (a) For low values of k_V/k_t, there is rocking mode and it may be observed that the natural frequency is 48% of the fixed-base frequency; (b) for high values of k_V/k_t, the mode is sway-bending of the tower with no rocking. In this case, the natural frequency is 93% of the fixed-base frequency.

Based on the above concept, design pointers are suggested. Essentially, there is a minimum vertical foundation stiffness necessary. Figure 6.25 presents a graph providing a guidance. Salient points are presented here:

(1) The parameter dictating whether the system vibrates in a rocking or sway bending mode is the ratio of foundation vertical stiffness (k_V) to superstructure stiffness (k_t). At low foundation stiffness, the structure is more susceptible to rocking, whilst at higher foundation stiffness values sway-bending vibration governs. It is important to note that in the rocking vibration region, any change in vertical stiffness results

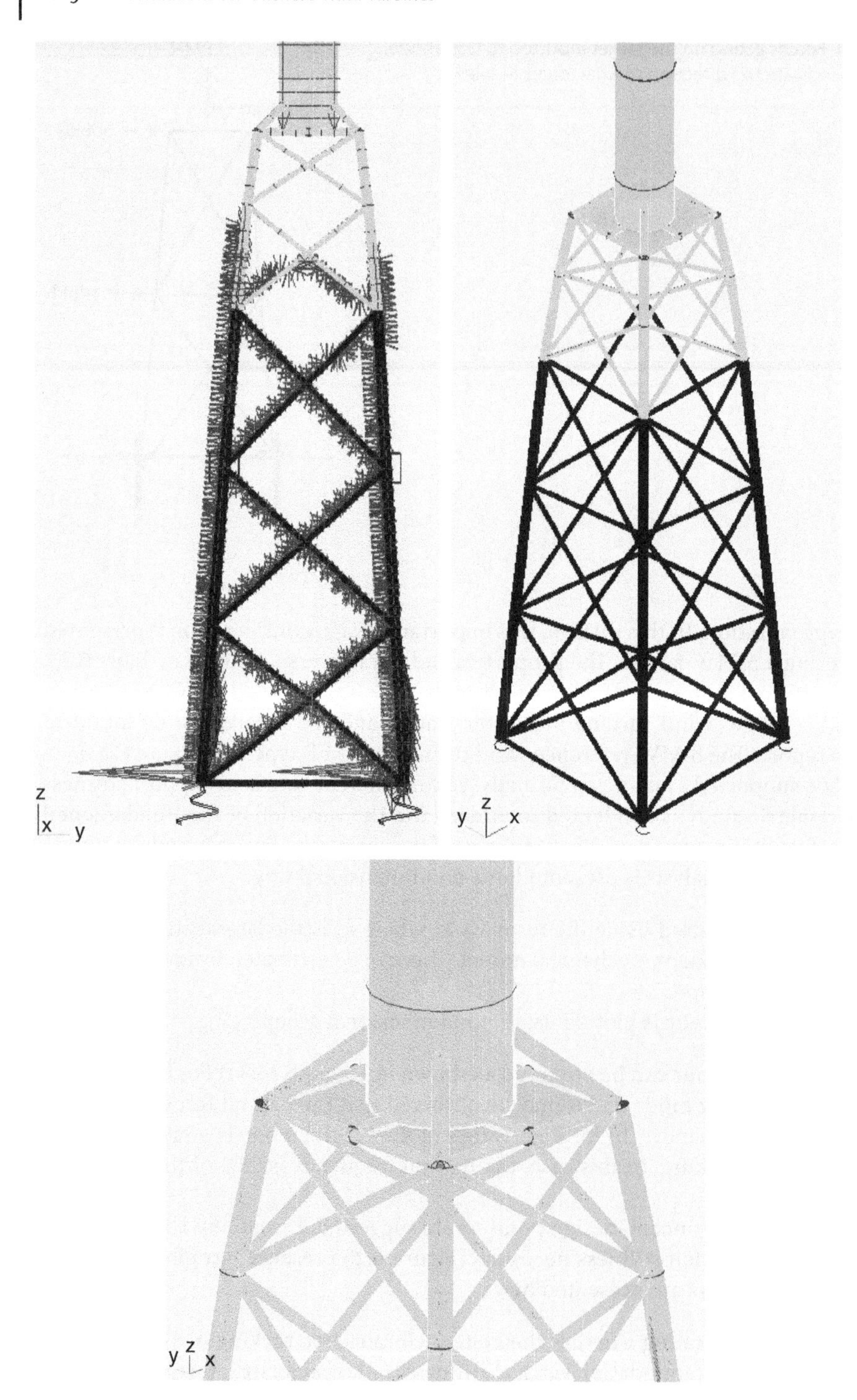

Figure 6.21 Modelling of jacket.

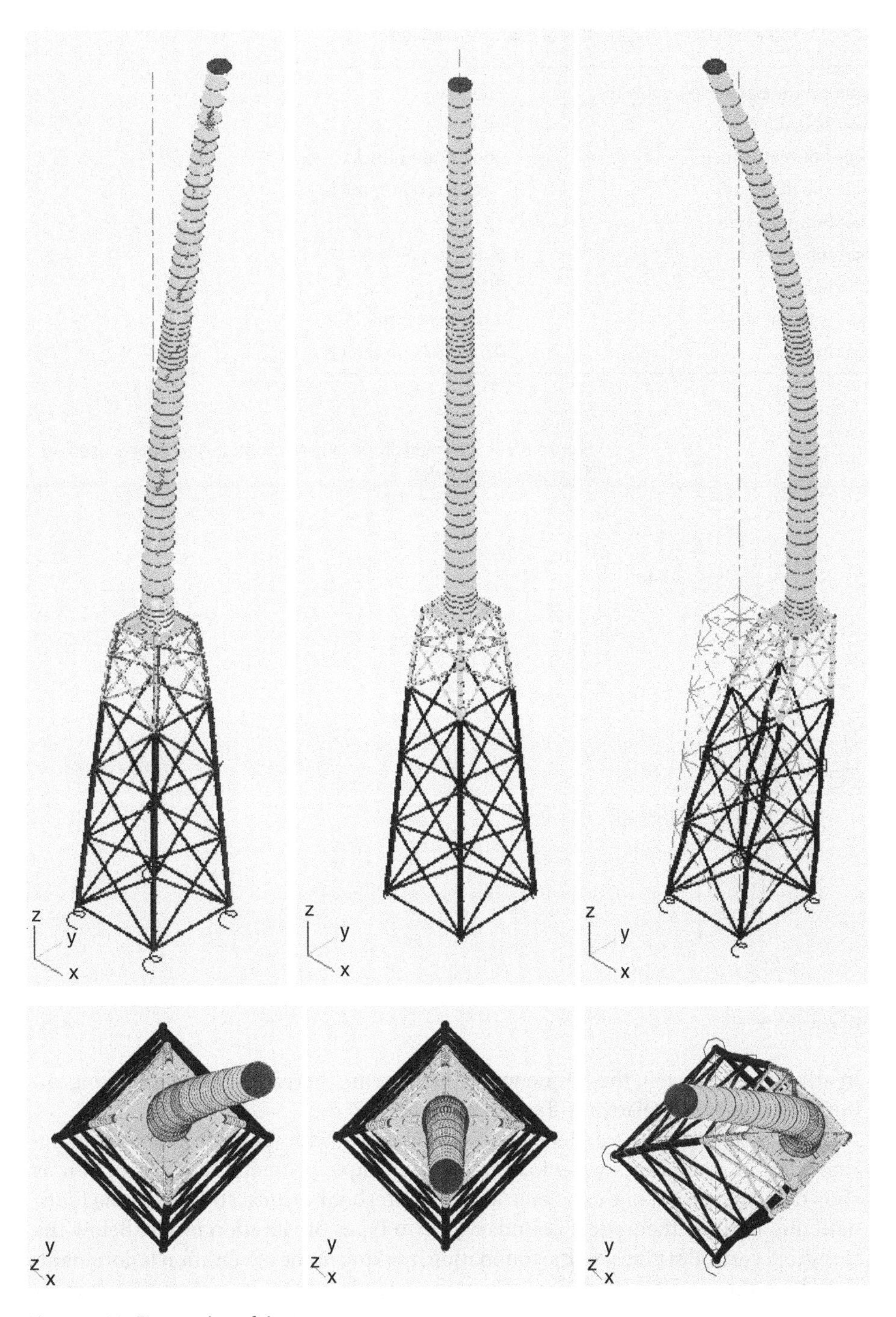

Figure 6.22 Eigen value of the system.

Table 6.23 Jacket and tower properties of example problem.

Mass of rotor-nacelle assembly (m_{RNA})	350 tons
Tower height	70.4 m
Tower bottom diameter	6 m (27 mm thick)
Tower top diameter	3.87 m (20 mm thick)
Jacket bottom width	12 m
Jacket top width	8 m
Jacket height	70.15 m
Jacket external legs	1 m (50 mm thick)
Jacket braces	0.5 m (50 mm thick)

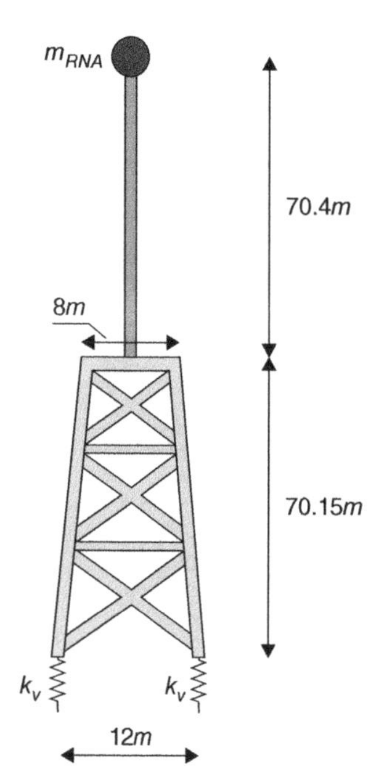

Figure 6.23 Schematic for example problem and details used for finite element model.

in an abrupt change in the frequency of the system. Therefore, to avoid rocking, an optimization of the relative stiffness may be carried out.

(2) Rocking modes are low frequency and may interfere with the 1P frequencies of the rotor and in some cases wave loading. Using simple geometrical construction as shown in Figure 6.25, one can determine the threshold vertical stiffness of the foundation to find the theoretical boundary of two types of vibration mode. Below the threshold vertical stiffness of the foundation, rocking mode of vibration is dominant.

(3) Based on the analysis carried out by Arany et al. (2016), it is shown that most monopile-supported wind turbine are close to the fixed-base frequency i.e. value of f_0/f_{fb} close to 0.9. In the absence of monitoring data of jacket supported on shallow foundations, vertical stiffness of the foundation is important, such that sway bending mode of vibration governs.

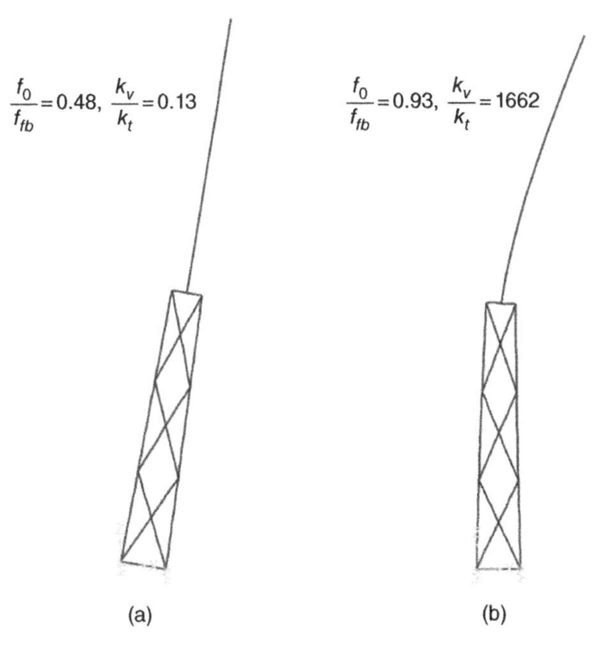

Figure 6.24 Typical output from finite element model showing rocking and sway-bending modes of vibration. (a) rocking mode of vibration for low k_v values; (b) Sway bending mode for high k_v values.

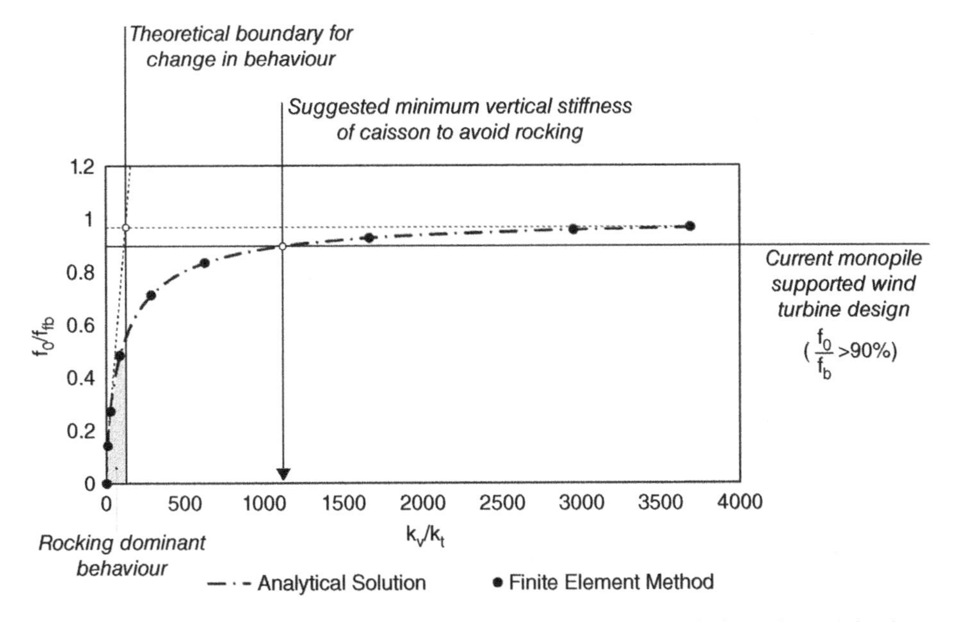

Figure 6.25 Variation of normalised 1st natural frequency of the system (f_0/f_{fb}) with normalised vertical stiffness of the foundation (k_v/k_t).

Example 6.10 ***Plastic Moment of Pile (M_P) and Moment at First Yield (M_Y)*** It is necessary to find the maximum moment that a pile section may withstand structurally. Two types of calculation can be carried out: moment at first yield (elastic) and fully plastic. The stress distribution for a solid section for the two conditions mentioned above can be visualised in Figure 6.26.

For a solid section:

Plastic Moment of a Solid Circular Section

The plastic moment of a solid section is given by the following equation:

$$\left(\sigma_y \times \frac{\pi R^2}{2}\right) \times 2 \times \frac{4R}{3\pi} = \frac{4}{3}R^3\sigma_y = \frac{4}{3}\frac{D^3}{8}\sigma_y = \frac{1}{6}D^3\sigma_y$$

$$M_p = \frac{1}{6}D^3\sigma_y \qquad (6.65)$$

Plastic Moment of a Hollow Pile Section

Properties of a hollow section can be computed by subtraction, i.e. outer diameter minus inside diameter, as shown in Figure 6.27. For a thin-walled section, it is assumed that thickness is much less as compared to the diameter.

If d_o is the outer diameter and d_i is the inside diameter, the plastic moment capacity is given by:

$$M_P = \frac{\sigma_y}{6}(d_o^3 - d_i^3) \qquad (6.66)$$

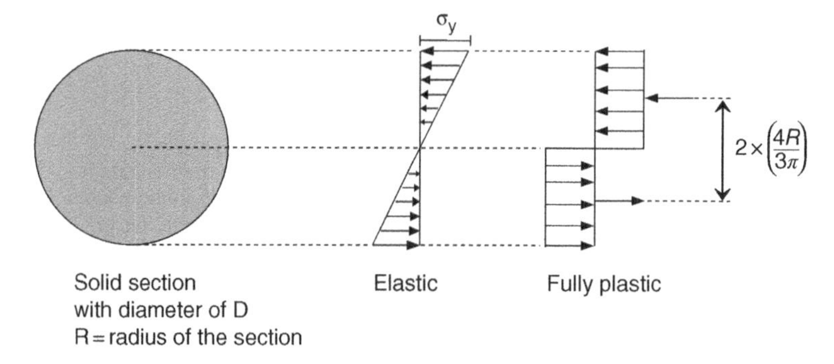

Figure 6.26 Stress distribution.

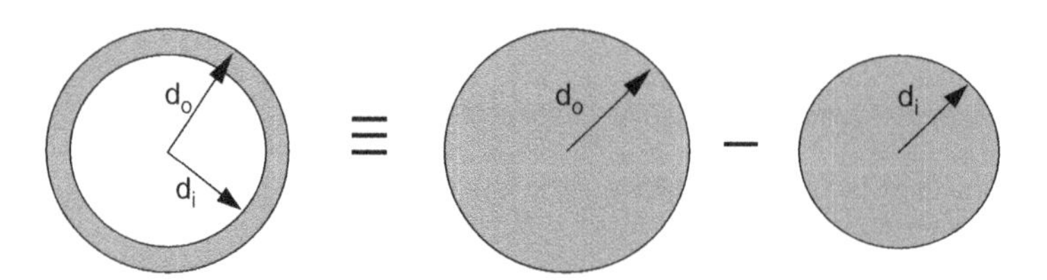

Figure 6.27 Plastic moment of section estimation.

Moment at First Yield

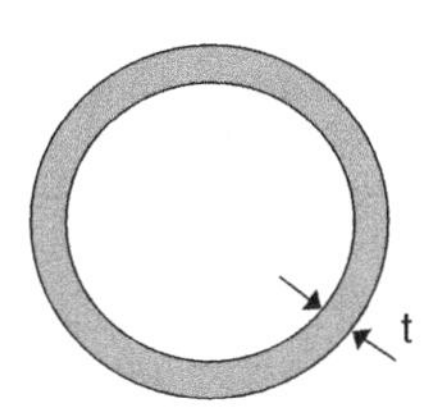

Figure 6.28 Thin-walled section.

Moment at first yield (M_Y) can be easily estimated for a hollow circular section using the concept of shape factor and making assumption of thin walls, i.e. $t << d$ (see Figure 6.28).

The area of the thin-walled section is given by $\pi.d.t$

The plastic moment is given by $\sigma_y.d^2.t$

The second moment of area is given by:

$$I = \frac{\pi}{64}[d^4 - (d-2t)^4] = \frac{\pi}{64}\left[d^4 - \left\{d^4\left(1 - \frac{2t}{d}\right)^4\right\}\right]$$

for

$t << d$

$$I = \frac{\pi}{64}\left[d^4 - d^4\left(1 - \frac{8t}{d}\right)\right] = \frac{\pi}{64}\left[d^4 - d^4 + \frac{8t}{d}.d^4\right] = \frac{\pi}{8}d^3t$$

The shape factor is the ratio of plastic moment to the elastic moment is given by:

$$\text{Shape factor}: \frac{Z_p}{Z_e} = \frac{d^2t}{\frac{\pi}{4}d^2t} = \left(\frac{4}{\pi}\right) \quad (6.67)$$

Example problem:

For a pile having outside diameter of 1.524 m and wall thickness of 63.5 mm, find the plastic moment capacity (M_P) and the moment capacity at first yield (M_Y). The yield strength may be taken as 500 MPa.

$$\text{Pile diameter (inside)} = \left(1.524 - \frac{2 \times 63.5}{1000}\right) = 1.398\text{ m}$$

Taking $\sigma_y = 500$ MPa

$$M_p = \left(\frac{1.524^3}{6} - \frac{1.398^3}{6}\right) m^3 \times 500\text{ MPa} = 67.2\text{ MN.m}$$

$$\text{Shape factor} = \left(\frac{M_p}{M_E}\right) = \left(\frac{4}{\pi}\right) \quad (6.68)$$

$$M_Y = \frac{M_P}{\frac{4}{\pi}} = \frac{67.2}{1.27}\text{MNm} = 52.9\text{ MNm}$$

Table 6.24 provides M_P and M_{es} for 11 field case records of wind turbines..

One point may be noted: These monopiles will have very high geotechnical moment capacity i.e. the capacity at which the soil surrounding the pile will fail. It is highly likely that the monopiles will fail by forming plastic hinge (structural failure) long before they will fail geotechnically i.e. soil fails first.

Table 6.24 Values of Plastic Moment and Yield Moment of monopiles of 11 Wind Farms.

Wind Farm	Blyth	Kentish Flats I	Barrow II	Thanet III	Belwind IV	Burbo Bank	Walney I	Gunfleet Sands	Horns rev	London Array 1	London Array 2
Dia (m)	3.5	4.3	4.75	4.7	5	4.7	6	5	4	5.7	4.7
WT (mm)	0.050	0.045	0.060	0.065	0.050	0.075	0.080	0.050	0.050	0.075	0.065
M_Y	163.6	224.8	363.3	384.0	338.2	440.2	771.3	338.2	214.8	652.9	384.0
M_P	211.3	289.2	468.5	495.8	434.9	569.6	995.4	434.9	277.0	842.5	495.8

NB: Yield stress is taken as 355 MPa; WT is wall thickness.

Example 6.11 ***Spar-Type Floating Wind Turbine*** A worked example presented here basically emulates the Hywind Pilot Park close to Peterhead in Scotland. Five 6 MW turbines are planned on a spar buoy platform, utilising suction caisson anchors and design data following (Statoil 2015) (see Table 6.25). The wind and wave loads are calculated in the following section.

Wind Load

Weibull distribution for long-term based on parameters in Table 6.25, gives the 50-year and 1-year return period 10-minutes mean wind speeds as

$$U_{10,50\ yr} = 35.7\left[\frac{m}{s}\right] U_{10,1\ yr} = 28.6\left[\frac{m}{s}\right]$$

The EOG wind speed is calculated as

$$u_{EOG} = 7.6\ \left[\frac{\mathrm{m}}{\mathrm{s}}\right]$$

and wind load due to the EOG at the rated wind speed is

$$F_{u,EOG} = 2.9[\mathrm{MN}]$$

The wind load on the shut-down structure in the 50-year extreme wind speed is

$$F_{u,U50} = 0.72[\mathrm{MN}]$$

of which the tower drag load and rotor drag load components are

$$F_{DT} = 0.18[\mathrm{MN}\ F_{DT,50\ yr} = 0.54[\mathrm{MN}]$$

The drag force on the tower at the rated wind speed is $F_{DT,U_R} = 0.02[\mathrm{MN}]$.

Wave and Current Loads

These loads are calculated by breaking up the spar into three sections as specified in Table 6.25. The bottom 58 m is modelled with a diameter of 14.4 m, the 15 m long-coned sections is modelled with the average diameter of 11.95 m, and the top section with a diameter of 9.5 m. This section can be modelled using an equivalent diameter of $D_D = 11.33m$ for drag load calculations, and $D_I = 12.89m$ for inertia load calculations.

Table 6.25 Parameters of the floating offshore wind turbine and the site.

Parameter	Symbol	Value	Unit
Turbine parameters			
Rotor diameter	D	154	m
Rated wind speed	U_R	12	m/s
Mass of the rotor-nacelle assembly	m_{RNA}	403	tons
Mass of the tower	m_T	626	tons
Drag coefficient of tower	C_{DT}	0.5	[−]
Tower bottom diameter	D_b	6.5	m
Tower top diameter	D_t	4.1	m
Hub height above sea level	z_{hub}	100	m
Spar and mooring parameters			
Spar diameter	D_S	14.4/9.5[a)]	m
Spar draft (depth below sea level)	B	85	m
Mass of the ballast	m_B	8000	tons
Mass of the spar buoy	m_S	1700–2500	tons
Centre of buoyancy below sea level	z_B	50	m
Mooring radius	r_m	600–1200	m
Unit weight of mooring chains	μ_C	200–550	kg/m
Mass of the mooring cables	m_C	120–660	tons
Wind parameters			
Mean wind speed at the site	κ	1.8	[−]
Weibull distribution shape parameter	λ	8	m/s
Wind profile exponent	γ	1/7	[−]
Integral length scale	L_k	340.2	m
Turbulence Intensity	I_{15}	20	%
Wave parameters			
Water depth	S	95–120	m
Significant wave height	H_S	10	m
Peak wave period	T_S	11.2	s
Density of sea water	ρ_w	1030	kg/m^3
Soil parameters			
Soil type		Medium sand	
Mooring chain friction on sand	μ	0.25	[−]
Internal angle of friction	ϕ'	30	°
Submerged unit weight	γ'	9	kN/m^3

a) The diameter of the lower 58 m of the spar is 14.4 m; then there is a coned section 15 m long, and the upper section has a diameter of 9.5 m.

The maximum of the drag load for the 50-year EWH is

$$F_{D,max,50yr} = 3.56[\mathrm{MN}]$$

and the maximum of the 50-year inertia load is

$$F_{I,max,50yr} = 20.04[\mathrm{MN}]$$

The peak loads occur at different time instants, therefore the 50-year extreme wave load is taken as

$$F_{w,50yr} \approx 20.1\ [\mathrm{MN}]$$

For the 1-year EWH scenario, the maximum of the drag load and inertia load are calculated as

$$F_{D,max,1yr} = 2.2[\mathrm{MN}]\ F_{I,max,50yr} = 16.67[\mathrm{MN}]$$

The total wave load is then

$$F_{w,1yr} \approx 16.7\ [\mathrm{MN}]$$

The current load is calculated as

$$F_C = \frac{1}{2}\rho_w D_P C_{DP} v_C^2 B \approx 2.32[\mathrm{MN}]$$

Anchor Load Combinations

The loads under the combined actions of wind and waves have to be considered. The two combinations of loads (E-1) and (E-2) are calculated as

$$F_{E-1} = 20.1 + 0.72 + 2.32 \approx 23.1\ [\mathrm{MN}]$$

$$F_{E-2} = 16.7 + 2.9 + 2.32 \approx 21.9[\mathrm{MN}]$$

As expected, the wave load dominates, and the scenario with the combination of the 50-year EWH and the 50-year extreme mean wind speed combination produces the ULS load. It should be noted here that this load is conservative for anchor design, as the load that acts on the anchor is reduced by the weight of the suspended section of the mooring line, the friction on the horizontal section (Touch Down Zone) of the mooring line, the soil reaction on the inverse catenary-shaped forerunner in the soil, and the weight of the forerunner. The vertical load acts on the spar at the instant when the surface elevation at the spar is at its highest point (wave crest), while the horizontal load is dominated by the inertia load, which is highest when the surface elevation is at the mean water level. Therefore, the ultimate load is taken as the horizontal load as calculated above.

Sizing the Anchor

In this section, a simple anchor sizing exercise is carried out, assuming a suction caisson anchor. The diameter of the caisson D and the embedment depth L are the two main independent parameters that govern the holding capacity of the caisson for a given soil profile. Formulations for both clayey and sandy soils are given in this section.

At the Hywind site, the top layer of the seabed soil is dominated by loose to medium sand. The sub-seabed soil within the embedment range of the anchor is dominantly soft clay with intermittent sand layers. In the worked example, three soil types are considered:

(1) Clay with constant undrained shear strength with depth, using an average value of $s_u = 30$ kPa.
(2) Clay with linearly increasing undrained shear strength with depth, using $s_{u0} = 15$ kPa and $\frac{ds_u}{dz} = 2$ kPa/m.
(3) Soft/medium sand with angle of internal friction of $\phi = 30^\circ$ and effective unit weight of $\gamma' = 9 \left[\frac{\text{kN}}{\text{m}^3}\right]$.

The holding capacity of suction caissons is typically determined in terms of an envelope based on the horizontal and vertical load components at the anchor. Following Randolph and Gourvenec (2011) and Supachawarote et al. (2004), the envelope is given as:

$$FP = \left(\frac{H_u}{H_m}\right)^a + \left(\frac{V_u}{V_m}\right)^b < 1 \tag{6.69}$$

where

$$a = \frac{L}{D} + 0.5 \quad b = \frac{L}{3D} + 4.5$$

An alternative formulation by Senders and Kay (2002) replaces a and b with $k = 3$. In Eq. (6.69), H_m is the horizontal capacity and V_m is the vertical capacity. On the other hand, H_u and V_u are the applied load. FP is the failure criterion and the maximum value can be 1 (limiting condition).

Suction Caisson Bearing Capacity in Clay

The horizontal capacity H_m in clay is given following Randolph and Gourvenec (2011) as

$$H_m = LD_eN_p\bar{s}_u \tag{6.70}$$

where

L penetration depth of the caisson.
D_e external caisson diameter.
N_p lateral bearing capacity factor (shown to depend only slightly on L/D_e in Randolph and Gourvenec (2011)); approximate values are given in Table 6.26.
$\bar{s}_u$ average undrained shear strength over the embedded length of the caisson.

The three formulations for the vertical capacity represent three failure modes:

Table 6.26 Lateral bearing capacity factor for clays with various strength profiles.

N_p	Linearly increasing s_u with depth	Uniform s_u with depth
Horizontal translation	~10.5	~10
Horizontal load at mudline	~2.5	~4

1) Presence of passive suction and reverse end bearing
2) No passive suction, caisson pullout
3) No passive suction, caisson, and soil plug pullout (internal soil plug failure)

Figure 6.29 shows the three failure modes, and the formulations are as follows:

V_{m1} = submerged weight of the caisson + external friction + reverse end bearing
V_{m2} = submerged weight of the caisson + external friction + internal friction
V_{m3} = submerged weight of the caisson + external friction + weight of the soil plug

Using the formulations of Randolph and Gourvenec (2011):

$$V_{m1} = W' + A_{se}\alpha_e\bar{s}_u + N_c s_u A_e$$

$$V_{m2} = W' + A_{se}\alpha_e\bar{s}_u + A_{si}\alpha_i\bar{s}_u$$

$$V_{m3} = W' + A_{se}\alpha_e\bar{s}_u + W'_{\text{plug}}$$

where

A_{se} external shaft surface area $\approx D_e\pi \times L$.
A_{si} internal shaft surface area $\approx D_i\pi \times L$.
A_e external cross-sectional area $= D_e^2\pi/4$.
α_e coefficient of external shaft friction between steel and soil.
α_i coefficient of internal shaft friction between steel and soil.
N_c reverse end bearing factor (~9).
s_u representative undrained soil shear strength at caisson tip level.
$\bar{s}_u$ average undrained soil shear strength over penetrated depth.
W' submerged caisson weight.
W'_{plug} effective weight of the soil plug.

Suction Caisson Bearing Capacity in Sand

The lateral capacity in sand can be calculated as follows:

$$H_{m,sand} = -LQ_{av} = 0.5A_b N_q \gamma' L^2$$

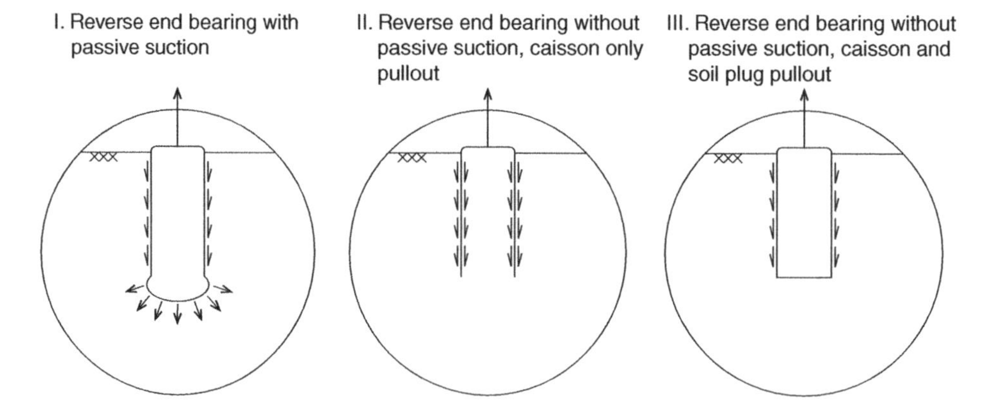

Figure 6.29 Pull-out failure modes of suction caissons.

where the average soil strength may be determined following Miedema et al. (2007):

$$LQ_{av} = D_e N_q \int_0^L \gamma' z dz = \frac{1}{2} D_e N_q \gamma' L^2$$

where

γ' submerged unit weight of the soil.

N_q bearing capacity factor, calculated based on DNV Classification Note 30.4 (DNV 1992) as

$$N_q = e^{\pi \tan\phi} \tan^2\left(45^\circ + \frac{\phi}{2}\right)$$

with ϕ being the internal angle of friction of the soil.

The vertical capacity in sand accounting for the effects of stress enhancement can be calculated following Houlsby et al. (2005a,b) as

$$V_{m,sand} = W' + \gamma' Z_e^2 y\left(\frac{h}{Z_e}\right)(K\tan\delta)_e(\pi D_e) + \gamma' Z_i^2 y\left(\frac{h}{Z_i}\right)(K\tan\delta)_i(\pi D_i) \tag{6.71}$$

where

$$y(x) = e^{-x} - 1 + x$$

$K\tan\delta$ factor that only appears together.

K the effective stress factor used to calculate the horizontal effective stress as constant times the effective vertical stress ($\sigma_H = K\sigma'_V$), δ is the mobilised angle of friction between the caisson wall and the soil, e represents the external and i is the internal circumference of the caisson.

Z $D/[4K\tan\delta]$ with e and i referring to external and internal values, respectively.

γ' submerged unit weight of the soil.

N_q bearing capacity factor as defined above.

Load Transfer from the Mudline to the Anchor

The actual load on the anchor is obtained by taking into account the load reduction on the inverse catenary forming at the anchor. This is important as not only the magnitude but also the angle at which the load is applied changes through the inverse catenary shape at the anchor. The anchor padeye tension T_a and angle θ_a can be determined by simultaneously solving the using two equations following Randolph and Neubecker (1995); see Figure 6.30 for definition of the terms:

$$\frac{T_a}{2}(\theta_a^2 - \theta_m^2) = z_a Q_{av}$$

$$\frac{T_m}{T_a} = e^{\mu(\theta_a - \theta_m)} \tag{6.72}$$

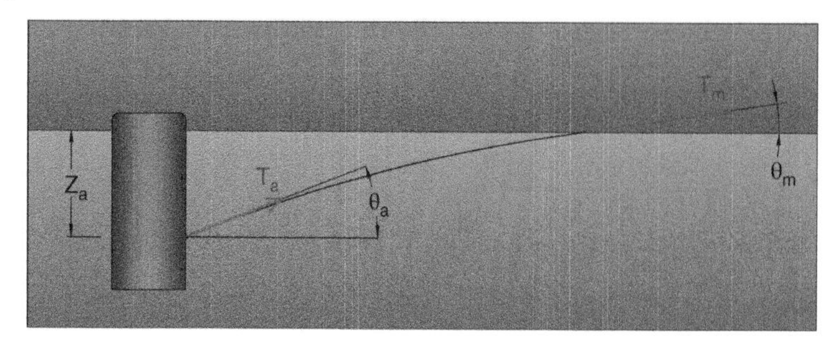

Figure 6.30 Loads on the anchor lines.

where

T_a tension at the anchor padeye.
T_m tension at the mudline.
θ_a angle of the tension at the anchor padeye to horizontal.
θ_m angle of the tension at the mudline to horizontal.
z_a depth of the anchor padeye below mudline.
μ friction coefficient between the forerunner (chain, rope, or wire) and the soil.
Q_{av} average soil resistance between the mudline and the padeye.

The average soil resistance can be determined for clay as

$$z_a Q_{av} = A_b N_c \int_0^{z_a} s_u(z) dz$$

where

A_b effective unit bearing area of the forerunner (equals the diameter of the rope or wire, and 2.5–2.6 times the bar diameter for a chain),
N_c bearing capacity factor (between 9 and 14 based on *DNVGL-RP-E301 Design and installation of fluke anchors (DNV GL 2017)*).
$s_u(z)$ distribution of the undrained shear strength with depth.

For sand, the average soil resistance is calculated following Miedema et al. (2007) as

$$z_a Q_{av} = A_b N_c \int_0^{z_a} \gamma' z dz$$

Required Caisson Dimensions

The required dimensions of the suction caisson necessary to anchor the floating platform are calculated using the ultimate load and the equations presented earlier. The caisson dimensions are determined for the three different soil types: clay with constant undrained shear strength (s_u), clay with linearly increasing s_u, and medium dense sand.

The anchor padeye is placed at z_a depth below mudline. This depth is determined based on moment balance such that soil resistance is mobilised due to horizontal translation of the anchor rather than rigid body rotation; see Randolph and Gourvenec (2011)

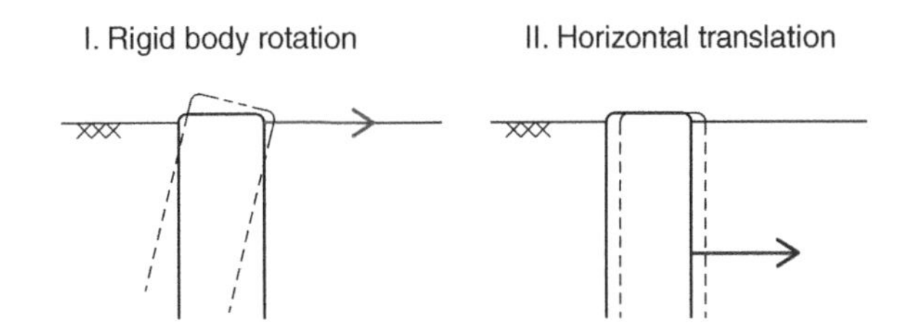

Figure 6.31 Failure modes of caissons under horizontal loading.

and Figure 6.31. For sand, where strength increases linearly from zero at the mudline, this depth is $z_a/L = 2/3$. For clay, the value varies between about $z_a = 1/2$ and $z_a = 2/3$.

Several values of the length-to-diameter ratio L/D are chosen for the analysis, and the required parameters are determined for each. This can be carried out by the following procedure:

1. An initial value D_0 of the caisson diameter is chosen.
2. The embedment depth of the caisson is calculated for the chosen length-to-diameter ratio.
3. The padeye depth is calculated from moment balance as described above as a portion of the embedment length.
4. The padeye tension T_a and forerunner angle θ_a are calculated using the soil data, the padeye depth, and equations presented above.
5. The horizontal and vertical force components are calculated from T_a and θ_a.
6. The wall thickness is simplistically estimated using a wall thickness to diameter ratio of 70. A sensitivity study showed that this has very limited effect as it only affects the internal diameter of the caisson used for calculating the internal friction and the weight of the caisson. By changing the value from 70 to 40 or 100 the required caisson diameter changes by less than 0.1 m.
7. The weight of the caisson and the horizontal and vertical capacities are calculated.
8. Equation (6.69) is used as failure criterion.
9. The process is repeated until the smallest diameter is found that satisfies the failure criterion.

Table 6.27 present the results of this analysis for soft clay with constant s_u. Other cases can be done similarly. The readers are referred to Arany and Bhattacharya (2018). The dimensions determined for the length to diameter ratio of 3.2 are as follows. For soft clay with constant undrained shear strength of $s_u = 30$ kPa the calculated diameter is identical to the actual dimensions of Hywind ($D = 5$ m, $L = 16$ m). For soft clay with undrained shear strength profile given by $s_u(z) = 15\,\text{kPa} + 2\frac{\text{kPa}}{\text{m}} \cdot z$, the dimensions are determined as ($D = 5.25$ m, $L = 16.8$ m). For the medium sand, the dimensions are higher at ($D = 6.9$ m, $L = 22.1$ m). The dimensions approximate the actual anchors very well, however, applying load factors would increase the required dimensions given by this methodology.

Table 6.27 Minimum caisson dimensions for various length to diameter ratios – soft clay with constant s_u with depth.

Length-to-diameter ratio of caisson	Lc/De	3.2	2	2.5	3	3.5	4
Minimum required caisson diameter for L/D [m]	D_{min}	5.00	6.35	5.65	5.20	4.80	4.50
Corresponding length [m]	L_{min}	16.0	12.7	14.1	15.6	16.8	18.0
Wall thickness [m]	t_w	0.05	0.063	0.056	0.052	0.048	0.045
Average shear strength of the soil [kPa]	s_{avg}	30.0	30.0	30.0	30.0	30.0	30.0
Shear strength of soil at caisson tip [kPa]	s_u	30.0	30.0	30.0	30.0	30.0	30.0
Submerged weight of the caisson [kN]	W_c	893	1183	1032	927	849	788
Submerged weight of the soil plug [kN]	W_p	2697	3422	3054	2785	2579	2416
External shaft friction [kN]	F_e	4878	4889	4881	4878	4879	4883
Internal shaft friction [kN]	F_i	4781	4791	4784	4781	4781	4786
Reverse end bearing [kN]	F_{reb}	4104	6581	5257	4378	3753	3287
Maximum vertical load – Mode I [kN]	V_I	9875	12 654	11 170	10 183	9481	8959
Maximum vertical load – Mode II [kN]	V_{II}	10 552	10 864	10 697	10 586	10 509	10 457
Maximum vertical load – Mode III [kN]	V_{III}	8468	9495	8967	8590	8307	8087
Maximum vertical load capacity [kN]	V_{max}	8468	9495	8967	8590	8307	8087
Maximum horizontal load capacity [kN]	H_{max}	21 500	21 548	21 514	21 500	21 503	21 523
Anchor padeye depth	z_a	7.98	6.32	7.06	7.73	8.35	8.93
Padeye location [%]	r_z	0.500	0.500	0.500	0.500	0.500	0.500
Angle at the padeye [deg]	ϑ_a	14.79	13.11	13.88	14.54	15.13	15.67
Tension at the padeye (variable) [kN]	T_a	21 657	21 816	21 743	21 680	21 624	21 574
Horizontal load on the anchor [kN]	H_a	20 940	21 248	21 109	20 985	20 874	20 772
Vertical load on the anchor [kN]	V_a	5527	4947	5214	5443	5645	5826
Horizontal load check exponent [–]	a	3.7	2.5	3	3.5	4	4.5
Vertical load check exponent [–]	b	5.57	5.17	5.33	5.50	5.67	5.83
Horizontal utilisation [–]	V_a/V_{max}	0.653	0.521	0.581	0.634	0.68	0.72
Vertical utilisation [–]	H_a/H_{max}	0.974	0.986	0.981	0.976	0.971	0.965

Appendix A

Natural Frequency of a Cantilever Beam with Variable Cross Section

The aim of the appendix is to provide the theoretical basis of the natural frequency estimation of monopile-supported wind turbine. Please also refer to Chapter 5 (see Sections 5.4.1 and 5.8) and publications by Arany et al. (2015a,b, 2016, 2017).

The motion of the cantilever beam can be described as a single degree of freedom mass-spring system. The free vibration of this system is given by

$$m\ddot{x}(t) + kx(t) = 0 \tag{A.1}$$

Assuming harmonic vibration, the following equation can be obtained:

$$-m\omega^2 + k = 0 \text{ with } k = \frac{3EI}{L_T^3} \text{ and } m = m_{RNA} + \frac{33}{140}m'_{tower} \tag{A.2}$$

and from the circular frequency the Hertz frequency is easily obtained:

$$\omega = \sqrt{\frac{k}{m}} \rightarrow f_1 = \frac{1}{2\pi}\sqrt{\frac{k}{m}} \tag{A.3}$$

The flexibility of the substructure expresses the dependence of the natural frequency on the water depth i.e. the flexibility of the monopile above the mudline and that of the transition piece. For the sake of simplicity, the model used in Chapter 5 assumes that the monopile's bending stiffness continues throughout the water depth and up to the root of the tower. For clarity, see Figure A.1.

$$\textit{Bending stiffness ratio}: \ \chi = \frac{E_T I_T}{E_P I_P} \tag{A.4}$$

$$\textit{Platform/tower length ratio}: \ \psi = \frac{L_S}{L_T} \tag{A.5}$$

Castigliano's second theorem for a linearly elastic 1 DoF structure can be written as

$$q = \frac{\partial U}{\partial Q}$$

where U is the strain energy, q is the generalised displacement, and Q is the generalised force. For the particular problem, the theorem can be used to calculate the top head

Design of Foundations for Offshore Wind Turbines, First Edition. Subhamoy Bhattacharya.

Companion website: www.wiley.com/go/bhattacharya/offshorewindturbines

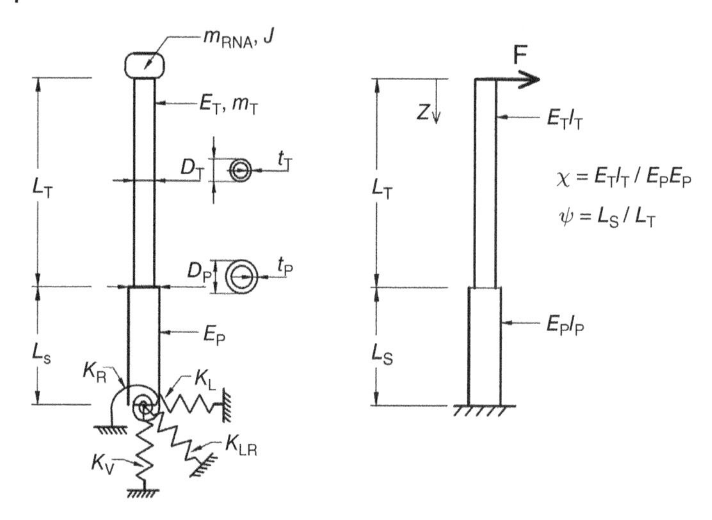

Figure A.1 Cantilever beam with variable cross section.

deflection (the total deflection at the hub $w(0)$) due to a horizontal force F acting at the hub as

$$w(0) = \frac{\partial U}{\partial F} = \frac{\partial}{\partial F}\int_0^{L_T+L_S}\frac{[M(z)]^2}{2EI(z)}dz = \frac{\partial}{\partial F}\int_0^{L_T}\frac{F^2z^2}{2E_TI_T}dz + \frac{\partial}{\partial F}\int_{L_T}^{L_T+L_S}\frac{F^2z^2}{2E_PI_P}dz$$
$$= \left[\frac{Fz^3}{3E_TI_T}\right]_0^{L_T} + \left[\frac{Fz^3}{3E_PI_P}\right]_{L_T}^{L_T+L_S} = \frac{FL_T^3}{3E_TI_T} + \frac{F(L_T+L_S)^3}{3E_PI_P} - \frac{FL_T^3}{3E_PI_P} \tag{A.6}$$

where the moment distribution along the structure caused by the horizontal force F is given as

$$M(z) = Fz \tag{A.7}$$

The stiffness of the 1DoF system is then given as

$$k = \frac{F}{w(0)} = \frac{1}{\frac{L_T^3}{3E_TI_T} + \frac{(L_T+L_S)^3}{3E_PI_P} - \frac{L_T^3}{3E_PI_P}} = \frac{3E_TI_T}{L_T^3 + (L_T^3 + 3L_T^2L_S + 3L_TL_S^2 + L_S^3)\chi - L_T^3\chi}$$
$$= \frac{3E_TI_T}{L_T^3(1+(1+\psi)^3\chi - \chi)} \tag{A.8}$$

From this the natural frequency is calculated as

$$f_{FB} = \frac{1}{2\pi}\sqrt{\frac{k}{m}} \tag{A.9}$$

where m is the generalised mass of the 1DoF system. The stiffness can be written as

$$k = \frac{1}{1+(1+\psi)^3\chi - \chi}k_T \tag{A.10}$$

where k_T is the stiffness of the tower without the substructure, given as

$$k = \frac{3E_T I_T}{L_T^3}$$

The fixed-base natural frequency of the tower-substructure system (excluding foundation stiffness) is given using the bending stiffness ratio χ and the platform/tower length ratio ψ as

$$f_{FB} = \sqrt{\frac{1}{1 + (1 + \psi)^3 \chi - \chi}} f_{FB,T} \tag{A.11}$$

Appendix B

Euler-Bernoulli Beam Equation

The readers are referred to Adhikari and Bhattacharya (2010), Bhattacharya and Adhikari (2011), and Arany et al. (2015a,b, 2016, 2017) for comprehensive understanding. The fundamentals are provided here.

The equation of motion using the Euler-Bernoulli beam model for a beam with an axial force is

$$\frac{\partial^2}{\partial z^2}\left(EI(z)\frac{\partial^2 w(z,t)}{\partial z^2}\right) + \mu(z)\frac{\partial^2 w(z,t)}{\partial t^2} + \frac{\partial}{\partial z}\left(P^* \frac{\partial w(z,t)}{\partial z}\right) = p(z,t) \tag{B.1}$$

where $EI(z)$ is the bending stiffness distribution along the axial coordinate z, $\mu(z)$ is the distribution of mass per unit length, P^* is the axial force acting on the beam due to the top head mass and the self-weight of the tower, $p(z, t)$ is the excitation of the beam, $w(z, t)$ is the deflection profile.

Using constant equivalent values for the axial force, bending stiffness and mass per length, and considering free harmonic vibration of the beam with separation of variables $w(z, t) = W(z) \cdot e^{i\omega t}$, the equation can be reduced to the following using the non-dimensional parameters of Table B.1 and the dimensionless axial coordinate $\xi = z/L$:

$$W'''' + \nu W'' - \Omega^2 W = 0 \tag{B.2}$$

where $\nu = P^* L_T^2 / EI_\nu$ is the nondimensional axial force and $\Omega = \omega/c_0 = \omega\sqrt{E_T I_T / m_T L^3}$ is the non-dimensional circular frequency).

Using the nondimensional numbers as defined above, the boundary conditions can be written for the bottom of the tower ($\xi = 0$):

$$W'''(0) + (\nu + \eta_{LR})W'(0) + \eta_L W(0) = 0 \tag{B.3}$$

$$W''(0) - \eta_R W'(0) + \eta_{LR} W(0) = 0 \tag{B.4}$$

and the top of the tower ($\xi = 1$):

$$W'''(1) + \nu W'(1) + \alpha\Omega^2 W(1) = 0 \tag{B.5}$$

$$W''(1) - \beta\Omega^2 W'(1) = 0 \tag{B.6}$$

The parameters used in the boundary conditions are defined in Table B.1. The characteristic equation for the equation of motion can be written as

$$\lambda^4 + \nu\lambda^2 - \Omega^2 = 0 \text{ or } \tilde{z}^2 + \nu\tilde{z} - \Omega^2 = 0 \text{ with } \tilde{z} = \lambda^2 \tag{B.7}$$

Design of Foundations for Offshore Wind Turbines, First Edition. Subhamoy Bhattacharya.

Companion website: www.wiley.com/go/bhattacharya/offshorewindturbines

Table B.1 Nondimensional variables.

Dimensionless group	Formula	Dimensionless group	Formula
Nondimensional lateral stiffness	$\eta_L = \dfrac{K_L L_T^3}{EI_\eta}$	Nondimensional axial force	$\nu = \dfrac{P^* L_T^2}{EI_\nu}$
Nondimensional rotational stiffness	$\eta_R = \dfrac{K_R L_T}{EI_\eta}$	Mass ratio	$\alpha = \dfrac{m_{RNA}}{m_T}$
Nondimensional cross stiffness	$\eta_{LR} = \dfrac{K_{LR} L_T^2}{EI_\eta}$	Nondimensional rotary inertia	$\beta = \dfrac{J}{\mu L^2}$

K_L, K_R, K_{LR} are the lateral, rotational, and cross stiffness of the foundation, respectively; EI_η is the equivalent bending stiffness of the tapered tower; L_T is the hub height above the bottom of the tower; P^* is the modified axial force, m_{RNA} is the mass of the rotor-nacelle assembly; m_T is the mass of the tower; J is the rotary inertia of the top mass; μ is the equivalent mass per unit length of the tower.

and

$$\tilde{z}_{1,2} = \frac{-\nu \pm \sqrt{\nu^2 + 4\Omega^2}}{2} = -\frac{\nu}{2} - \sqrt{\left(\frac{\nu}{2}\right)^2 + \Omega^2}, -\frac{\nu}{2} + \sqrt{\left(\frac{\nu}{2}\right)^2 + \Omega^2}$$

The four solutions are then

$$r_1 = i\sqrt{\tilde{z}_1} \qquad r_2 = -i\sqrt{\tilde{z}_1} \qquad r_3 = \sqrt{\tilde{z}_2} \qquad r_4 = -\sqrt{\tilde{z}_2} \tag{B.8}$$

with which the solution is in the form

$$W(\xi) = C_1 e^{r_1\xi} + C_2 e^{r_2\xi} + C_3 e^{r_3\xi} + C_4 e^{r_4\xi} \tag{B.9}$$

which can be transformed using Euler's identity to

$$W(\xi) = P_1 \cos(\lambda_1\xi) + P_2 \sin(\lambda_1\xi) + P_3 \cosh(\lambda_2\xi) + P_4 \sinh(\lambda_2\xi) \tag{B.10}$$

with

$$\lambda_1 = \sqrt{|\tilde{z}_1|} \text{ and } \lambda_2 = \sqrt{|\tilde{z}_2|}. \tag{B.11}$$

Substituting this form of the solution into the boundary conditions, one obtains four equations, written in matrix form as

$$\boldsymbol{M} \cdot \boldsymbol{p} = \underline{0} \tag{B.12}$$

with

$$\boldsymbol{p}^T = [\boldsymbol{P}_1 \quad \boldsymbol{P}_2 \quad \boldsymbol{P}_3 \quad \boldsymbol{P}_4] \tag{B.13}$$

and

$$\boldsymbol{M} = \begin{bmatrix} \eta_L & \lambda_1^3 + (\nu + \eta_{LR})\lambda_1 \\ \lambda_1^2 - \eta_{LR} & -\eta_R\lambda_1 \\ (\lambda_1^3 + \nu\lambda_1)\sinh(\lambda_1) + \alpha\Omega^2\cosh(\lambda_1) & (\lambda_1^3 + \nu\lambda_1)\cosh(\lambda_1) + \alpha\Omega^2\sinh(\lambda_1) \\ \lambda_1^2\cosh(\lambda_1) - \beta\Omega^2\lambda_1\sinh(\lambda_1) & \lambda_1^2\sinh(\lambda_1) - \beta\Omega^2\lambda_1\cosh(\lambda_1) \end{bmatrix}$$

$$\begin{array}{cc} \eta_L & -\lambda_2^3 + (\nu + \eta_{LR})\lambda_2 \\ -\lambda_2^2 - \eta_{LR} & -\eta_R\lambda_2 \\ (\lambda_2^3 - \nu\lambda_2)\sin(\lambda_2) + \alpha\Omega^2\cos(\lambda_2) & (-\lambda_2^3 + \nu\lambda_2)\cos(\lambda_2) + \alpha\Omega^2\sin(\lambda_2) \\ -\lambda_2^2\cos(\lambda_2) + \beta\Omega^2\lambda_2\sin(\lambda_2) & -\lambda_2^2\sin(\lambda_2) - \beta\Omega^2\lambda_2\cos(\lambda_2) \end{array}\right] \quad \text{(B.14)}$$

Looking for nontrivial solutions of this equation one obtains

$$det(M) = 0 \quad \text{(B.15)}$$

from which one can obtain the nondimensional circular frequency Ω, and from that the natural frequency using

$$f_1 = \frac{\omega}{2\pi} = \Omega c_0 = \Omega\sqrt{\frac{m_T L_T^3}{EI_\eta}} \quad \text{(B.16)}$$

The equation that has to be solved is transcendental and therefore solutions can only be obtained numerically.

Appendix C

Tower Idealisation

The towers of offshore wind turbines are tapered towers with diameters decreasing from the bottom to the top. Typically, the wall thickness of the tower also decreases with height. However, some small and medium-sized turbines have constant wall thickness. The formulation presented in this book (Chapter 5 – Sections 5.4.1 and 5.8) replaces this tower shape with an equivalent constant diameter, constant wall thickness tower. The average tower diameter

$$D_T = \frac{D_b + D_t}{2} \tag{C.1}$$

is used in combination with an equivalent tower wall thickness t_T. Note that if the average wall thickness is determined from a range of wall thicknesses of the tower or from the mean of the top and bottom wall thicknesses, then the tower mass, as calculated from the idealised tower geometry,

$$m'_T = D_T \pi t_T \rho_T L_T \tag{C.2}$$

may not be the same as the actual tower mass m_T. In the case that information about tower wall thickness is not available, the equivalent thickness can be chosen such that the actual tower mass is maintained, that is,

$$t_T = \frac{m_T}{L_T D_T \pi \rho_T} \tag{C.3}$$

The nondimensional stiffness parameters are normalised with the length L_T and the bending stiffness $E_T I_T$ of the tower. When calculating these nondimensional stiffness parameters, the equivalent bending stiffness is calculated such that the deflection at the tower top due to a force acting perpendicular to the tower at the tower top is the same for the equivalent constant diameter tower as that of the tapered tower.

The derivation is given here.

$$D_b = q D_t \tag{C.4}$$

The diameter varies along the structure as

$$D(z) = \frac{D_t}{L_T}[L_T + (q-1)z] \tag{C.5}$$

Design of Foundations for Offshore Wind Turbines, First Edition. Subhamoy Bhattacharya.

Companion website: www.wiley.com/go/bhattacharya/offshorewindturbines

with $z = 0$ at the top of the tower and positive downwards. The second moment of area is then given as

$$I_T(z) = \frac{1}{8}\pi D_T^3(z)t_T = \frac{1}{8}\pi \frac{D_t^3}{L_T^3} t_T[L_T + (q-1)z]^3 = I_t(1+az)^3 \tag{C.6}$$

where

$$a = \frac{q-1}{L_T}. \tag{C.7}$$

Moment curvature relation is written as

$$E_T I_T \frac{d^2w}{dz^2} = Fz \tag{C.8}$$

where F is the horizontal force at the hub. One can write

$$E_T I_t (1+az)^3 \frac{d^2w}{dz^2} = Fz \tag{C.9}$$

and from that w is obtained via integration

$$\frac{d^2w}{dz^2} = \frac{Fz}{E_T I_t (1+az)^3} \tag{C.10}$$

$$\frac{dw}{dz} = \frac{F}{E_T I_t a^2}\left[\frac{1}{2(1+az)^2} - \frac{1}{1+az}\right] + C_1 \tag{C.11}$$

$$w = \frac{F}{E_T I_t a^2}\left[-\frac{1}{2a(1+az)^2} - \frac{ln(1+az)}{a}\right] + C_1 z + C_2 \tag{C.12}$$

The boundary conditions are used to calculate the constants.

$$z = L_T, \frac{dw}{dz} = 0\ C_1 = \frac{F}{E_T I_t a^2}\left[\frac{1}{1+aL_T} - \frac{1}{2(1+aL)^2}\right] \tag{C.13}$$

$$z = L_T, w = 0\ C_2 = \frac{F}{E_T I_t a^2}\left[\frac{ln(1+aL_T)}{a} + \frac{1}{2a(1+aL_T)} - \frac{L_T}{1+aL_T} + \frac{L_T}{2(1+aL_T)^2}\right] \tag{C.14}$$

The deflection at the end of the column is obtained by substituting $z = 0$ into the equation of deflection

$$w_{free} = \frac{H}{E_T I_b a^2}\left[\frac{ln(1+aL_T)}{a} + \frac{1}{2a(1+aL_T)} - \frac{L_T}{1+aL_T} + \frac{L_T}{2(1+aL_T)^2} - \frac{1}{2a}\right] \tag{C.15}$$

$$w_{free} = \frac{FL^3}{E_T I_t}\left[\frac{q^2(2\,ln\,q - 3) + 4q - 1}{2q^2(q-1)^3}\right] \tag{C.16}$$

The stiffness is then given as

$$k = \frac{E_T I_b}{L_T^3}\left[\frac{2q^2(q-1)^3}{q^2(2\,ln\,q-3)+4q-1}\right] \tag{C.17}$$

Verification: for a cantilever beam of constant diameter:

$$lim_{q\to 1}\left[\frac{2q^2(q-1)^3}{q^2(2\,ln\,q-3)+4q-1}\right] = 3 \tag{C.18}$$

In the book the following notification is used:

$$E_T I_t = EI_t$$

Appendix D

Guidance on Estimating the Vertical Stiffness of Foundations

Methods to estimate the vertical stiffness of the foundation in literature:

	Source	Formulae and their applications
Surface foundation	Poulos and Davis (1974) Spence (1968)	Circular rigid footing on surface of homogenous elastic half-space $$k_v = \frac{2G_s D}{1 - \nu_s}$$ To account for the roughness of the footing base that allow full transmission of shear stress, Spence (1968) proposed; $$k_v = \frac{2G_s D \ln(3 - 4\nu_s)}{1 - 2\nu_s}$$ The results from this analytical solution showed up to 10% increase in stiffness values at low ν_s values.
Shallow embedded foundation	Gazetas (1983) DNV (2014)	Circular footing on stratum over half space $$k_v = \frac{2G_s D}{1 - \nu_s}\left(1 + 1.28\frac{D}{2H}\right)$$
	Gazetas (1991) DNV (2014)	Circular rigid footing embedded in homogenous stratum over bedrock. Developed for machine-type inertial loading. Range of validity: $L/D < 1$ $$k_v = \frac{2G_s D}{1 - \nu_s}\left(1 + 1.28\frac{D}{2H}\right)\left(1 + \frac{L}{D}\right)\left[1 + \left(0.85 - \frac{0.28L}{0.5D}\right)\left(\frac{L}{H - L}\right)\right]$$
	Wolf (1988)	General prismatic footing embedded in a linear elastic half space $$k_v = \frac{2G_s b}{1 - \nu_s}\left(3.1\left(\frac{l}{b}\right)^{0.75} + 1.6\right)\left(1 + \left(0.25 + 0.25\frac{b}{l}\right)\left(\frac{e}{b}\right)^{0.8}\right)$$ Where $2l$, $2b$ are the base dimensions of circumscribed rectangle and e is the embedment depth.
Deep foundation	Fleming et al. (1992)	Embedded piles considering shaft friction only $$k_v = \frac{2\pi L G_s}{\varsigma}$$ where ς is between 3 and 5
	Sharma and El Naggar (2015)	Single pile under axial load for seismic design of highway bridges $$k_v = \frac{1.25 E_p A}{L}$$

Design of Foundations for Offshore Wind Turbines, First Edition. Subhamoy Bhattacharya.

Companion website: www.wiley.com/go/bhattacharya/offshorewindturbines

N.B: K_V is vertical stiffness of the foundation, G_s is shear modulus of the soil, ν_s is Poisson's ratio of the medium, D is diameter, L is embedment depth, H is thickness of the soil layer, and E_p is modulus of elasticity of pile material.

In practice, t-z type of analysis or calibrated FEA (finite element analysis) can be carried out to obtain the axial stiffness of the piles.

Appendix E

Lateral Stiffness K_L of Piles

A literature review of formulas for stiffness are provided in Tables 5.3 and 5.4. As an example, the latest formulation (Shadlou and Bhattacharya 2016) of K_L is given in the Table E.1. Shadlou and Bhattacharya (2016) formulation considers the base shearing of the pile as well as the shaft resistance of the pile. This is in contrast to the conventional p-y formulation where only compressional resistance (i.e. lateral pile-soil interaction) is taken.

Table E.1 Lateral stiffness of piles.

Ground profile/pile type	Rigid pile $\frac{K_L}{DE_{SO}f(\upsilon_s)}$	Flexible pile
Homogeneous	$3.2\left(\frac{L}{D}\right)^{0.62}$	$1.45\left(\frac{E_P}{E_{SO}}\right)^{0.186} f_{(vs)}E_{SO}D_P$
Parabolic	$2.65\left(\frac{L}{D}\right)^{1.07}$	$1.015\left(\frac{E_P}{E_{SO}}\right)^{0.27} f_{(vs)}E_{SO}D_P$
Linear	$2.35\left(\frac{L}{D}\right)^{1.53}$	$0.79\left(\frac{E_P}{E_{SO}}\right)^{0.34} f_{(vs)}E_{SO}D_P$

L: Pile embedded length; D: Pile diameter; E_{SO}: Soil Young's modulus at 1 diameter depth; E_p: Equivalent Young's modulus of the pile; υ_s: Soil Poisson's ratio; $f_{(v_s)} = 1 + |v_s - 0.25|$.

Homogeneous soils are soils that have a constant stiffness with depth such as overconsolidated clays. On the other hand, a linear profile is typical for normally consolidated clays (or 'Gibson Soil') and parabolic behaviour can be used for sandy soils.
Eurocode 8 (Part 5) also provides stiffness of flexible piles. Expert judgement is required for choosing the formulas given in Table 5.3.

Conversely, one may carry out beam on nonlinear Winkler foundation model using state-of-the-art *p-y* curves to obtain the pile head stiffness values. The method to convert *p-y* analysis to the pile head stiffness values is provided in Examples 6.2 and 6.3 of Chapter 6.

Design of Foundations for Offshore Wind Turbines, First Edition. Subhamoy Bhattacharya.

Companion website: www.wiley.com/go/bhattacharya/offshorewindturbines

Appendix F

Lateral Stiffness K_L of Suction Caissons

Complete formulation for skirted foundations (K_L, K_R, and K_{LR}) for three types of grounds are shown in Tables 5.5 and 5.6 (Chapter 5). As an example, Table F.1 shows the K_L value for lateral stiffness of rigid suction caissons.

Table F.1 Stiffness of suction caissons for 0.5 < L/D < 2.

Ground profile/pile type	Rigid $\frac{K_L}{DE_{SO}f(v_s)}$
Homogeneous	$2.91\left(\frac{L}{D}\right)^{0.56}$
Parabolic	$2.7\left(\frac{L}{D}\right)^{0.96}$
Linear	$2.53\left(\frac{L}{D}\right)^{1.33}$

$f(v_s) = 1.1 \times \left(0.096\left(\frac{L}{D}\right) + 0.6\right) v_s^2 - 0.7v_s + 1.06$

Design of Foundations for Offshore Wind Turbines, First Edition. Subhamoy Bhattacharya.

Companion website: www.wiley.com/go/bhattacharya/offshorewindturbines

Bibliography

4C Offshore Limited. Global offshore wind farms database. In: 4COffshore.com web page. http://www.4coffshore.com/windfarms/; 2015a [accessed 12.10.17].

4C Offshore Limited (2016) Offshore Wind Turbine Database. In: http://4COffshore.com web page. http://www.4coffshore.com/windfarms/turbines.aspx. Accessed 19 Mar 2015.

Abbs, A.F. (1983). Lateral pile analysis in weak carbonate rocks. In: *Proceedings of the Conference on Geotechnical Practice in Offshore Engineering*, 546–556. Austin: ASCE.

Abed, Y., Bouzid, D.A., Bhattacharya, S. et al. (2016). Static impedance functions for monopiles supporting offshore wind turbines in nonhomogeneous soils – emphasis on soil/monopile interface characteristics. *Earthquakes and Structures* 10 (5): 1143–1179. https://doi.org/10.12989/eas.2016.10.5.1143.

Abhinav, K.A. and Saha, N. (2015). Coupled hydrodynamic and geotechnical analysis of jacket offshore wind turbine. *Soil Dynamics and Earthquake Engineering* 73: 66–79.

Abhinav, K.A. and Saha, N. (2018). Nonlinear dynamical behaviour of jacket supported offshore wind turbines in loose sand. *Marine Structures* 57: 133–151.

Achmus M, Abdel-Rahman K. (2005). Finite Element Modelling of Horizontally Loaded Monopile Foundations for Offshore Wind Energy Converters in Germany. Proc, Int Symp Front Offshore Geotech ISFOG 2005:391–6. doi:10.1201/NOE0415390637.ch38

Achmus, M., Kuo, Y.-S., and Abdel-Rahman, K. (2009). Behavior of monopile foundations under cyclic lateral load. *Computers and Geotechnics* 36: 725–735. https://doi.org/10.1016/j.compgeo.2008.12.003.

Adedipe, O., Brennan, F., and Kolios, A. (2016). Review of corrosion fatigue in offshore structures: present status and challenges in the offshore wind sector. *Renewable Energy and Sustainable Energy Reviews* 61: 141–154.

Adhikari, S. and Bhattacharya, S. (2011). Vibrations of wind-turbines considering soil-structure interaction. *Wind and Structures – An International Journal* 14: 85–112.

Adhikari, S. and Bhattacharya, S. (2012). Dynamic analysis of wind turbine towers on flexible foundations. *Shock and Vibration* 19: 37–56.

Aissa, M., Bouzid, D.A., and Bhattacharya, S. (2017). Monopile head stiffness for servicibility limit state calculations in assessing the natural frequency of offshore wind turbines. *International Journal of Geotechnical Engineering* (6): 1, 267–17. https://doi.org/10.1080/19386362.2016.1270794, 283.

Aldridge, T.R., Carrington, T.M., and Kee, N.R. (2005). Propagation of pile tip damage during installation. In: *Frontiers in Offshore Geotechnics ISFOG 2005* (ed. S. Gourvenec and M. Cassidy), 823–827. London: Taylor and Francis Group. ISBN: 0415 39063 X.

Design of Foundations for Offshore Wind Turbines, First Edition. Subhamoy Bhattacharya.

Companion website: www.wiley.com/go/bhattacharya/offshorewindturbines

Alexander, N.A. (2010). Estimating the nonlinear resonant frequency of a single pile in nonlinear soil. *Sound and Vibration, Issue* 329: 1137–1153.

Alexander, N., Chanerley, A., Crewe, A. et al. (2014). Obtaining spectrum matching time series using a reweighted volterra series algorithm (RVSA). *Bulletin of the Seismological Society of America* 104 (4): 1663–1673.

American Petroleum Institute (1993). *Recommended Practice for Planning, Designing and constructing Fixed Offshore Platforms - Load and Resistance Factor Design.* Washington DC: American Petroleum Institute.

American Petroleum Institute. Recommended Practice for Planning, Designing and Constructing Fixed Offshore Platforms – Working Stress Design. API Recommended Practice 2A-WSD (RP 2A-WSD), 2000, 21st edition.

Andersen, O.J. and Løvseth, J. (2006). The Frøya database and maritime boundary layer wind description. *Marine Structures* 19.

API (2005) Recommended Practice for Planning, Designing and Constructing Fixed Offshore Platforms—Working Stress Design. 68–71.

API (2007). Recommended Practice for Planning, Designing and Constructing Fixed Offshore Platforms – Working Stress Design. API Recommenced Practice, 2A-WSD (RP 2A-WSD)

Arany (2017) A methodology for simplified integrated design of offshore wind turbine foundations; PhD thesis, University of Bristol (UK)

Arany, L. and Bhattacharya, S. (2018). Simplified load estimation and sizing of suction anchors for spar buoy type floating offshore wind turbines. *Ocean Engineering* 159: 348–357.

Arany L, Bhattacharya S, Hogan SJ, et al. (2014) Dynamic soil-structure interaction issues of offshore wind turbines. Proc. 9th Int. Conf. Struct. Dyn. EURODYN 2014. pp 3611–3618.

Arany, L., Bhattacharya, S., Macdonald, J. et al. (2015a). Simplified critical mudline bending moment spectra of offshore wind turbine support structures. *Wind Energy* 18: 2171–2197. https://doi.org/10.1002/we.1812.

Arany, L., Bhattacharya, S., Adhikari, S. et al. (2015b). An analytical model to predict the natural frequency of offshore wind turbines on three-spring flexible foundations using two different beam models. *Soil Dynamics and Earthquake Engineering* 74: 40–45. https://doi.org/10.1016/j.soildyn.2015.03.007.

Arany, L., Bhattacharya, S., Macdonald, J.H.G. et al. (2016). Closed form solution of Eigen frequency of monopile supported offshore wind turbines in deeper waters incorporating stiffness of substructure and SSI. *Soil Dynamics and Earthquake Engineering* 83: 18–32. ISSN: 0267-7261, https://doi.org/10.1016/j.soildyn.2015.12.011.

Arany, L., Bhattacharya, S., Macdonald, J. et al. (2017). Design of monopiles for offshore wind turbines in 10 steps. *Soil Dynamics and Earthquake Engineering* 92: 126–152. https://doi.org/10.1016/j.soildyn.2016.09.024.

ASCE/SEI 7-2016: *Minimum Design Loads and Associated Criteria for Buildings and Other Structures.*

Ashour, M. and Norris, G. (2000). Modeling lateral soil-pile response based on soil pile interaction. *Journal of Geotechnical and Geoenvironmental Engineering* 126 (5): 420–428.

ASTM (2005) E1049–85 Standard Practices for Cycle Counting in Fatigue Analysis.

Augustesen, A.H., Brødbæk, K.T., Møller, M. et al. (2009). Numerical modelling of large-diameter steel piles at Horns Rev. *Civil Comp.[CD-ROM].*

AxanyShama, A.A. and El Naggar, H. (2015). Bridge Foundations. In: *Encyclopedia of earthquake engineering*, vol. 1, 298–317.

Axelsson, G. Long-term set-up of driven piles in sand, PhD thesis; Royal Institute of Technology, Stockholm, 2000.

Ayyar, S., Jansson, J., and Sorensen, R. (2014). Cathodic protection design for offshore wind turbine foundations. *Materials Performance* 53: 26–29.

Barber ES (1953) Discussion to paper by S. M. Gleser. ASTM Spec Tech Publ 154:96–99.

Betz, A. (1919). Maximum der theoretisch möglichen Ausnützung des Windes durch Windmotoren. *Zeitschrift für das gesamte Turbinenwesen, Heft* 26: 1920.

Bhattacharya, S. (2014). Challenges in design of foundations for offshore wind turbines. *IET Journal, Engineering and Technology Reference* 9: https://doi.org/10.1049/etr.2014.0041, ISSN: 2056-4007.

Bhattacharya, S. (2017). Chapter 12: Civil engineering aspects of a wind farm and wind turbine structures. In: *Wind Energy Engineering: A Handbook for Onshore and Offshore Wind Turbines* Hardcover. Elsevier. ISBN: 9780128094518.

Bhattacharya, S. and Adhikari, S. (2011). Experimental validation of soil–structure interaction of offshore wind turbines. *Soil Dynamics and Earthquake Engineering* 31 (5–6): 805–816.

Bhattacharya, S. and Goda, K. (2013). Probabilistic buckling analysis of axially loaded piles in liquefiable soils. *Soil Dynamics and Earthquake Engineering* 45: 13–24.

Bhattacharya, S., Carrington, T.M., and Aldridge, T.R. (2005). Buckling considerations in pile design. In: *Frontiers in Offshore Geotechnics ISFOG 2005* (ed. S. Gourvenec and M. Cassidy), 815–821. London: Taylor and Francis Group.

S. Bhattacharya, T.M. Carrington and T.R. Aldridge (2006): Design of FPSO piles against Storm Loading, OTC 17861

Bhattacharya, S., Adhikari, S., and Alexander, N.A. (2009a). A simplified method for unified buckling and free vibration analysis of pile-supported structures in seismically liquefiable soils. *Soil Dynamics and Earthquake Engineering* 29 (8): 1220–1235.

Bhattacharya, S., Carrington, T., and Aldridge, T. (2009b). Observed increases in offshore pile driving resistance. *Proceedings of the Institution of Civil Engineers-Geotechnical Engineering* 162: 71–80.

Bhattacharya, S., Hyodo, M., Goda, K. et al. (2011a). Liquefaction of soil in the Tokyo Bay area from the 2011 Tohoku (Japan) earthquake. *Soil Dynamics and Earthquake Engineering* 31 (11): 1618–1628.

Bhattacharya, S., Lombardi, D., and Muir Wood, D.M. (2011b). Similitude relationships for physical modelling of monopile-supported offshore wind turbines. *International Journal of Physical Modelling in Geotechnics* 11 (2): 58–68.

Bhattacharya, S., Cox, J., Lombardi, D. et al. (2013a). Dynamics of offshore wind turbines supported on two foundations. *Geotechnical Engineering: Proceedings of the ICE* 166 (2): 159–169. https://doi.org/10.1680/geng.11.00015.

Bhattacharya, S., Nikitas, N., Garnsey, J. et al. (2013b). Observed dynamic soil–structure interaction in scale testing of offshore wind turbine foundations. *Soil Dynamics and Earthquake Engineering* 54: 47–60.

Bhattacharya, S. and Goda, K. (2016). Use of offshore wind farms to increase seismic resilience of Nuclear Power Plants. *Soil Dynamics and Earthquake Engineering* 80 (2016): 65–68.

Bhattacharya, S., Nikitas, G., Arany, L. et al. (2017). Soil-structure interactions for offshore wind turbines. *Engineering and Technology.*

Bhattacharya, S., Wang, L., Liu, J. et al. (2017a). Civil engineering challenges associated with design of offshore wind turbines with special reference to China. In: *Chapter 13 of Wind Energy Engineering: A Handbook for Onshore and Offshore Wind Turbines,* 243–273. Academic Press (Elsevier).

Bhattacharya, Subhamoy, Georgios Nikitas, and Saleh Jalbi (2018): On the Use of Scaled Model Tests for Analysis and Design of Offshore Wind Turbines. *Geotechnics for Natural and Engineered Sustainable Technologies.* Springer, Singapore, 2018. 107–129.

Bhattacharya, S, Orense R and Lombardi, D (2019): *Seismic Design of Foundations: Concepts and Applications,* ICE Publishing, ISBN 978-0-7277-6166-8

Bhattacharya, S., Nikitas, G., and Vimalan, N. (2019). *Dynamic SSI of Monopile-Supported Offshore Wind Turbines: Geotechnical Design and Practice.,* 113–123. Singapore: Springer.

Bisoi, S. and Haldar, S. (2014). Dynamic analysis of offshore wind turbine in clay considering soil–monopile–tower interaction. *Soil Dynamics and Earthquake Engineering* 63: 19–35.

Black, A. (2015). Corrosion risks and mitigation strategies for offshore wind turbine foundations. *Materials Performance* 54.

Blazquez, R. and Lopez-Querol, S. (2006). Generalized densification law for dry sand subjected to dynamic loading. *Soil Dynamics and Earthquake Engineering* 26 (9): 888–898. https://doi.org/10.1016/j.soildyn.2005.09.001.

Blevins, R.D. (1984). *Formulas for Natural Frequency and Mode Shape.* Krieger Publishing Company.

Bond, A.J., Jardine, R.J., and Dalton, C.P. (1992). Design and performance of the Imperial College Instrumented Pile. *ASTM Geotechnical Testing Journal* 14 (4): 413–425.

Bouzid, D.A., Bhattacharya, S., and Dash, S.R. (2013). Winkler Springs (p-y curves) for pile design from stress-strain of soils: FE assessment of scaling coefficients using the mobilized strength design concept. *Geomechanics and Engineering* 5 (5): 379–399. https://doi.org/10.12989/gae.2013.5.5.379.

Bowles, J.E. (1996). *Foundation Analysis and Design,* 5e. New York: The McGraw-Hill Companies, Inc.

Brandenberg, S. (2005). Behavior of pile foundations in liquefied and laterally spreading ground, PhD thesis, University of California, Davis.

Brennan, F. and Tavares, I. (2014). Fatigue design of offshore steel monopile wind substructures. *Inst Civ Eng* 167: 1–7.

Breusers, H.N.C. and Raudkivi, A.J. (1991). *Scouring.* Rotterdam: A.A. Balkema.

Breusers, H.N.C., Nicollet, G., and Shen, H.W. (1977). Local scour around cylindrical piers. *Journal of Hydraulic Research* 15: 211–252.

Broms, B.B. (1964). Lateral resistance of pile in cohesionless soils. *Journal of the Soil Mechanics and Foundations Division, ASCE* 90: 123–156.

Budhu, M. (2011). *Soil Mechanics and Foundations,* 3e. Wiley.

Burton, T., Jenkins, N., Sharpe, D. et al. (2001). *Wind Energy Handbook.* Chichester, West Sussex, England: Wiley.

Burton, T., Sharpe, D., Jenkins, N. et al. (2011). *Wind Energy Handbook,* 2e. Chichester, UK: Wiley.

Butterfield, R., Houlsby, G.T., and Gottardi, G. (1997). Standardized sign conventions and notation for generally loaded foundations. *Geotechnique* 47: 1051–1054.

Byrne, B.W. (2000). *Investigations of suction caissons in dense sand.* Doctoral dissertation: University of Oxford.

Byrne, B.W. and Houlsby, G.T. (2003). Foundations for offshore wind turbines. *Philosophical Transactions of the Royal Society of London, Series A: Mathematical, Physical and Engineering Sciences* 361: 2909–2930. https://doi.org/10.1098/rsta.2003.1286.

Byrne, B.W., Burd, H.J., Houlsby, G.T. et al. (2015a). *New Design Methods for Large Diameter Piles under Lateral Loading for Offshore Wind Applications.* ISFOG.

Byrne BW, Mcadam RA, Burd HJ, et al. (2015b) Field testing of large diameter piles under lateral loading for offshore wind applications. Proc XVI ECSMGE Geotech Eng Infrastruct Dev 1255–1260. doi: 10.1680.60678

Cambou, B. and Hicher, P.Y. (2010). Elastoplastic modeling of soils: cyclic loading. In: *Constitutive Modeling of Soils and Rocks* (ed. P.-.Y. Hicher and J.-.F. Shao), 143–186. Hoboken, NJ: Wiley https://doi.org/10.1002/9780470611081.ch4.

Camp TR, Morris MJ, van Rooij R, et al. (2004) Design Methods for Offshore Wind Turbines at Exposed Sites (Final Report of the OWTES Project EU Joule III Project JOR3-CT98-0284). Bristol

Carter, J. (2007). North Hoyle offshore wind farm: design and build. *Energy: Proceedings of the Institution of Civil Engineers EN1* 160 (1): 21–29.

Carter, J. and Kulhawy, F. (1992). Analysis of laterally loaded shafts in rock. *Journal of Geotechnical Engineering* 118: 839–855.

Carter, J.P., Randolph, M.F., and Wroth, C.P. (1980). Some aspects of the performance of open and closed ended piles. In: *Numerical Methods in Offshore Piling*, 165–170. London: ICE.

Chow, F.C. Investigation into the behaviour of displacement piles for offshore foundations, PhD thesis; University of London, Imperial College, 1996.

Cox, J (2014): Long-term serviceability behaviour of suction caisson supported offshore wind turbines, PhD thesis (University of Bristol).

Cox, J.A., O'Loughlin, C.D., Cassidy, M. et al. (2014). Centrifuge study on the cyclic performance of caissons in sand. *International Journal of Physical Modelling in Geotechnics* 14: 99–115. https://doi.org/10.1680/ijpmg.14.00016.

Cox, James A., and Subhamoy Bhattacharya (2017). "Serviceability of suction caisson founded offshore structures." Proceedings of the Institution of Civil Engineers: Geotechnical Engineering170.3 (2016): 273–284.

Coyle, H.M. Bartoskewitz, R.E and Lowery, L.L. Prediction of static bearing capacity from wave equation analysis, Offshore Technology Conference, 1970, Paper number OTC 1202.

Cuéllar P. Pile foundations for offshore wind turbines: numerical and experimental investigations on the behaviour under short-term and long-term cyclic loading. Thesis 2011a. Technischen Universität Berline.

Cuéllar, P., Georgi, S., Baeßler, M. et al. (2012). On the quasi-static granular convective flow and sand densification around pile foundations under cyclic lateral loading. *Granular Matter* 14 (1): 11–25.

Cui L (2006) Developing a virtual test environment for granular materials using discrete element modelling. PhD. Thesis, University College Dublin, Ireland.

Cui, L. and Bhattacharya, S. (2016). Soil-monopile interactions for offshore wind turbine. *Engineering and Computational Mechanics, Proceedings of the Institution of Civil Engineers* 169: 171–182.

Cui, L., O'Sullivan, C., and O'Neill, S. (2007). An analysis of the triaxial apparatus using a mixed boundary three-dimensional discrete element model. *Geotechnique* 57 (10): 831–844.

Cundall, P. and Strack, O. (1979). A discrete numerical model for granular assemblies. *Geotechnique* 29 (1): 47–65.

Dammala, P.K., Krishna, A.M., Bhattacharya, S. et al. (2017). Dynamic soil properties for seismic ground response studies in Northeastern India. *Soil Dynamics and Earthquake Engineering* 100: 357–370.

Damgaard, M., Zania, V., Andersen, L. et al. (2014). Effects of soil–structure interaction on real time dynamic response of offshore wind turbines on monopiles. *Engineering Structures* 75: 388–401. https://doi.org/10.1016/j.engstruct.2014.06.006.

Dash SR (2010) Lateral pile soil interaction in liquefiable soils, PhD thesis. (University of Oxford).

Dash, S.R., Bhattacharya, S., and Blakeborough, A. (2010). Bending–buckling interaction as a failure mechanism of piles in liquefiable soils. *Soil Dynamics and Earthquake Engineering* 30 (1-2): 32–39.

Dash, S., Rouholamin, M., Lombardi, D. et al. (2017). A practical method for construction of py curves for liquefiable soils. *Soil Dynamics and Earthquake Engineering* 97: 478–481.

Davenport, A. (1961). The spectrum of horizontal gustiness near the ground in high winds. *Quarterly Journal of the Royal Meteorological Society* 87 (372): 194–211.

Davisson MT (1970) Lateral Load Capacity of Piles. Highw Res Rec 104–112.

Davisson, M.T. and Salley, J.R. (1970). Model study of laterally loaded piles. *Journal of the Soil Mechanics and Foundations Division* 96: 1605–1627.

De Risi, R., Bhattacharya, S., and Goda, K. (2018). Seismic performance assessment of monopile-supported offshore wind turbines using unscaled natural earthquake records. *Soil Dynamics and Earthquake Engineering* 109: 154–172.

De Vries, W., Vemula, N.K., Passon, P. et al. (2011). Support structure concepts for deep water sites. *Delft University of Technology, Delft, The Netherlands, Technical Report No. UpWind Final Report WP4* 2: 167–187.

Dean, E.T.R. (2010). *Offshore geotechnical engineering: Principles and practice*. ICE Publication.

Debnath, K. and Chaudhiri, S. (2009). Laboratory experiments on local scour around cylinder for clay and clay-sand mixture beds. *Engineering Geology* 111: 51–61.

DNV (2005) DNV-RP-C203- Fatigue design of offshore steel structures. Recomm Pract DNV-RPC203 126.

Design of Offshore Wind Turbine Structures, October 2010a, Offshore Standard DNV-OS-J101.

DNV (2010b) Recommended Practice DNV-RP-C205 - Environmental conditions and environmental loads.

DNV (2014) Offshore Standard DNV-OS-J101 Design of Offshore Wind Turbine Structures. Høvik, Norway.

DNV (Det Norske Veritas) (2002). *Guidelines for Design of Wind Turbines*, 2e. London, UK: DNV.

Doyle, E. H., Dean, E. T. R., Sharma, J. S., Bolton, M. D., Valsangkar, A. J. & Newlin, J. A. (2004). Centrifuge Model Tests on Anchor Piles for Tension Leg Platforms. Proc. Ann. Technol. Conf., Houston, TX, paper:16845

Drnevich, V.P., Hall, J.R., and Richart, F.E. (1967). Effects of amplitude of vibration on the shear modulus of sand. Proceedings. In: *International Symposium on Wave Propagation and Dynamic properties of Earth Materials*. Albuquerque: NM.

Durning, P.J., Rennie, I.A., Thompson, J.M et al. Installing a piled foundation in hard, over-consolidated North Sea clay for the Heather platform, 1978, Proceedings of European Offshore Petroleum Conference, London, Vol. 1, pp 375–382.

Dyson, G.J. (1999) Laterally loaded piles in calcareous sediments, PhD thesis, The University of Western Australia.

Dyson, G.J. and Randolph, M.F. (1997) Load Transfer Curves for Piles in Calcareous Sand, 8th International Conference on the Behaviour of Offshore Structures, Delft, The Netherlands, Elsevier Science Ltd. 2 245-258.

Dyson, G.J. and Randolph, M.F. (2000). *Monotonic Lateral Loading of Piles in Calcareous Sediments, Research report G: 1472, Department of Civil and Resource Engineering*. The University of Western Australia.

Ebelhar, R.J., Drnevich, V.P., and Kutter, B.L. (eds.) *Dynamic Geotechnical Testing II*. Philadelphia, PA: ASTM.

Eide, O. and Andersen, K.H. (1984). Guest Lecture – Foundation Engineering for Gravity Structures in the Northern North Sea. In: *International Conference on Case Histories in Geotechnical Engineering. 1*. http://scholarsmine.mst.edu/icchge/1icchge/1icchge-theme4/1.

Encyclopedia of Maritime and Offshore Engineering homepage Encyclopedia of Maritime and Offshore Engineering, First published: 29 September 2017, Print ISBN: 9781118476352| Online, ISBN: 9781118476406| DOI: 10.1002/9781118476406

EWEA (European Wind Energy Association)

European Committee for Standardization (2003) Eurocode 8: Design of Structures for earthquake resistance - Part 5: Foundations, retaining structures and geotechnical aspects.

European Committee for Standardization (CEN) (2004): EN 1998-5:2004 - Eurocode 8: Design of structures for earthquake resistance, Part 5: Foundations, retaining structures and geotechnical aspects.

Farouki, O.T. (1986). *Thermal properties of soils*. Rockport, MA: Trans Tech.

Fleming, W.G.K., Weltman, A.J., Randolph, M.F. et al. (1992). *Piling Engineering*. Glasgow: Blackie.

Forehand, P.W. and Reese, J.L. (1964a). Prediction of pile capacity by wave equation. *Journal of Soil Mechanics and Foundation Engineering, ASCE* (SM2): 1–25.

Forehand, P.W. and Reese, J.L. (1964b). Prediction of pile capacity by wave equation. *Journal of ASCE* 90 (2): 1–26.

Fragio, A.G., Santiago, J.L., and Sutton, V.J.R. (1985). Load tests on grouted piles in rock, Proceedings 17th Annual Offshore Technology Conference. *Houston, OTC* 4851: 93–104.

Frohboese P, Schmuck C (2010) Thrust coefficients used for estimation of wake effects for fatigue load calculation. Eur. Wind Energy Conf. 2010, Warsaw, Pol. pp 1–10.

Gazetas, G. (1984). Seismic response of end-bearing single piles. *International Journal of Soil Dynamics and Earthquake Engineering* 3: 82–93. https://doi.org/10.1016/0261-7277(84)90003-2.

Gazetas, G., (1991). Formulas and charts for impedances of surface and embedded foundations. *Journal of geotechnical engineering* 117 (9): 1363–1381.

Georgiannou, V.N., Rampello, S., and Silvestri, F. (1991). Static and dynamic measurements of undrained stiffness on natural overconsolidated clays. *Proc. X ECSMFE, Florence.* 1: 91–96.

Germanischer, L.L.O.Y.D. (2012). *Rules and Guidelines Industrial Services: Guideline for the Certification of Offshore Wind Turbines.* Germany: Hamburg.

Gibson, G.C. and Coyle, H.M. Soil damping constants related to common soil properties in sand and clays, Research report 125-1, Texas Transportation Institute, September 1968.

Goble G.G., Rausche F., Likins G. and Associates Inc. GRLWEAP manual, 2003.

Gottardi, G. & Houlsby, G.T. (1995). Model Tests of Circular Footings on Sand Subjected to Combined Loads, Report No. 2071/95. Oxford.: Department of Engineering Science, University of Oxford.

Goda, K., Pomonis, A., Chian, S. et al. (2012). Ground motion characteristics and shaking damage of the 11th March 2011 M w9.0 Great East Japan earthquake. *Bulletin of Earthquake Engineering* 11 (1): 141–170.

Guang-Yu, Z. Wave equation Application for piles in soft ground, Proceedings of the 3rd International Conference of Stress-Wave theory to piles, Ottawa, Ontario, Canada, 1988, pp 831–836.

Guo, Z., Yu, L., Wang, L. et al. (2015). Model tests on the long-term dynamic performance of offshore wind turbines founded on monopiles in sand. *ASME Journal of Offshore Mechanics and Arctic Engineering*; Paper No: OMAE-14-1142 137 (4): https://doi.org/10.1115/1.4030682.

Hall, J.R. and Richart, F.E. (1963). Dissipation of Elastic Wave Energy in Granular Soils. *Journal of the Soil Mechanics and Foundations Division, ASCE* 6: 27–56.

Hald T, Mørch C, Jensen L, et al. (2009) Revisiting monopile design using p-y curves. Results from full scale measurements on Horns Rev. Eur. Wind Energy Conf.

Hamming RW. Digital filters (Prentice-Hall signal processing series). 1977.

Hansteen, O.E. (1980), Dynamic performance' Shell Brent B Instrumentation Project, Seminar, London 1979, Proceedings, London, Society for Underwater Technology, pp. 89- Also pub!. in: Norwegian Geotechnical Institute, Publication, 137

Hardin, B.O. and Black, W.L. (1966). Sand Stiffness Under Various Triaxial Stresses. *Journal of the Soil Mechanics and Foundations Division, ASCE* 94: 353–369.

Hardin, B.O. and Blandford, G.E. (1989). Elasticity of particulate materials. *Journal of Geotechnical Engineering, ASCE* 115: 788–805.

Hardin, B.O. and Drnevich, V.P. (1972). Shear Modulus and Damping in Soils: Design equations and curves. *Journal of the Soil Mechanics and Foundations Division, ASCE* 98: 667–692.

Hardin, B.O. and Music, J. (1965). Apparatus for vibration of soil specimens during the triaxial test. *ASTM, STP* 392: 55–74.

Hardin, B.O. and Richart, F.E.J. (1963). Elastic wave velocities in granular soils. *Journal of the Soil Mechanics and Foundations Division, ASCE* 89: 33–65.

Hashash, Y.M.A., Phillips, C., and Groholski, D.R. (2010). Recent advances in nonlinear site response analysis. In: *Fifth International Conference in Recent Advances in Geotechnical Earthquake Engineering and Soil Dynamics*, vol. CD-Volume. San Diego, CA: OSP 4.

Head, K.H. (2006). *Manual of Soil Laboratory Testing: Soil Classification and Compaction Tests*, 3rd Revised. Whittles Publishing.

Hearn E. Finite Element Analysis of an Offshore Wind Turbine Monopile. Master's Thesis, Dep Civ Environ Eng Tufts Univ Medford, MA 2009. https://doi.org/10.1061/41095(365)188.

Hettler A (1981) Verschiebungen starrer und elastischer Gründungskörper in Sand bei monotoner und zyklischer Belastung. Univ. Fridericiana, Karlsruhe.

Higgins W, Basu D (2011). Fourier finite element analysis of laterally loaded piles in elastic media - Internal Geotechnical Report 2011–1. Storrs, Connecticut.

Hjorth, P. (1975). *Studies on the Nature of Local Scour*. Lund, Sweden: Institute of Technology, Dept. of Water Resources Engineering.

Hoffmans, G.J.C.M. and Verheij, H.J. (1997). *Scour Manual*. Rotterdam: A.A. Balkema.

Houlsby, G.T. and Byrne, B.W. (2005a). Design procedures for installation of suction caissons in sand. *Proceedings of the Institution of Civil Engineers-Geotechnical Engineering* 158 (3, 2005): 135–144.

Houlsby, Guy T., and Byron W. Byrne (2005b). "Design procedures for installation of suction caissons in clay and other materials". *Proceedings of the Institution of Civil Engineers-Geotechnical Engineering* 158.2 (2005): 75–82.

Houlsby, G.T. and Martin, C.M. (1992) "Modelling of the Behaviour of Foundations of Jack up Units on Clay", *Proc. of the Wroth Memorial Symp. Predictive Soil Mechanics*, Oxford, July 27–29, pp 339–358.

Houlsby, G., Kelly, R., and Byrne, B. (2005a). The tensile capacity of suction caissons in sand under rapid loading. *Front. Offshore Geotech* 405–410. https://doi.org/10.1201/NOE0415390637.ch40.

Houlsby, G.T., Ibsen, L.B., and Byrne, B.W. (2005b). Suction caissons for wind turbines. In: *Front. Offshore Geotech. ISFOG 2005-Gourvenec Cassidy* (ed. C. Gourvenec), 75–94. London, UK: Taylor and Francis Group.

Hu WH, Thöns S, Said S, et al. (2014) : Resonance phenomenon in a wind turbine system under operational conditions; Proceedings of the 9th International Conference on Structural Dynamics, EURODYN 2014, Porto, Portugal, 30 June - 2 July 2014, A. Cunha, E. Caetano, P. Ribeiro, G. Müller (eds.)

Huang, S. Application of Dynamic Measurements on long H-Pile driven into soft ground in Shanghai. Proceedings of the 3rd International Conference of Stress-Wave theory to piles, Ottawa, Ontario, Canada, 1988, pp 635–643.

Idriss, I.M. and Boulanger, R.W. (2008). *Soil Liquefactions during Earthquakes*. Oakland, California, USA: Earthquake Engineering Research Institute (EERI).

IEC (2005) International Standard IEC-61400-1 Wind Turbines - Part 1: Design requirements, Third Edition. 2005:

IEC (2009a) International Standard IEC 61400-3 Wind turbines - Part 3: Design requirements for offshore wind turbines, 1.0 ed. International Electrotechnical Commission (IEC).

IEC (2009b) International Standard IEC-61400-1 Amendment 1 - Wind turbines - Part 1: Design requirements, 3rd Edition.

Iida, K. (1938). The velocity of elastic waves in sand. *Bulletin of Earthquake Research Institute, University of Tokyo* 16: 131–145.

Imai, T. and Tonouchi, K. (1982): Correlation of N value with S-wave velocity, Proc., 2nd European Symposium on Penetration Testing, Amsterdam, pp. 67–72.

Ishimoto, M. and Lida, K. (1937). Determination of Elastic Constants of Soils by Means of Vibration Methods. Part 2. Modulus of Rigidity and Poisson's Ratio. Bulletin of the Earthquake Research Institute. *Tokyo Imperial University* 15: 67–86.

Ishihara, K. (1996). *Soil Behaviour in Earthquake Geotechnics.* Oxford: Oxford Science Publication. Clarendon Press.

Itasca (2008). PFC2D manual 4.0.

Jalbi, S., Arany, L., Salem, A. et al. (2019). A method to predict the cyclic loading profiles (one-way or two-way) for monopile supported offshore wind turbines. *Marine Structures* 63: 65–83.

Jalbi, S. and Bhattacharya, S. (2018). Closed form solution for the first natural frequency of offshore wind turbine jackets supported on multiple foundations incorporating soil-structure interaction. *Soil Dynamics and Earthquake Engineering* 113: 593–613.

Jalbi, S., Shadlou, M., and Bhattacharya, S. (2017). Chapter 16: Practical method to estimate foundation stiffness for design of offshore wind turbines. In: *Wind Energy Engineering: A Handbook for Onshore and Offshore Wind Turbines* Hardcover. Elsevier. ISBN: 9780128094518.

Jalbi, S., Shadlou, M., and Bhattacharya, S. (2018). Impedance functions for rigid skirted caissons supporting offshore wind turbines. *Ocean Engineering* 150: 21–35.

James K. Mitchell, Kenichi Soga (2005): *Fundamentals of Soil Behavior*, 3rd Edition, Wiley, ISBN: 978-0-471-46302-3

Jamieson Peter (2018) *Innovation in Wind Turbine Design*, Wiley, 2nd Edition ISBN: 978-1-119-13790-0

Jardine, R.J. (1992). Some observations on the kinematic nature of soil. *Soils and Foundations* 32 (2): 111–124.

Jardine, R., Chow, F., Overy, R. et al. (2005). *ICP Design Methods for Driven Piles in Sands and Clays.* London: Imperial College Press.

Jardine, R.J., Yang, Z.X., Foray, P. et al. (2013). Interpretation of stress measurements made around closed-ended displacement piles in sand. *Géotechnique* 63: 613–627. https://doi.org/10.1680/geot.9.P.138.

Jeanjean P (2017): A Framework for Monotonic p-y Curves in Clays, SUT conference Proceedings London

Jonkman, J., Butterfield, S., Musial, W. et al. (2009). *Definition of a 5-MW reference wind turbine for offshore system development (No. NREL/TP-500-38060).* Golden, CO: National Renewable Energy Laboratory (NREL).

L. Junfeng, C. Fengbo, Q. Liming, et al. 2012, China wind energy outlook. Greenpeace, GWEC and Chinese Wind Power Association.

Kaimal, J.C., Wyngaard, J.C., Izumi, Y. et al. (1972). Spectral characteristics of surface-layer turbulence. *Quarterly Journal of the Royal Meteorological Society* 98: 563–589.

Kallehave, Dan, et al. (2014) "Optimization of monopiles for offshore wind turbines." *Philosophical Transactions of the Royal Society A: Mathematical, Physical and Engineering Sciences373.*2035: 20140100.

Kallehave D, Thilsted CL (2012) Modification of the API p-y Formulation of Initial Stiffness of Sand. Offshore Site Investig. Geotech. Integr. Geotechnol. - Present Futur.

Kitazawa, G., Kitayama, K., Suzuki, K., et al. (1959): Tokyo ground map, Gihodo

Klinkvort RT, Leth CT, Hededal O (2010) Centrifuge modelling of a laterally cyclic loaded pile. 7th Int Conf Phys Model Geotech 959–964.

Kokusho, T., Yoshida, Y., and Esashi, Y. (1982). Dynamic properties of soft clay for wide strain range. *Soils and Foundations.* 22 (4): 1–18.

Kramer, S.L. (1996). *Geotechnical Earthquake Engineering*, Prentice-Hall Civil Engineering and Engineering Mechanics Series. Upper Saddle River, NJ: Prentice Hall.

Kucharczyk, P., Rizos, A., Münstermann, S. et al. (2012). Estimation of the endurance fatigue limit for structural steel in load increasing tests at low temperature. *Fatigue and Fracture of Engineering Materials and Structures* 35: 628–637. https://doi.org/10.1111/j.1460-2695.2011.01656.x.

Kuhn, M. (2000). *Dynamics of offshore wind energy converters on mono-pile foundation experience from the Lely offshore wind turbine.* OWEN Workshop.

Kühn M (1997) Soft or stiff: A fundamental question for designers of offshore wind energy converters. Proc. Eur. Wind Energy Conf. EWEC '97.

Kuhn, M. (2002). *Offshore wind farms.* Wind Power Plants: Fundamentals, desin, construction and operation, Gash and Twelve eds, 365–384.

Kuo, Y.-S., Achmus, M., and Abdel-Rahman, K. (2012). Minimum embedded length of cyclic horizontally loaded monopiles. *Journal of Geotechnical and Geoenvironmental Engineering* 138: 357–363. https://doi.org/10.1061/(ASCE)GT.1943-5606.0000602.

Lanzo, G., Vucetic, M., and Doroudian, M. (1997). Reduction of shear modulus at small strains in simple shear. *Journal of Geotechnical and Geoenviromental Engineering ASCE.* 123 (11): 1035–1042.

Leblanc, C (2009): Design of Offshore Wind Turbine Support Structures - Selected topics in the field of geotechnical engineering, Aalborg University, Denmark, 2009.

Leblanc, C., Houlsby, G.T., and Byrne, B.W. (2010). Response of stiff piles in sand to long-term cyclic lateral loading. *Géotechnique* 60: 79–90. https://doi.org/10.1680/geot.7.00196.

Lesny, K. and Wiemann, J. (2005). Design aspects of monopiles in German Offshore Wind Farms. In: *Frontiers in Offshore Geotechnics* (ed. S. Gourvenec and M. Cassidy), 383–389. ISFOG.

Li Z, Haigh SK, Bolton MD (2010) Centrifuge modelling of mono-pile under cyclic lateral loads. 7th Int Conf Phys Model Geotech 2:Vol 2 pg 965–970.

Lin, S.-S. and Liao, J.-C. (1999). Permanent strains of piles in sand due to cyclic lateral loads. *Journal of Geotechnical and Geoenvironmental Engineering* 125: 798–802.

Lin, C., Bennett, C., Han, J. et al. (2010). Scour effects on the response of laterally loaded piles considering stress history of sand. *Computers and Geotechnics* 37: 1008–1014.

Lin, C., Han, J., Bennett, C. et al. (2014). Behaviour of laterally loaded piles under scour conditions considering the stress history of undrained soft clay. *Journal of Geotechnical and Geoenvironmental Engineering* 140 (6): https://doi.org/10.1061/(ASCE)GT.1943-5606.0001111.

Lindoe Offshore Renewables Center (2011) Offshore Wind Farm Knowledge Base. In: lorc.dk web page. http://www.lorc.dk/knowledge. Accessed 19 Mar 2015.

Little RL, Briaud J-L (1988) Full scale cyclic lateral load tests on six single piles in sand (No. TAMU-RR-5640). College Station, Texas.

Lowe, J. (2012). Hornsea Met Mast - A Demonstration of the 'Twisted Jacket' Design. In: *Power Point Presentation, 2012. Presented at the Future Offshore Wind Turbine*

Foundation Conference, Bremen, Germany 2012; see the link. http://www.windpowermonthlyevents.com/events/future-offshore-wind-foundations-2012/.

Lopez-Querol, S., Liang, C., and Bhattacharya, S. (2017). Numerical Methods for SSI Analysis of Offshore Wind Turbine Foundations. *Wind Energy Engineering.* 275–297.

Lunne, T., Robertson, P.K. and Powell, J.J.M. (1997), Cone Penetration Testing in Geotechnical Practice, Blackie Academic/Routledge Publishing, New York

G. Lloyd (2005) Guideline for the certification of offshore wind turbines. Uetersen, Germany.

Lombardi, D. and Bhattacharya, S. (2016). Evaluation of seismic performance of pile-supported models in liquefiable soils. *Earthquake Engineering and Structural Dynamics* 45 (6): 1019–1038.

Lombardi, D., Bhattacharya, S., and Muir Wood, D. (2013). Dynamic soil-structure interaction of monopile supported wind turbines in cohesive soil. *Soil Dynamics and Earthquake Engineering* 49: 165–180. https://doi.org/10.1016/j.soildyn.2013.01.015.

Lombardi, D., Dash, S.R., Bhattacharya, S. et al. (2017). Construction of simplified design py curves for liquefied soils. *Géotechnique* 67 (3): 216–227.

Long, J.H. and Vanneste, G. (1994). Effects of cyclic lateral loads on piles in sand. *Journal of Geotechnical Engineering* 120: 225–244.

Long, J.H., Kerrigan, J.A., and Wysockey, M.H. (1999). Measured time effects for axial capacity of driven piling. *Journal of the Transportation Research Board* 1663, paper number 99-1183.

Massarsch, K.R. (2004). Deformation properties of fine-grained soils from seismic tests. Keynote Lecture, Intern. Conf. Site Characterization, Porto, September 2004. 133–146.

Matlock, H. (1970). Correlations of design of laterally loaded piles in soft clay. In Proceedings of the II Annual Offshore Technology Conference, Houston, Texas (OTC 1204): 577–594.

Matlock H, Foo SHC, Bryant LM (1978) Simulation of lateral pile behavior under earthquake motion. Proc. ASCE Spec. Conf. Earthq. Engrg. Soil Dyn. American Society of Civil Engineers, 19–21 June, Pasadena, California, pp 600–619.

Matsuishi M, Endo T (1968) Fatigue of metals subjected to varying stress. Japan Soc. Mech. Eng.

Matutano, C., Negro, V., López-Gutiérrez, J.S. et al. (2013). Scour prediction and scour protections in offshore wind farms. *Renewable Energy* 57: 358–365.

May, R.W.P., Ackers, J.C. and Kirby, A.M. 2002. Manual on scour at bridges and other hydraulic structures, London, Ciria.

Mayne, P.W. and Rix, G.J. (1993). G_{max}-qc relationships for clays. *Geotechnical Testing Journal* 16 (1): 54–60.

McClelland, B. and Focht, J.A. (1958). Soil modulus for laterally loaded piles. *Transactions of the American Society of Civil Engineers* 123 (2954): 1049–1086.

Melville, B.W. and Colema, S.E. (2000). *Bridge Scour.* Highlands Ranch, CO: Water Resources Publications.

Melville, B.W. and Sutherland, A.J. (1988). Design method for local scour at bridge piers. *Journal of Hydraulic Engineering* 114.

Meyerhof, G. G. (1957): Discussion, Proc. 4th Int. Conf. on Soil Mechanics and Foundation Engineering, Vol. 3, p. 110.

Michael Rattley, Richard Salisbury, Timothy Carrington, Carl Erbrich & Gary Li (2017): Marine site characterisation and its role in wind turbine geotechnical engineering;

Proceedings of TC 209 Workshop - 19th ICSMGE, Seoul 20 September 2017 Foundation design of offshore wind structures
Mindlin, R. and Deresiewicz, H. (1953). Elastic spheres in contact under varying oblique forces. *ASME Journal of Applied Mechanics* 20: 327–344.
Miner, M.A. (1945). Cumulative damage in fatigue. *The American Society of Mechanical Engineers (ASME) Journal of Applied Mechanics* 12: 159–164.
Mitchell, J.K. and Kao, T.C. (1978). Measurement of soil thermal resistivity. *Journal of the Geotechnical Engineering Division* 104: 1307–1320.
Morison, J.R., Johnson, J.W., and Schaaf, S.A. (1950). The force exerted by surface waves on piles. *Journal of Petroleum Technology* 2: 149–154. https://doi.org/10.2118/950149-G.
den Boon, J.H., Sutherland, J., Whitehous, R.J.S., et al. Scour Behaviour and Scour Protection for Monopile Foundations of Offshore Wind Turbines. Proceedings for the European Wind Energy Conference, 2004.
Neubecker, S.R., Randolph, M.F. 1995, Performance of embedded anchor chains and consequences for anchor design. Offshore. Technol. Conf. https://doi.org/10.4043.
Nikitas, G., Vimalan, N., and Bhattacharya, S. (2016). An innovative cyclic loading device to study long term performance of offshore wind turbines. *Journal of Soil Dynamics and Earthquake Engineering* 82: 154–160.
Nikitas, G., Arany, L., Aingaran, S. et al. (2017). Predicting long term performance of offshore wind turbines using cyclic simple shear apparatus. *Soil Dynamics and Earthquake Engineering* 92: 678–683.
Eide, O. and Andersen, K.H. (1984). Guest Lecture – Foundation Engineering for Gravity Structures in the Northern North Sea. In: *International Conference on Case Histories in Geotechnical Engineering. 1.* http://scholarsmine.mst.edu/icchge/1icchge/1icchge-theme4/1.
Ochi, M. and Shin, V. (1988). *Wind turbulent spectra for design consideration of offshore structures*, 461–467. In: Offshore Technology Conference. Offshore Technology Conference.
Osman, A.S. and Bolton, M.D. (2005). Simple plasticity-based prediction of the undrained settlement of shallow circular foundations on clay. *Géotechnique.* 55 (6): 435–447.
Offshore Limited. Offshore wind turbine database. In: 4COffshore.com web page. http://www.4coffshore.com/windfarms/turbines.aspx; 2015b [accessed 12.10.17].
Okur, D.V. and Ansal, A. (2007). Stiffness degradation of natural fine grained soils during cyclic loading. *Soil Dynamic Earthquake Engineering* 27: 843–854.
O'Neill MW, Murchinson JM (1983) An Evaluation of p-y Relationships in Sands.
O'Sullivan, C., Cui, L., and O'Neil, S. (2008). Discrete element analysis of the response of granular materials during cyclic loading. *Soils and Foundations* 48 (4): 511–530.
Parra, E. (1996). Numerical Modeling of Liquefaction and Lateral Ground Deformation including Cyclic Mobility and Dilative Behavior in Soil Systems. PhD dissertation, Dept. of Civil Engineering, Rensselaer Polytechnic Institute.
Parry, R.H.G (1978): A study of pile capacity for the Heather platform, European offshore petroleum conference and Exhibition (EUR 49), 24-27th October, London.
Parsons, J.D. (1966). Piling difficulties in the New York area. *Journal of Soil mechanics and Foundation Engineering, ASCE* 92 (1): 43–64.
Pender, M.J. (1993). Aseismic pile foundation design analysis. *Bulletin of the New Zealand National Society for Earthquake Engineering* 26: 49–161.

Pierson, W.J. and Moskowitz, L. (1964). A proposed spectral form for fully developed wind seas based on the similarity theory of S.A. Kitaigordskii. *Journal of Geophysical Research* 69: 5181–5190.

Potts, D.M. (2003). Numerical analysis: a virtual dream or practical reality? *Géotechnique* 53 (6): 535–573.

Poulos, H. and Davis, E. (1980). *Pile Foundation Analysis and Design*. Rainbow-Bridge Book Co.

Poulos, H.G. and Davis, E.H. (1974). *Elastic solutions for soil and rock mechanics*. New York: John Wiley & Sons.

Prakash, S. (1962). *Behaviour of Pile Groups Subjected to Lateral Loads*. University of Illinois at Urbana-Champaign.

Randolph, M.F. (1981). The response of flexible piles to lateral loading. *Géotechnique* 31: 247–259.

Randolph M, Gourvenec MRS (2011) Offshore Geotechnical Engineering. Text book CRC Press Published April 16, 2017, Reference - 530 Pages,ISBN 9781138074729.

Reese, C.L. and Van Impe, F.W. (2001). *Single Piles and Pile Groups under Lateral Loading*, 2e. CRC Press/Balkema.

Reese LC, Cox WR, Koop FD (1974) Analysis of Laterally Loaded Piles in Sand. Pap. No. 2080, Sixth Annu. Offshore Technol. Conf.

Reese LC, Cox WR, Koop FD (1975) Field Testing and Analysis of Laterally Loaded Piles in Stiff Clay. Seventh Annu. Offshore Technol. Conf.

Richardson, E. V. and Davis, S. R. 1995. Evaluating Scour at Bridges, Hydraulic Engineering Circular no.18 (HEC-18). Washington DC, USA: Federal Highway Administration.

Rix, GJ and Stokoe, KHII. Correlation of initial tangent modulus and cone penetration resistance, Proc, 1 st International Symposium on Calibration Chamber Testing/ ISOCCT1, Postdam, New York, A.-B. Huang, ed., 351–362, 1991.

Robertson. P.K., Campanella, R.G., Gillespie, D., and Greig, J., (1986). Use of Piezometer Cone data. *In-Situ'86 Use of In-situ testing in Geotechnical Engineering, GSP 6 , ASCE*, Reston, VA, Specialty Publication, pp 1263–1280.

Rouholamin, M., Bhattacharya, S., and Orense, R.P. (2017). Effect of initial relative density on the post-liquefaction behaviour of sand. *Soil Dynamics and Earthquake Engineering* 97: 25–36.

RTRI (1999). *Design Standard for Railway Facilities – Seismic Design*. Japan: Railway Technical Research Institute.

Sa'don, N.M., Pender, M.J., Abdul Karim, A.R. et al. (2014). Pile head cyclic lateral loading of single pile. *Geotechnical and Geological Engineering* 32: 1053–1064. https://doi.org/10.1007/s10706-014-9780-5.

Salgado, R. (2006). *The Engineering of Foundations*. McGraw-Hill Education.

Senders, M. and Kay, S. (2002). Geotechnical suction pile design in deep water soft clay. In: *London, International Conference on Deep Riers Moorings Anchorings*.

Schmertmann, J.H. (1991). The mechanical ageing of soils. *Journal of Geotechnical Engineering, ASCE* 117 (9): 1289–1329.

Seed, H.B. (1979). Soil liquefaction and cyclic mobility evaluation for level ground during earthquakes. *Journal of Geotechnical Engineering, ASCE* 201–255.

Seed, H.B., Tokimatsu, K., Harder, L.F. et al. (1985). Influence of SPT procedures in soil liquefaction resistance evaluations. *Journal of Geotechnical Engineering, ASCE* 111 (12): 1425–1445.

Shadlou, M. and Bhattacharya, S. (2014). Dynamic stiffness of pile in a layered elastic continuum. *Geotechnique* 64 (4): 303–319.

Shadlou, M. (2016). Contribution to the static and dynamic response of piles in liquefiable ground, PhD thesis. In: *University of Bristol (UK).*

SCHAKENDA, B., NIELSEN, S. A. & IBSEN, L. B. 2011. Foundation structure. United States of America patent application 12/226,255. 22nd February 2011.

Shadlou, M. and Bhattacharya, S. (2016). Dynamic stiffness of monopiles supporting offshore wind turbine generators. *Soil Dynamics and Earthquake Engineering* 88: 15–32.

Siemens Wind Turbine SWT-3.6-120 Design Specification Document, 2015, http://www.siemens.com/energy

Simiu, E. and Leigh, S. (1984). Turbulent wind and tension leg platform surge. *Journal of Structural Engineering* 110 (4): 785–802.

Skov R. and Denver, H. Time dependence of bearing capacity of piles, Proceedings of the 3rd International Conf. App. Stress-wave Theory to piles, Ottawa, 1988, pp 879–888.

Smith, E.A.L. (1960). Pile-driving analysis by the wave equation. *Journal of the Soil Mechanics and Foundation Division, ASCE* 86 (SM4): 35–61.

de Sonneiveille, B., Rudolph, D. and Raaijmakers, T. Scour Reduction by Collars Around Monopiles. International Conference on Scour Erosion, 2010 San Francisco, California.

Spence, D.A. (1968). Self-similar solutions to adhesive contact problems with incremental loading. *Proc. R. Soc. London A* 305: 55–80.

Squorch 2006, *Ground Resonance: Rear View*, [accessed 1.10.17], https://www.youtube.com/watch?v=D2tHA7KmRME

Statoil, 2015. Hywind Scotland Pilot Park Environmental Statement.

Stroescu, I.-E., Frigaard, P., and Fejeskov, M. (2016). Scour development around bucket foundations. *International Journal of Offshore and Polar Engineering* 26: 57–64.

Stroud MA (1988). The standard penetration test – its application and interpretation. Proceedings of the Geotechnology Conference, Penetration Testing in the UK, ICE, Birmingham, 29-49.

Sumer, B.M. and Fredsøe, J. (2001). Scour around pile in combined waves and current. *Journal of Hydraulic Engineering* 127: 403–411.

Sumer, B.M. and Fredsøe, J. (2002). *Mechanics of Scour in the Marine Environment*. World Scientific Press.

Sumer, B.M. and Nielsen, A.W. (2013). Sinking failure of scour protection at wind turbine foundation. *Energy* 166. Proceedings of the Institution of Civil Engineers Energy 166 November (EN4): 170–188. http://dx.doi.org/10.1680/ener.12.00006.

Sumer, B.M., Fredsøe, J., and Christiansen, N. (1992). Scour around vertical pile in waves. *Journal of Waterway, Port, Coastal, and Ocean Engineering* 118: 15–31.

Sumer, B.M., Peterson, T.U., Locatelli, L. et al. (2013). Backfilling of a scour hile around a pile in waves and current. *Journal of Waterway, Port, Coastal, and Ocean Engineering* 139: 9–23.

SUT (2014): Guidance notes for the planning and execution of geophysical and geotechnical ground investigations for offshore renewable energy development; Society for Underwater Development, Editor: Mike Cook; ISBN 0 906940 54 0 ISBN 13 978 0 906940 54 9

Sun, X., Diangui, H., and Guoqing, W. (2012). The current state of offshore wind energy technology development. *Energy* 41 (1): 298–312.

Supachawarote, C., Randoph, M., and Gourvenec, S. (2004, January 1). *Inclined pull-out capacity of suction caissons*. Offshore Polar Eng: Int. Soc.

Svinkin, M. (1996). Discussion on set-up and relaxation in glacial sand. *Journal of Geotechnical Engineering, ASCE* 122 (4): 319–321.

Swane IC, Poulos H (1982) A theoretical study of the cyclic shakedown of laterally loaded piles. Sydney.

Taylor, G.I. (1938) The Spectrum of Turbulence. Proceedings of the Royal Society of London, A164, 476-490. http://dx.doi.org/10.1098/rspa.1938.0032

Te Kamp, W. G. B.. Cone Penetration tests and piled foundations, Second Fugro-Cesco CPT Symposium, 1977, 119–132, Holland.

Terzaghi, K. (1955). Evaluation of coefficients of subgrade modulus. *Geotechnique* 5: 297–326.

Terzaghi, K. and Peck, R.B. (1948). *Soil Mechanics in Engineering Practice*. Wiley, 566pp.

Ting, F.C.K., Briaud, J.-L., Chen, H.C. et al. (2001). Flume tests for scour in clay at circular piers. *Journal of Hydraulic Engineering* 127: 969–978.

Towhata, I. (2008). *Geotechnical Earthquake Engineering*. Berlin Heidelberg: Springer-Verlag.

Vaitkunaite, E., Molina, S.D. and Ibsen, L.B. 2012. Comparison of Calculation Models for Bucket Foundation in Sand. *DCE Technical Memorandum*. Aalborg: Aalborg Universitet. Institut for Byggeri og Anlæg.

Vattenfall (2016) – the Youtube Link: https://www.youtube.com/watch?v=7vftUaViSQ0

Verdure, L., Garnier, J., and Levacher, D. (2003). Lateral cyclic loading of single piles in sand. *International z of Physical Modelling in Geotechnics* 3: 17–28.

Villalobos, F.A. (2006). *Model Testing of Foundations for Offshore Wind Turbines*. Doctor of Philosophy: University of Oxford.

Villalobos, F.A., Byrne, B.W., and Houlsby, G.T. (2005). Moment loading of caissons installed in saturated sand. In: *Frontiers in Offshore Geotechnics: ISFOG 2005* (ed. G. Cassidy). Taylor and Francis Group, London.

Villalobos, F.A., Byrne, B.W., and Houlsby, G.T. (2009). An experimental study of the drained capacity of suction caisson foundations under monotonic loading for offshore applications. *Soils and Foundations* 49: 477–488.

Vucetic, M. and Dobry, R. (1988). Degradation of marine clays under cyclic loading. *Journal of Geotechnical Engineering ASCE*, 117, No. 1, 89–107.

Vucetic, M. and Dobry, R. (1991). Effect of soil plasticity on cyclic response. *Journal of Geotechnical Engineering (ASCE)*. Vol. 117, No. 1, pp. 89–117.

Vucetic, M. (1994). Cyclic threshold shear strains in soils. *Journal of Geotechnical Engineering* 120 (12): 2208–2228.

Wei, K., Myers, A.T., and Arwade, S.R. (2017). Dynamic effects in the response of offshore wind turbines supported by jackets under wave loading. *Engi\neering Structures* 142: 36–45.

Wei-Hua, H., S. Thöns, S. Said, et al. (2014) : Resonance phenomenon in a wind turbine system under operational conditions; Proceedings of the 9th International Conference on Structural Dynamics, EURODYN 2014, Porto, Portugal, 30 June - 2 July 2014, A. Cunha, E. Caetano, P. Ribeiro, G. Müller (eds.); ISSN: 2311-9020; ISBN: 978-972-752-165-4.

Wesselink, B.D., Murff, J.D., Randolph, M.F., Nunez, I.L., and Hyden, A.M. (1988) Analysis of centrifuge model test data from laterally loaded piles in calcareous sand, Proc. Int. Conf. Calcareous Sediments, 1, Balkema, 261–270.

Whitehouse, R. J. S. 2004. Marine Scour at At Large Foundations. Second International Conference on Scour and Erosion (ICSE-2). Singapore.

Whitehouse, R.J.S., Harris, J.M., Sutherland, J. et al. (2011). The nature of scour development and scour protection at offshore windfarm foundations. *Marine Pollution Bulletin* 62: 73–88.

Whittle, A.J. and Sutabutr, T. (1991). Prediction of pile set-up in clay. *Transportation Research Record* 1663, Paper number 99-1152: 33–40.

Whittle, A.J. and Sutabutr, T. (1999). Predic\tion of pile set-up in clay. *Transportation Research Record* 1663, Paper number 99-1152: 33–40.

Wichtmann T. Explicit accumulation model for non-cohesive soils under cyclic loading. Inst Für Grundbau Und Bodenmechanik 2005; PhD:274.

Wichtmann, T., Niemunis, A., and Triantafyllidis, T. (2005). Strain accumulation in sand due to cyclic loading: drained triaxial tests. *Soil Dynamics and Earthquake Engineering* 25 (12): 967–979. https://doi.org/10.1016/j.soildyn.2005.02.022.

Williams MO (2008) Wave mapping in UK waters. Yarmouth, Isle of Wight.

Williams, A.F., Dunnavant, T.W., Anderson, S. et al. (1988). The performance and analysis of lateral load tests on 356 mm dia piles in reconstituted calcareous sand. In: *Proc. Int. Conf. Calcareous Sediments, 1, Balkema*, 271–280.

Winkler, E. (1867a). Die Lehre von der Elasticitaet und Festigkeit mit besonderer Rucksicht auf ihre Anwen- dung in der Technik fur polytechnische Shulen, Bauakademien, Ingenieure, Maschinenbauer, Architek- ten, etc. Verlag von H. Domenicus, Prag.

Winkler E (1867b) Die Lehre Von der Elasticitaet und Festigkeit. H. Dominicus, Prag.

Wolf, J.P. (1988). *Soil-Structure-Interaction Analysis in Time Domain (Prentice-Hall).* Englewood Cliffs: N J.

Yamada, S., Hyodo, M., Orense, R.P., and Dinesh, S.V. (2008). Initial shear modulus of remolded sand-clay mixtures. *Journal of Geotechnical and Geoenviromental Engineering ASCE.* 134 (7): 960–971.

Yang, N.C. (1970). Relaxation of piles in sand and inorganic silt. *Journal of Soil Mechanics and Foundation Engineering Division ASCE* 96 (2): 395–409.

Yang, Z. (1999). Identification and Numerical Modeling of Earthquake Ground Motion and Liquefaction,' Ph.D. dissertation, Dept. of Civil Engineering and Engineering Mechanics, Columbia University, New York, NY

Yang, M., Ge, B., Li, W. et al. (2016). Dimension effect on P-y model used for design of laterally loaded piles. *Procedia Engineering* 143: 598–606. https://doi.org/10.1016/j.proeng.2016.06.079.

York, D.L., Brusey, W.G., Clemente, F.M. et al. (1994). Set-up and relaxation in glacial sand. *Journal of Geotechnical Engineering, ASCE* 120 (9): 1498–1513.

Youhu Zhang, Knut H. Andersen (2017), Scaling of lateral pile p-y response in clay from laboratory stress-strain curves, *Marine Structures* 53 (2017) 124 to 135

Youhu Zhang, Knut H Andersen, Rasmus T. Klinkvort, Hans Petter Jostad, Nallathamby Sivasithamparam, Noel P. Boylan Thomas Langford (2016) Monotonic and Cyclic p-y Curves for Clay based on Soil Performance Observed in Laboratory Element Tests; OTC-26942-MS

Yoshida, N. (2015). *Seismic Ground Response Analysis.* Japan: Tohoku Gakuin University.

Yoshida, I. and Yoshinaka, R. (1972). A method to estimate modulus of horizontal subgrade reaction for a pile. *Japanese Society of Soil Mechanics and Foundation Engineering* 12.

Yoshimi, Y., Richart, F.E., Prakash, S. et al. (1977). Soil dynamics and its application to foundation engineering. *9th International Conference on Soil Mechanics and Foundation Engineering. Tokyo.*

Yu, L., Wang, L., Guo, Z. et al. (2015). Long-term dynamic behavior of monopile supported offshore wind turbines in sand. *Theoretical and Applied Mechanics Letters* 5 (2): 80–84.

Zaaijer, M.B. (2006). Foundation modelling to assess dynamic behaviour of offshore wind turbines. *Applied Ocean Research* 28: 45–57. https://doi.org/10.1016/j.apor.2006.03.004.

Zaaijer, M.B. (2009). Review of knowledge development for the design of offshore wind energy technology. *Wind Energy* 12: 411–430.

Zania, V. (2014). Natural vibration frequency and damping of slender structures founded on monopiles. *Soil Dynamics and Earthquake Engineering* 59: 8–20. https://doi.org/10.1016/j.soildyn.2014.01.007.

Zanke, U.C., Hsu, T.-W., Roland, A. et al. (2011). Equilibrium scour depths around piles in noncohesive sediments under currents and waves. *Coastal Engineering* 58: 986–991.

Zdravkovic, L., Taborda, D.M.G., Potts, D.M. et al. (2015). Numerical modelling of large diameter piles under lateral loading for offshore wind applications. In: *Frontiers in Offshore Geotechnics III* (ed. V. Meyer), 759–764. London: Taylor and Francis Group.

Zhang, J.H., Zhang, L.M., and Lu, X.B. (2007). Centrifuge modeling of suction bucket foundations for platforms under ice-sheet-induced cyclic lateral loadings. *Ocean Engineering* 34: 1069–1079.

Zhang, P., Ding, H., Le, C. et al. 2011. Test on the Dynamic Response of the Offshore Wind Turbine Structure with the Large-Scale Bucket Foundation. Procedia Environmental Sciences, 12, Part B, 856-863.

Zhang, P., Ding, H., and Le, C. (2013). Installation and removal records of field trials for two mooring dolphin platforms with three suction caissons. *Journal of Waterway, Port, Coastal, and Ocean Engineering* 139: 502–517.

Zhang, P., Ding, H., and Le, C. (2014). Seismic response of large-scale prestressed concrete bucket foundation for offshore wind turbines. *Journal of Renewable and Sustainable Energy* 6: 013127.

Zhu, B., Byrne, B., and Houlsby, G. (2012). Long-term lateral cyclic response of suction caisson foundations in sand. *Journal of Geotechnical and Geoenvironmental Engineering* 73–83. https://doi.org/10.1061/(ASCE)GT.1943-5606.0000738.

Zienkiewicz, O.C., Chang, C.T., and Bettess, P. (1980). Drained, undrained, consolidating and dynamic behaviour assumptions in soils. *Geotechnique* 30: 385–395.

Index

Design of Foundations for Offshore Wind Turbines, First Edition. Subhamoy Bhattacharya.

Companion website: www.wiley.com/go/bhattacharya/offshorewindturbines

u

v

w

y

Printed and bound by CPI Group (UK) Ltd, Croydon, CR0 4YY
16/04/2025
14050395-0004